THE NAVIGATORS

The Navigators

A History Of NASA's Deep-Space Navigation

Andrew J. Butrica   2014

© Andrew J. Butrica

All rights reserved. No part of this book may be reproduced in any form by any electronic or mechanical means (including photocopying, recording, or information storage and retrieval) without permission in writing from the author.

Andrew J. Butrica is the author and publisher of this work.

Andrew J. Butrica, PhD, is an independent scholar and historian of technology and science who researches and publishes in nineteenth-century French science, technology, engineering, and industrialization. His "day job" is the researching and writing of history for NASA, the Department of Defense, and other public and private organizations.

ISBN (cloth): 1-492-77783-8

Government Rights Notice
This work was authored by Andrew J. Butrica under Contract N° NNH07CC14C with the National Aeronautics and Space Administration. The United States Government retains and the publisher, by accepting this manuscript for publication, acknowledges that the United States Government retains a non-exclusive, paid-up, irrevocable, worldwide license to reproduce, prepare derivative works, distribute copies to the public, and perform publicly and display publicly, or allow others to do so, for United States Government purposes. All other rights are reserved by the copyright owner.

"Science is much more than a body of knowledge. It is a way of thinking. This is central to its success. Science invites us to let the facts in, even when they don't conform to our preconceptions. It counsels us to carry alternative hypotheses in our heads and see which ones best match the facts. It urges on us a fine balance between no-holds-barred openness to new ideas, however heretical, and the most rigorous skeptical scrutiny of everything—new ideas and established wisdom. We need wide appreciation of this kind of thinking. It works. It's an essential tool for a democracy in an age of change. Our task is not just to train more scientists but also to deepen public understanding of science."

Carl Edward Sagan

"Why We Need To Understand Science," *The Skeptical Inquirer*, vol. 14, no. 3, Spring 1990

Contents

	Preface	*i*
	Introduction	1

Part One — Origins
One	Home on the Range	7
Two	Applied Celestial Mechanics	21
Three	Vanguard	31

Part Two — Pioneering Space Navigation
Four	Pioneers of Space Navigation	43
Five	Adapting Astronomy to Navigation	53
Six	Modeling Earth	61
Seven	Navigator Scientists	71
Eight	The First Navigators	85

Part Three — The Golden Age of Exploration
Nine	Navigators in the Golden Age	99
Ten	Navigation Becomes Astronomy	113
Eleven	Scientists and Explorers	127
Twelve	To Be More Precise	139
Thirteen	The Quest	147
Fourteen	A Page from Radio Astronomy	161
Fifteen	Optical Navigation	171

Part Four — Back to Earth
Sixteen	Multi-Mission, Multinational Navigation	187
Seventeen	Earth and Space Joined	199
Eighteen	Opening the ΔDOR to VLBI Navigation	209
Nineteen	The New Scientists of Navigation	223

Part Five — Things Go Awry
Twenty	Navigating with a Bull in the China Shop	235
Twenty-One	The Angry Red Planet	251
Twenty-Two	Autonomous Navigation	267

Part Six — A New Normal
Twenty-Three	Navigating in a New Era	283
Twenty-Four	Cassini, Navigation, and Science	293
Twenty-Five	Mars Redux	303
	Conclusion	315
	Oral Histories	323
	List of Abbreviations	327
	Index	361

Preface

This book is organized roughly chronological and presents an overarching history of space navigation in three phases. The first traces the origins of space navigation. The second, running from the late 1950s to about 1980, deals with the evolution of navigation within the context of the national security crisis, defined as the period of World War II and the Cold War. The third phase, the period from about 1980 to the present, has its own peculiarities, including the revolution in computing—the replacement of mainframes by minicomputers—new management philosophies, and persistent budget and personnel reductions that continue to shape navigation today. The chapters are arranged in six parts each of which, with the exception of the first, more or less represents a decade of time during the roughly half century of JPL deep-space navigation. Readers may detect a rather sprawling narrative, one occasioned by the necessity to produce a comprehensive readable book of a genre I can describe only as a hybrid of a contract "deliverable" and scholarly work.

* * * * *

This history was researched and written under a contract with NASA Headquarters. It would not have come into existence without the efforts and enthusiasm of Barry Geldzahler, who secured the funding to make it happen. It is also to his credit that he found additional support to complete the project, whose completion exceeded the scheduled deadline for numerous reasons. I hope this work meets and surpasses his expectations.

I benefited significantly from the research, writing, and oral histories that I undertook previously for a history of planetary radar astronomy published by NASA in 1996, and I wish to thank again the late Nick Renzetti and Steve Ostro. Without them, that history never would have been written and this one far less rich.

The location of NASA's deep-space navigators at JPL made numerous and at times lengthy research trips to the laboratory's library and archives essential. My warmest thanks go to those librarians and archivists who helped: Barbara J. Amago, Julie A. Cooper, Charlene Nichols, and especially Mickey Honchell. My thanks also go to JPL historian Erik Conway.

Among the most helpful on this project were the navigators themselves. I am grateful to all who agreed to be interviewed either formally in an oral history or informally: Shyam Bhaskaran, Jim Border, Al Cangahuala, Dave Curkendall, Sam Dallas, Len Efron, Barry Geldzahler, Tom Hamilton, Frank Jordan, Earl Maize, Tomás Martin-Mur, Neil Mottinger, Bill Sjogren, Myles Standish, Steve Synnott, Catherine Thornton, Lincoln Wood, and Don Yeomans. Additionally, I owe a debt to Frank Jordan and Judy Greenberg for their generous cooperation, documents, and even a place to work for a day as well as to Bill Zielenbach and Lincoln Wood for providing various documents and expediting them through security processing.

I also thank Lincoln Wood, George Null, Bill Sjogren, and especially Len Efron, who helped to find navigators willing to read an earlier, and even longer, version of this manuscript—or at least parts of it. They included, in alphabetical order, John Anderson, James Border, Pete Breckheimer, Al Cangahuala, Carl Christenson, Elliott Cutting, Sam Dallas, Len Efron, Jordan Ellis, Barry Geldzahler, Don Gray, Jerry Hintz, Frank Jordan, Bill Kirhofer, James McDanell, Bill Melbourne, James Miller, Neil Mottinger, George Null, Bill Purdy, Kara Reiter, Bill Sjogren, Myles Standish, Rhody Stephenson, Tony Taylor, Catherine Thornton, Don Trask, Bobby Williams, Pamela Wolken, and Lincoln Wood. Jim Border's chapter edits and comments as well as the editorials on the entire manuscript provided by Lincoln Wood, who has written his own three-part history of JPL navigation, were especially useful. I also want to thank Dave Curkendall and Tom Hamilton for being so helpful, but who passed before the manuscript became available for review.

Additional thanks go to those who helped in the various libraries and archives along the way. Sally Bosken and Greg Shelton aided in making available the books and oral histories of the James Melville Gilliss Library, Naval Observatory, and in particular the papers of Paul Herget. I thank Mark Kahn, Marilyn Graskowiak, David Schwartz, and Larry Wilson at the Smithsonian Air and Space archives, where much of the early research took place. Additionally, I am grateful to Susie F. Cook, Special Collections Department, University Libraries, Virginia Polytechnic Institute, Blacksburg, Virginia, for access to the Samuel Herrick papers; to Fordyce Williams, Archives and Special Collections, Clark University, Worcester, Massachusetts, for Robert Goddard documents; to Sarah Hambleton for access to the E. Dorrit Hoffleit papers at the Arthur and Elizabeth Schlesinger Library on the History of Women in America, Radcliffe Institute for Advanced Study, Harvard University Library, Cambridge, Massachusetts; to Nancy Honeyford, Niels Bohr Library and Archives, and Gregory A. Good, director, American Institute of Physics Center for History of Physics, College Park, Maryland; to Eric Conway, then in charge of the archives at NASA's Langley Research Center, Hampton, Virginia; and to Elizabeth Hauser for transcribing interview tapes and MP3 files. My thanks equally go to the staffs of the Library of Congress and the National Archives and Records Administration in College Park, Maryland.

At NASA Headquarters, I thank Steve Dick, retired chief historian, and Barry Geldzahler, who supervised the contract; Nadine Andreassen, program support specialist; Jane Odom, archivist; the knowledgeable and ever professional staff of the Historical Reference Collection—Colin Fries, John Hargenrader, and Liz Suckow—and Rick Spencer, manager of the Dr. T. Keith Glennan Memorial Library as well as Matthew Levine and the rest of the library team.

I also am appreciative of the discussions I have had regarding navigation history and related topics with Hamilton Cravens (Iowa State University and the University of Minnesota); Allan Olley (University of Toronto), who also read manuscript sections and supplied correspondence relating to Neil Block; Michael Osborne (Oregon State University); the late Craig Waff; and David Whalen (University of North Dakota).

*****

Over the years, I have enjoyed researching and writing for the NASA History Office. I have been fortunate and privileged to produce publications that have won awards from the Organization of American Historians, the National Council on Public History, and the American Institute of Aeronautics and Astronautics, among others. This project offered a rather different experience, however.

The Statement of Work gave little hint as to the absolute vastness and complexity of the subject, nor did the contract allot sufficient time to complete the project as directed in that document. These obstacles in themselves were not insurmountable. The first serious impediment encountered was the requirement that each and every JPL document undergo both proprietary and export control security processing. It was over a year before the first documents cleared. The resulting delays impacted the delivery of chapters and even altered the narrative structure. Further impediments arose when the History Office canceled the e-mail account that facilitated locating and communicating with navigators and required that records collected in the course of this project be delivered before the manuscript had been prepared for publication (both the usual practice and a contract stipulation). Then there was the lack of civility and professionalism that came in various forms. The author overcame these adversities and invested a not insignificant number of unpaid hours in completing this project.

Normally, a NASA history manuscript such as this one would be published by the government or an academic press. A different route was taken at the behest of Barry Geldzahler, who wanted to see this book in the hands of NASA navigators as quickly as possible, faster than was possible with an academic press (my personal preference). With the space agency's permission, I accommodated him by turning the "final deliverable" manuscript version into this book. The work was done entirely at my expense on my own time.

Developing and testing Cold War missiles launched JPL into space navigation (Author)

Introduction

Science since World War II has taken on a primarily "applied" character. Paul Forman argues that "the convergence of science and engineering that characterizes our modern world was effectively launched in its primarily military direction with the mobilization of U.S. scientists especially physicists by the Manhattan Project and the Office of Scientific Research and Development."[1] Hamilton Cravens extends this analysis to the social sciences.[2] Lillian Hoddeson also identifies the Manhattan Project as the origin, but specifically of the model of the hierarchically-organized, multidisciplinary, government-supported research laboratory that spread into the organization of national laboratories and the design of large-scale federally-funded projects in such areas as space, microwaves, and lasers.[3]

Deep-space navigation exemplifies this model of pragmatic, multidisciplinary, government-supported research. William Corliss, decades ago, suggested that one can think of navigation as an applied or engineering version of celestial mechanics.[4] The multiplicity of institutions with which navigators interacted—the Bureau International de l'Heure, the International Polar Motion Service, the International Union of Geodesy and Geophysics, the International Astronomical Union, to name a few—reflects the gamut of disciplines—astronomy and radio astronomy, geodesy and geophysics, cartography and meteorology, ionospheric physics and radio science—that are a necessary basis for space navigation.

It goes without saying that NASA's deep-space navigation takes place in a government-supported research laboratory: the Jet Propulsion Laboratory. The Goddard Space Flight Center has been active in space navigation since its founding, but it has focused on Earth-circling satellites. The origins of JPL navigation predate the establishment of NASA. World War II united scientists and engineers with testing ranges and radars in the development and testing of missiles. Those radars, of course, were a product of World War II. JPL, then operating as GALCIT, engaged in rocket research on behalf of the Army Air Forces and the Army's Ordnance Department during World War II.

That war was only the beginning; the Cold War continued the national security crisis until the collapse of the Soviet Union. The role of the Cold War in shaping navigation is one of this book's leitmotifs. Army projects introduced JPL to missile ranges and tracking Earth-orbiting satellites. The ARPA lunar program initiated the lab in

---

[1] Paul Forman, "Behind Quantum Electronics: National Security as Basis for Physical Research in the United States, 1940-1960," *Historical Studies in the Physical and Biological Sciences* 18 (1987): 149-229.
[2] Hamilton Cravens, "The Social Sciences as Social Technologies: How the National Security Crisis of 1941–91 Transformed the American Social Disciplines," American Historical Association, January 2, 2014, Washington, DC; Mark Solovey and Hamilton Cravens, ed., *Cold War Social Science: Knowledge Production, Liberal Democracy, and Human Nature* (New York: Palgrave Macmillan, 2012).
[3] Lillian Hoddeson, *Critical Assembly: A Technical History of Los Alamos during the Oppenheimer Years, 1943-1945* (New York: Cambridge University Press, 1993), 407, 413 & 415.
[4] Corliss, *Space Probes*, 11.

tracking and navigating deep-space probes. NASA, like Vanguard, attempted to put a civilian face on the nation's space endeavors, but its missions served to project the Cold War into space. Although the greatest attention has been paid to the attempt to put an American on the Moon by 1970, that race to the Moon appears to have been one-sided. Meanwhile, probes launched by both sides to the Moon and beyond fought in the real Cold War in space. Navigators were under massive pressure to ensure the often elusive goal of success.

Following the entente cordiale of the 1970s, the national security crisis inspired the Strategic Defense Initiative. That grandiose space project exerted a weighty influence on NASA. Its approach to project management became that of NASA ("Faster, Better, Cheaper"), just as the Manhattan Project had served earlier as a model. Both the Strategic Defense Initiative and the influence of its management philosophy outlasted the national security crisis, because they were part and parcel of the conservative space agenda. The conservative agenda and "reinventing government" meshed seamlessly with government budget and personnel cuts.

The country's space program turned back toward Earth. NASA had but three planetary missions in the works, none of which was on a particularly solid financial footing. The Space Shuttle embodied the shifting emphasis in the country's space efforts as exploration gave way to exploitation. Deep-space navigation also reflected this terrestrially-centered emphasis.

Intimately tied to the nurturing of rocketry and space exploration by the national security crisis is the proliferation of institutions and firms with expertise in either tracking craft or determining their trajectories fostered by the armed forces to support its space efforts. As a result, by the time JPL began navigating NASA missions, they were not alone. For the most part, JPL maintained its hegemony over NASA's deep-space missions, but navigation of several flights did end up in the hands of private enterprise. The business-friendly culture of government from the 1980s favored contracting firms for navigation services which, by and large, were furnished by former JPL navigators. Indeed, the navigators themselves were now "for hire" by foreign space agencies.

These factors shaped the milieu in which navigators worked and both directly and indirectly shaped their idiosyncratic work culture. The core question of this study, though, is the relationship between science and navigation. Maritime navigation, like space navigation, depends on applying science (the branch of astronomy known as celestial mechanics) to determining one's position at sea with the aid of instruments and ephemerides published in the form of books (almanacs).

This ongoing improvement of their ephemerides was one way in which navigators became scientists. Celestial mechanics was the vital science at the heart of their software. After initially borrowing and adapting their ephemerides data and practices from astronomers at the Naval Observatory's almanac office, they began offering their ephemerides to academic and commercial users. In the 1970s, they became the scientists in charge of the ephemerides—its values and models included—for astronomers and almanac offices. How they accomplished this feat illustrates one of the paths taken as navigators became scientists.

Navigators also became scientists as mission project scientists. Not surprisingly, they began by doing celestial mechanics experiments. Next, in collaboration with Stanford University, they undertook radio occultations from spacecraft, and later they performed gravitational research. Like the celestial mechanics experiments, these studies yielded scientific results that served to improve navigational accuracy for future missions. The trials that navigators conducted of the Fourth Test of general relativity, on the other hand, contributed to the ongoing assessment of Albert Einstein's theories. Navigators also made a number of scientific breakthroughs in their own right, including the discoveries of mascons on the Moon, volcanism on Io, and numerous satellites around the outer planets.

A key part of the link between navigation and science in this multidisciplinary application-oriented setting is how navigators made and used experimental observations, in a sense, their scientific method. The instruments of observation were the antennas of NASA's global tracking network. The experiments, as I use the term, were the repeated observations and orbit determinations performed during flights to estimate spacecraft positions and velocities.

In preparation for each flight, navigators made mathematical models of the conditions that would impact the probe's course as well as the tracking data. They hoped to realize a fit between the predicted and observed values, so that the residuals—the differences between the predicted and the observed values—approached zero. The goal, in a word, was to match this "fitter's universe" with the "real universe."[5]

Modeling the tracking data entailed accounting for known error sources in media effects (for example, the influence of charged particles in the ionosphere) and tracking station locations (the impacts of polar motion, clock inaccuracies, and other factors). While gravity was the main force acting on probes, modeling also required that navigators account for non-gravitational forces acting on spacecraft, such as solar radiation pressure—the force exerted on a craft by solar photons ejected into space by the Sun—and the attitude control system. The unexpected is the bane of navigators. A small amount of force acting over a long planetary trek can have a major impact. To the degree possible, navigators attempted to collect data on spacecraft positions and velocities that allowed them to determine the contributions from all variables, including error sources.

Errors are eternal. The question is: how does one deal with them? In what I am calling the fitter's approach, one measures and develops mathematical formulas for their impact. In other words, one models them. A different method was to arrange the experiment (spacecraft observation) in a different way so as to try to cancel out or at least mitigate error sources from, say, media effects or station locations. This "direct" technique replaced making numerical "corrections" with experimental design.[6]

The "direct" method was a peculiar characteristic of French experimental physics during the late nineteenth century inspired by Victor Regnault. The chemist Jean-Baptiste Dumas illustrated the approach by describing how Regnault measured the weight of a given volume of gas contained in a sizeable glass balloon with a balance and small metal weights. The difference in the amount of air displaced by the balloon and by

---

[5] This language is from Jay Light, "An Investigation of the Orbit Redetermination Process Following the First Midcourse Maneuver," 8-17 in JPLSPS 37-33:IV.
[6] Daniel Jon Mitchell, "Measurement in French Experimental Physics from Regnault to Lippmann. Rhetoric and Theoretical Practice," *Annals of Science* 69,4 (2012): 453-482, discusses making "direct" measurements versus "corrections" to measured values in French physics.

the weights required an uncertain correction to the measured value because of the surrounding air's buoyancy. Rather than attempt to model that correction, Regnault eliminated the need for the correction by altering the experiment. He suspended an identical but evacuated "compensatory" glass balloon from the opposite side of the balance. The only remaining buoyancy correction—for the metallic weights balancing the weight of the gas—was negligible.[7]

In navigation, "direct" measurements—alterations to the observational method—have taken many forms and, in general, were motivated by preparation for future flights to the outer planets as well as comets and asteroids. Many varieties entailed differencing data from two separated antennas. One might difference Doppler or range or even take the difference between Doppler and range readings. One could difference signals of different frequencies. This search for "self-calibrating" tracking data led logically to attempts, often frustrated, to adapt the radio astronomy method known as Very Long Baseline Interferometry (VLBI).

A very different "direct" approach, equally adapted to the special challenges of outer-planet exploration, was to acquire and process images taken by the spacecraft's own science camera to obtain positional information about the craft and its target. This optical navigation, as it is known, circumvented error sources originating in media effects and station locations and provided greater navigational exactitude, but it was not entirely unproblematic. Nonetheless, advances in computing hardware and software eventually enabled probes to perform onboard navigational calculations under certain circumstances and more recently to interact with the craft's guidance and control systems. As a result, the very definition of deep-space navigation seemed to be changing away from an entirely Earth-based endeavor.

---

[7] For a brief summary of Dumas' description of Regnault's "direct method," see Matthias Doërries, "Easy Transit: Crossing Boundaries Between Physics and Chemistry in Mid-Nineteenth Century France," 246-262 in Crosbie Smith and Jon Agar, ed., *Making Space for Science: Territorial Themes in the Shaping of Knowledge* (New York: St. Martin's Press, 1998).

Rocket pioneer Robert Goddard observing a launch (NASA GRIN 74-H-1245)

Part One

Origins

Chapter 1

Home on the Range

NASA's deep-space navigation began at JPL with the country's first attempts to launch probes toward the Moon. However, the roots of JPL navigation lay in the laboratory's earlier role as a developer of military missiles during the Cold War. Missile development required conducting trials at a test range. The methods for tracking those launches and determining their flight paths later became the foundation for deep-space navigation. Optical observation and recording were commonplace, but radar, thanks to its accelerated improvement to meet wartime needs, was the launch tracker's sine qua non whether at White Sands or Wallops Island. The missile test ranges also saw the earliest use of a mainframe computer to calculate flight paths as well as the hiring of astronomers as consultants, a marriage of civilian scientists and military projects that had become commonplace. By the end of the 1950s, as the country prepared to launch its first rockets to the Moon and beyond, JPL was just one of many organizations, civilian and military, possessed of expertise in determining flight paths.

Tracking the origins. Scant little is known about launch tracking methods between the world wars, especially for the Gas Dynamics Laboratory in the Soviet Union and the German *Heeresversuchsstelle Peenemünde* (the Peenemünde Research Center) program which led to, among others, the infamous V-2 rocket.[1] One exception is the missile testing of Robert Goddard. His descriptions of the trials he carried out near Roswell, New Mexico, from 1930 to 1932 provide some insights. One observer stationed 3,000 feet from the launch site had a recording telescope. Equipped with two pencils and a clockwork mechanism, the telescope recorded the altitude and azimuth of the rocket's flight on a paper strip. The device worked satisfactorily, Goddard reported, except when the rocket's trajectory was in the plane of the launch tower and the telescope. He suggested the use of short-wave radio direction finders to track the rocket's descent from greater heights in lieu of the telescope. Goddard also referred to the use of a 16-mm motion picture camera to record test flights,[2] an approach taken later at White Sands.

During the interwar period, anyone wishing to calculate a rocket trajectory could draw on long-standing knowledge of the mathematics and geometry of ballistic trajectories. Under ideal conditions, they trace a parabolic path described poetically by Thomas Pynchon as "gravity's rainbow"[3] and mathematically by Galileo Galilei (1564-

---

[1] For a scholarly discussion of the development of the V-2 and other rockets by the Nazis, see Michael J. Neufeld, *The Rocket and the Reich: Peenemünde and the Coming of the Ballistic Missile Era* (New York: Free Press, 1995).
[2] Goddard, "Liquid-Propellant Rocket Development," *Smithsonian Miscellaneous Collections* 95 (March 16, 1936), 5-6 & 9-10 in Goddard, *Rockets* (New York: American Rocket Society, 1946).
[3] Pynchon, *Gravity's Rainbow* (New York: Viking Press, 1973).

1642)[4] and later by the English savant Isaac Newton (1643-1727).[5] Newtonian mechanics provided the mathematical basis for computing basic rocket trajectories under ideal conditions. Missiles flying in the real world experienced a myriad of perturbing forces. The same was true of ordnance racing through the atmosphere.

During World War I, the need to predict ballistic trajectories with greater accuracy led the U. S. Army's ordnance branch to produce so-called range tables. They took into account such factors as air pressure, temperature, velocity, and mass. The Army performed these calculations using numerical integration, a mathematical technique that consists of a broad range of algorithms (mathematical solutions) for calculating the position and velocity of a moving object (missile, satellite, or spacecraft) as it passes over a particular location.[6]

Radar. The introduction of radar was a watershed moment in the history of tracking rockets that presaged the end of the domination of optical instrumentation. The development of radar largely by the armed forces plus the high cost of the equipment meant that civilians could not avail themselves of this technological luxury, leaving them dependent on simpler, cheaper optical systems. Radar became available for test ranges largely as a result of World War II, although its advancement as a defensive technology predated that war. The new technology made its mark during the war as an integral and necessary instrument of offensive and defensive warfare. World War II was the first electronic war, and radar was its prime agent.[7]

---

[4] A large body of scholarship has established and elaborated upon Galileo's contributions in this area: Stillman Drake and James MacLachlan, "Galileo's Discovery of the Parabolic Trajectory," *Scientific American* 232,3 (1975): 102-110; Ronald H. Naylor, "An Aspect of Galileo's Study of the Parabolic Trajectory," *Isis* 66 (1975): 394-396; Naylor, "Galileo: The Search for the Parabolic Trajectory," *Annals of Science* 33 (1976): 153-172; Naylor, "Galileo's Theory of Projectile Motion," *Isis* 71 (1980): 550-570; Naylor, "Galileo's Early Experiments on Projectile Trajectories," *Annals of Science* 40 (1983): 391-394; Pierre Thuillier, "De l'art à la science: La découverte de la trajectoire parabolique," *Recherche* 18 (1987): 1082-1089; David K. Hill, "Dissecting Trajectories: Galileo's Early Experiments on Projectile Motion and the Law of Fall," *Isis* 79 (1988): 646-668; most recently, Jim Bennett and Stephen Johnston, *The Geometry of War, 1500-1750: Catalogue of the Exhibition* (Oxford: Museum of the History of Science, 1996).
[5] Schorn, 33-35. For a more complete discussion of Newton and his contributions, see, among others, Richard S. Westfall, *Never at Rest* (New York: Cambridge University Press, 1980); Westfall, *Force in Newton's Physics: The Science of Dynamics in the Seventeenth Century* (New York: American Elsevier, 1971); Westfall, *The Life of Isaac Newton* (New York: Cambridge University Press, 1993); Betty Jo Teeter Dobbs and Margaret C. Jacob, *Newton and the Culture of Newtonianism* (Atlantic Highlands, NJ: Humanities Press, 1995); Dobbs, *The Foundations of Newton's Alchemy: or, "The Hunting of the Greene Lyon"* (New York: Cambridge University Press, 1975); Isaac Newton, *Papers and Letters in Natural Philosophy*, ed. I. Bernard Cohen (Cambridge: Harvard University Press, 1958).
[6] For example, Office of the Chief of Ordnance, *The Method of Numerical Integration in Exterior Ballistics: Ordnance Textbook* (Washington: GPO, 1921).
[7] Charles Süsskind, "Who Invented Radar?" *Endeavour* 9 (1985): 92-96; Henry E. Guerlac, "The Radio Background of Radar," *Journal of the Franklin Institute* 250 (1950): 284-308; Sean S. Swords, *A Technical History of the Beginnings of Radar* (London: Peter Peregrinus Press, 1986), 270-271; Alfred Price, *Instruments of Darkness: The History of Electronic Warfare*, 2d ed. (London: MacDonald and Jane's, 1977); Tony Devereux, *Messenger Gods of Battle, Radio, Radar, Sonar: The Story of Electronics in War* (Washington: Brassey's, 1991); Davie E. Fisher, *A Race on the Edge of Time: Radar—The Decisive Weapon of World War II*

Illustrative of the rise of tracking radar is the case of the National Advisory Committee for Aeronautics (NACA), the civilian institutional predecessor to NASA. Created to address the military concerns of World War I, the NACA collaborated in the military-driven rocket effort of World War II that encompassed all branches of the defense establishment plus the civilian Office of Scientific Research and Development (OSRD). The NACA undertook guided missile research as early as June 1941—in response to a request from the Army Air Forces—and continued until 1946, by which time the NACA had undertaken more than 18 specific guided missile projects for the Army, Navy, and OSRD. After the war, the Army Air Forces again provided the impetus and funding for the NACA to develop its own multiple-stage guided missile, the Tiamat (named for a Babylonian goddess), to study general missile-guidance problems.[8] This defense work put the NACA in the business of tracking launches.

When the NACA began testing the Tiamat on June 27, 1945, at its Wallops Island facility on the Delmarva Peninsula,[9] the range equipment consisted of so-called tracking and Doppler radars. These two radar varieties were standard equipment at contemporary missile test ranges. One provided the missile's position, while the other measured its velocity using that positional information. Position and velocity were and remain the sine qua non for determining a missile's flight path and a spacecraft's trajectory.

The tracking radar was an SCR-584 unit that the NACA's Langley laboratory already had acquired for a different flight program and which its Instrument Research Division had adapted with an optical sight, a computer, and a plot board. Developed by the Army Signal Corps for World War II, the SCR-584 was a mobile unit designed to follow artillery targets automatically in azimuth and elevation, so that searchlights and antiaircraft guns could be aimed at them, a high priority during the London Blitz. In June 1945, Wallops also acquired its first model AN/TPS-5 (commonly known as the "tipsy") Doppler radar to ascertain rocket velocities from their Doppler shift. The AN/TPS-5 had been developed during the war for use in sentry duty, especially to detect movements across bridges. By May 1947, Wallops had four SCR-584 radars, two TPS-5 Doppler radars, and one Sperry Model 10 Velocimeter, an advanced form of Doppler radar.[10]

Using these radars on the test range proved to be tricky. Langley personnel described Doppler radars to be "very temperamental and difficult to keep in operating condition." The search for better radars was ongoing through either procurement or internal innovations.[11] As rocket models proven at Wallops Island flew increasingly faster, higher, and farther, they pushed the NACA radars to the limits of their

---

(New York: McGraw-Hill, 1988).
[8] NACA, *Thirty-Second Annual Report of the National Advisory Committee for Aeronautics* (Washington: NACA, 1946), 36; materials in "Missile 1946-47," B35, Bonney Files; Shortal, 12. Shortal, 12-14 & 49-54, provides a history of Tiamat.
[9] Shortal, 46.
[10] Shortal, 38, 48, 57, 97 & 98.
[11] Minutes of Langley Committee on Guided Missiles, November 9, 1945, August 2, 1946, November 22, 1946, and February 27, 1948, all in M.A.C. 65-2, LaRCA.

capabilities and provided a strong impetus to acquire newer, more powerful equipment from either the armed services or directly from military suppliers.[12]

If radar power could not be boosted satisfactorily, Wallops Island engineers reasoned, a feasible alternative approach would be to develop a device to be mounted on test rockets that would receive, amplify, and retransmit radio signals, thereby overcoming the need for increasingly more powerful radars to provide vehicle velocity readings. Furthermore, they argued, if the device operated at the same frequency as the telemetry system, no special tracking radar would be needed, as the telemetry receivers would function in their stead.[13] This approach became an integral part of JPL tracking techniques.

Tracking at GALCIT. JPL began as a small group of rocket enthusiasts at Caltech under Theodore von Kármán, a professor of aerodynamics, helped along with funding from the recently founded Daniel Guggenheim Fund for the Promotion of Aeronautics. Their initial labors (1936-1938) worked toward the development of rockets for high-altitude scientific research. Nevertheless, from 1939 and the onset of World War II, the Caltech group operated as the Guggenheim Aeronautical Laboratory of the California Institute of Technology (GALCIT) and engaged in rocket research on behalf of the Army Air Forces.[14]

In June 1944, toward the end of the war, GALCIT began the so-called ORDCIT Project for the Army's Ordnance Department on the technical problems associated with long-range missiles. Experiments took place in a nearby arroyo (dry riverbed) ad hoc test range. Later in 1944, GALCIT acquired even more rocket projects, and the procurement of a launch tracking site, equipment, and personnel began in earnest. At the same time, at the invitation of von Kármán, William H. Pickering in Caltech's Electrical Engineering Department started working on the ORDCIT Project.[15]

Soon after joining, Pickering traveled to the East Coast to study radar and optical tracking techniques at MIT's Radiation Laboratory in Massachusetts and the Army's Aberdeen Proving Ground in Maryland. He concluded that GALCIT should develop its own range equipment. Returning in the fall of 1944, Pickering activated the Remote Control Section of what had been known since July 1, 1944, as the Jet Propulsion Laboratory. The section's responsibilities included rocket test instrumentation, telemetry, and tracking. A 1950 JPL reorganization by Director Louis Dunn, a Caltech professor and protégé of von Kármán, restructured Pickering's section into the Guided Missile Electronics division and created the Guided Missile Engineering division under Paul Meeks.[16]

---

[12] Shortal, 38, 292 & 302; Harold D. Wallace, Jr., *Wallops Station and the Creation of an American Space Program* (Washington: NASA, 1997), 16.
[13] Minutes of Langley Group on Guided Missiles, February 27, 1948.
[14] Koppes, 2-17; Malina, "Memoir," 154, 156, 158 & 159-180; Neufeld, 39.
[15] Malina, "Memoir," 159-181; Neufeld, 18, 36 & 42; Pickering, 386; Pickering, interview, 36.
[16] Pickering, 387 & 400; Koppes, 19, 20, 21 & 31-32.

In 1944, JPL had two major rocket development projects underway—the Private and the Corporal—both of which required new test range equipment. The Private requirements were simpler than those of the Corporal, as it lacked guidance systems and radio telemetry. Range personnel deduced its flight path from radar and movies. The Corporal was by far a more sophisticated missile and for several years was the focus of most of JPL research. The initial version of the Corporal, tested in the fall of 1945 at White Sands, New Mexico, was one of the first missiles proven there. Indeed, early maps of the range indicate the area as the ORDCIT extension of Fort Bliss and only later as the White Sands Proving Ground.[17]

Although White Sands had tracking equipment, JPL chose to procure its own and, from mid-1951 onward, stationed a fulltime group of JPL engineers there. Starting with the first Corporal E launch on May 22, 1947, the laboratory deployed its tracking equipment and plotted flight paths from radar position and velocity measurements acquired by tracking an onboard flight transponder. JPL utilized a modified SCR-584 tracking radar (the same model as Wallops Island) plus a continuous-wave Doppler unit. The SCR-584, one of six such tracking radars reengineered by a private company expressly for the Signal Corps, indicated the missile's position and velocity and sent commands to it as well. A change made to the unit in 1950 allowed it to relay rocket position observations to a computer and, in turn, to transmit computer commands to the missile, thereby keeping it on course if necessary. An auxiliary radar tracked the missile in both azimuth and elevation and plotted its trajectory automatically. A Doppler radar measured the Corporal's velocity and sent the data to a computer, which calculated precise velocity and acceleration values for a number of points in the flight path.[18] The practice of tracking launches by monitoring an onboard flight transponder was the same solution that Wallops Island engineers had envisioned.

White Sands. Although JPL relied on its own equipment and personnel for testing Private and Corporal missiles, it lacked its own range and relied on the White Sands facility operated by the Army's Aberdeen Proving Ground. The Aberdeen Ballistic Research Laboratory was in charge of coordinating all tracking and data reduction, functioning as a service organization to all test-range users who had their own launch sites there.[19] The White Sands complex marked a milestone in the evolution of space navigation. For the first time, astronomers were involved in tracking rocket launches. White Sands also featured the application of computers to the determination of missile trajectories. The digital computer later was a commonplace tool for calculating spacecraft flight paths.

Louis A. Delsasso, a physicist from Princeton University, oversaw White Sands tracking. The laboratory had recruited him in 1943 as its chief physicist. Delsasso enlarged his staff at least in part by hiring local talent, including Clyde W. Tombaugh, the astronomer known for discovering Pluto. Tombaugh developed optical tracking systems

---

[17] Pickering, 388; Pickering, interview, 48; Thomas, 113.
[18] Koppes, 45 & 47; Pickering, 392-393 & 401; JPLCBS 15, 29 & 30; JPLCBS 19A, 7; JPLCBS 20A, 7 & 8; JPLCBS 21A, 5.
[19] Thomas, 133.

for the range under the supervision of Dirk Reuyl, an astronomer who led the Ballistic Research Laboratory's optical measurements branch.[20]

In addition to lending his expertise, perhaps Tombaugh's best known range contribution was his recommendation to paint the Viking test rockets flat white in order to facilitate optical observation of them. The unpainted aluminum vehicles were just too glinty. The white paint allowed telescope photographs to show the true shape of the vehicle, thereby improving the tracking process. Other noteworthy astronomers that Delsasso hired included William Bidelman, E. Dorrit Hoffleit, Edwin Hubble, Rebecca Karpov, Kaj Strand, and Richard Zug.[21]

These astronomers were in their element. White Sands employed a variety of optical instruments to follow and record launches. They included high-speed domestic (Mitchell) and foreign cine-theodolites—instruments used to triangulate positions by measuring both horizontal and vertical angles—captured in Germany as well as high-speed Bowen-Knapp fixed-motion cameras and specialized tracking telescopes. Recording on 35-mm film, they followed missiles up to 30 miles and sometimes higher.[22]

Optical equipment was critical at other Army-operated test ranges. At Eglin Field, for example, ballistic cameras followed flights of captured German V-1 rockets launched over the Gulf of Mexico after the end of the war, while Navy dirigibles obtained range information. Triangulating the position of a missile necessitated a pair of synchronized optical instruments. In contrast, one radar station alone could determine an entire trajectory. Radar had a further advantage over optical instruments because of its ability to "see" missiles regardless of lighting conditions. On the other hand, the early V-2 launches during the summer of 1946, which relied on both optical and radar units, saw some advantages of tracking telescopes over radar, which, for example, disclosed that the V-2 tumbled midcourse.[23]

The advantages and disadvantages of optical versus radar equipment were subjects of ongoing debates. The astronomer Dorrit Hoffleit concluded that: "No one system has yet proved 'best' for all the information that is wanted on a missile's entire flight path and behavior (including velocity, deceleration, spin, yaw, etc.)."[24] Frank Malina, one of the founders of JPL, recalled that radar tracking at White Sands "gave a minimum of data to permit the altitude to be calculated. On most rounds it failed completely."[25] Even so, Milt Rosen, who was in charge of developing the Viking rocket from 1947 to 1955, recalled: "We didn't dare fly without radar, and they knew it."[26]

---

[20] DeVorkin, 112 & 129; Newell, 414; "Electronic Computers within the Ordnance Corps: Chapter IV: ORDVAC," http://ftp.arl.mil/~mike/comphist/61ordnance/chap4.html (accessed September 11, 2007).
[21] Rosen, 154; Kaj Strand, interview, 49 & 50; DeVorkin, 112.
[22] A theodolite is an instrument used in triangulation work to measure both horizontal and vertical angles. Cine-theodolites were motion-picture instruments that picked up and followed launches until the booster reached the limit of visibility. DeVorkin, 111; Thomas, 121; Green and Lomask, 99; Hoffleit, "DOVAP," 172.
[23] Strand, interview, 48-51; Hoffleit, "DOVAP," 172; Thomas, 114.
[24] Hoffleit, "DOVAP," 172.
[25] Frank J. Malina, "America's First Long-Range Missile and Space Exploration Program: The ORDCIT Project of the Jet Propulsion Laboratory (1943-1946): A Memoir," 340 in Hall, Essays.
[26] Rosen, 201.

Analyzing missile trajectories—and spacecraft flight paths—required calculating their probable course in advance of launch and, with the aid of tracking data, computing their actual trajectories as accurately as possible. The Ballistic Research Laboratory computed missile trajectories before launches and was the Army Ordnance's main calculator and predictor of trajectories. For the reduction of optical data, Delsasso contracted with the New Mexico College of Agriculture and Mechanical Arts (today New Mexico State University).[27] The analysis of radar data took place both immediately at White Sands and subsequently on automated processing equipment.

The earliest attempt to analyze radar readings with such machinery involved a new relay multiplier developed by IBM. The estimated time to process 800 trajectory points was about 4 weeks, assuming that the machine ran 40 hours per week and including machine down time. Two relay multipliers wired in tandem cut processing time in half, with each trajectory point taking 5 to 8 minutes.[28]

Another relay machine, designed by Bell Telephone Laboratories' George Stibitz, was not as fast, but it did process trajectory point data in 5 minutes without any major preparatory setup time. The advantage of the Bell multiplier was its ability to run unattended all night. As a result, it could analyze a trajectory of 800 points in only 3 days. The machine's efficiency, though, was not up to that of the Ballistic Research Laboratory's ENIAC (Electronic Numerator Integrator Analyzer and Computer).[29] ENIAC marked the dawn of data reduction by computer at White Sands.

Among the first data that ENIAC tackled were measurements from the DOVAP (Doppler Velocity and Position)[30] radars at White Sands, some of which JPL had procured from the Army's Aberdeen Proving Grounds. ENIAC quickly won great acclaim for its speed in resolving trajectory points. The computer calculated the coordinates of a missile in a time comparable to the missile's flight time, so that in 10 minutes it could do a 10-minute trajectory consisting of data taken every half second. On the other hand, preparing ENIAC for these computations took about two days. Moreover, with all the demands for ENIAC time, obtaining a full reduction of flight data might take several weeks.[31]

---

[27] DeVorkin, 111-112; Strand, interview, 49 & 50.
[28] Hoffleit, "DOVAP," 175. Relay multipliers were so named because they used electromechanical relays. The IBM fast relay multiplier was able to divide faster than any other machine available at the time. Aiken, interview, 25.
[29] Hoffleit, "DOVAP," 175; Franz L. Alt, "Operating Characteristics of the Aberdeen Machines," December 29, 1947, F13.2, MC 529, Hoffleit Papers.
[30] JPLCBS 15, 30; JPLCBS 16, 23; JPLCBS 17, 24. DeVorkin, 113, 130 & 307-308, discusses the DOVAP.
[31] Hoffleit, "DOVAP," 175; Pickering, 389 & 391; DeVorkin, 69, 110, 111, 113 & 129; Koppes, 23-24. Although the ENIAC was designed specifically to calculate trajectories, the literature treats it as the first large-scale digital electronic general-purpose computer: John G. Brainerd, "Genesis of the ENIAC," *Technology and Culture* 17 (1976): 482-488; Arthur W. Burks and Alice R. Burks, "The ENIAC: First General-Purpose Electronic Computer," *IEEE Annals of the History of Computing* 3 (1981): 310-389; W. Barkley Fritz, "ENIAC: A Problem Solver," *IEEE Annals of the History of Computing* 16 (1994): 25-45; Scott McCartney, *ENIAC: The Triumphs and Tragedies of the World's First Computer* (New York: Walker and Co., 1999); Thomas Parke Hughes, "ENIAC: Invention of a Computer," *Technikgeschichte* 42 (1975): 148-165. On the role of female technicians in the development of ENIAC, see Jennifer S. Light, "When Computers were Women," *Technology and Culture* 40 (1999): 455-483; W. Barkley Fritz, "The Women of ENIAC," *IEEE Annals of the History of Computing* 18 (1996): 13-28.

Counting Doppler. Quicker results were available at White Sands by analyzing tracking data in a more labor-intensive manner. The nature of the recorded observations closely resembled those of space navigators in that they were Doppler measurements with an associated time tag. Milt Rosen described the route taken by data collected from the "dozens of radio, radar, and optical tracking stations . . . scattered over the desert and perched on the surrounding mountains." The readings flowed to the "ballistic laboratory, a huge block of concrete without windows," which was "headquarters for the hundred radio, radar and optical tracking stations that dot the range." The building's basement housed "the fabulous electronic computers" that "print out the long columns of figures that record point by point the motions of every rocket fired at White Sands."[32]

During rocket tests, the blockhouse sent to each tracking station "a continuous series of timing signals, the close-in stations by wire, the more distant ones by radio," which originated from the Naval Observatory in Washington, DC. The time was imprinted on the same film, frame by frame, that recorded the position of the rocket. Subsequently, each data record was synchronized with every other one in order to facilitate the comparison of data obtained by radio and optical instruments.[33]

The creation of a mathematical means for determining the position of a missile at a given point in its trajectory via Doppler alone was the work of Dorrit Hoffleit. Better known for her Bright Star Catalog, Hoffleit already held a Ph.D. in astronomy from Radcliffe College, when she joined the Ballistic Research Laboratory in 1943. After the war, Hoffleit put off going back to Harvard and remained at Aberdeen for three more years, computing the trajectories of captured V-2 rockets being tested at White Sands. She returned to Harvard in 1948, but continued as a laboratory consultant until 1961.[34]

During World War II, Hoffleit supervised as many as 16 Women's Army Corps (WAC) computers in the Aircraft Fire Section. The WAC computers were people, not equipment, who performed calculations. All computers held ranks from private to corporal. Traditionally, college-educated women were observatory computers, the most noted example of which was Pickering's Harem at the Harvard College Observatory named after Edward Charles Pickering, the observatory's director from 1877 to 1919.[35]

---

[32] Rosen, 43 & 226.
[33] Rosen, 45-46 & 226.
[34] Harvard University Library, "Hoffleit, Dorrit. Papers, 1906-2005: A Finding Aid," http://oasis.lib.harvard.edu/oasis/deliver/~sch00362 (accessed October 29, 2007).
[35] Barbara L. Welther, "Pickering's Harem," *Isis* 73 (1982): 94, provides a photograph of Pickering and his computers along with their names. There is a rich literature on women in astronomy. See, for example, the extensive bibliography posted at: University of Toronto, Department of Astronomy and Astrophysics, Library, "Women in Astronomy," http://www.astro.utoronto.ca/AALibrary/womenbib.html; Astronomical Society of the Pacific, "Women in Astronomy: An Introductory Resource Guide," http://www.astrosociety.org/education/resources/womenast_bib.html, among others. A far less explored area is that of women as computers in astronomy. The best treatment to date is John Lankford, *American Astronomy: Community, Careers, and Power, 1859-1940* (Chicago: University of Chicago Press, 1997), 287-359, while M. T. Brück, "Lady Computers at Greenwich in the early 1890s," *Quarterly Journal of the Royal*

By 1949, Hoffleit's expertise in the field was such that the Naval Ordnance Test Station (NOTS) at China Lake, California, hired her to work for its new Ballistics Laboratory because of her "considerable experience with Doppler radar measurement systems."[36]

In that same year, Hoffleit described how the Ballistic Research Laboratory determined missile positions from DOVAP radar information. White Sands had one DOVAP radar system and four receivers arranged roughly at the corners of a diamond (see Figure). The radar transmitter and one of the receivers (B) were close together and about two miles south of the launch pad. The distance to each of the other receivers (F, K, G) was about 14 miles. Although the data from all four stations was not necessary to calculate rocket velocities, the additional receivers provided a safety net in case one of the stations failed and helped to evaluate the accuracy of the results obtained.[37]

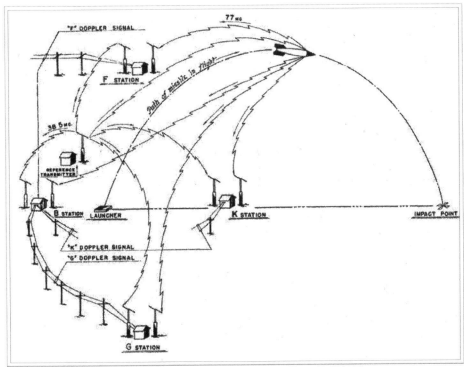

A Typical DOVAP Field Setup (Hoffleit, DOVAP)

The signals acquired at the four stations traveled to a single building via ground wire where they were recorded simultaneously side by side on 35-mm movie film

---

Astronomical Society 36 (1995): 83-95, recounts the short-lived (1890-1895) experiment of employing women on the staff of the Greenwich Royal Observatory.
[36] Hoffleit, informal memorandum to Lt. Marks, March 14, 1944, F13.3, Hoffleit Papers; "Description of Duties for E. Dorrit Hoffleit," [March 1949], F13.2, Hoffleit Papers.
[37] Rosen, 46; Hoffleit, "DOVAP," 173.

together with a time tag. Radar headquarters was a few miles south of the blockhouse. It contained the plotting room where, as Milt Rosen recalled, one found "a long row of automatic plotting tables which trace out continuously the rocket's position in flight. As the rocket speeds away from the launching site each pen moves forward, controlled by radars or theodolites, and draws the rocket's track on a map of the proving ground."[38]

By continuously recording the missile's speed, each ground station measured its distance; three or more stations in unison could fix its position. The DOVAP radar sent signals simultaneously to the ground stations and to the missile. The missile's transceiver received the signals, doubled their frequency, and sent the doubled frequency to the ground receiving stations. Those stations, in turn, "mixed" the doubled missile signals and the signals from the DOVAP transmitter. The result was recorded as Doppler waves along a time scale. Basically, a count of the number of recorded cycles from launch to any moment in time (and the signal's wavelength) indicated the distance flown by the rocket.[39] JPL space navigation similarly used signals modulated by a spacecraft transponder, but Doppler processing took place electronically and digitally.

Turning the cycle count into a measure of the distance that the missile flew started with the number of cycles (N) and the Doppler wavelength (λ). One derived the wavelength from the frequency (f) using a long-standing formula, namely, $\lambda=c/2f$, where c is a known constant, namely the velocity of light and radio waves. Ground surveys provided the distances from the DOVAP transmitter to the launcher (TL) and from the launcher to the receiver (LR). The distance from the transmitter to the missile to a given receiver was equal to $TL + LR_1 + N_1\lambda$ (the number of counted cycles at that station times the Doppler wavelength). Calculations of the distance from the transmitter to the missile to a second and third receiver were made in the same fashion, that is, $TL + LR_2 + N_2\lambda$ for the second receiver and $TL + LR_3 + N_3\lambda$ for the third receiver.[40]

The missile's course, Hoffleit concluded, described a prolate spheroid whose foci were the transmitter and the receiver. A prolate spheroid is one in which the polar diameter is longer than the equatorial diameter, resembling the shape of a rugby football, but not as pointed at the ends as an American or Canadian football. Looking at the mathematical results derived from all three receivers, Hoffleit pointed out, "we know that the missile is somewhere on an ellipse that is the intersection of two prolate spheroids having one focus, the transmitter, in common."[41]

Hoffleit also observed an important discrepancy between what mathematics predicted and the Viking rockets' actual flight paths. In principle, because the DOVAP transmitter frequency (around 38.5 MHz) corresponded to a wavelength of approximately 12.5 feet (about 3.8 meters), and the time signals were accurate to better than 1 part in 100,000, the distances computed from the counts of recorded cycles should be accurate to within a foot (0.3 meters). In reality, though, "most of the records suffer[ed] blemishes" created by "peculiarities in the missile behavior, interference of radio waves, and atmospheric effects," Hoffleit reported. Typical errors amounted to

---

[38] Hoffleit, "DOVAP," 173; Rosen, 46.
[39] Rosen, 47; Hoffleit, "DOVAP," 172.
[40] Hoffleit, "DOVAP," 172-173.
[41] Hoffleit, "DOVAP," 172-173.

around 50 feet (about 15 meters), but were much larger at higher altitudes. Still, she concluded, these errors aside, the results obtained surpassed the accuracy of those derived "from any of the other present types of position determination."[42]

These calculations did not represent the complete reduction of DOVAP data for any one launch. The process often required several weeks of reading film, interpreting data, and computing. As Hoffleit recalled, "the determination of missile coordinates requires some forty arithmetic operations for each point on the trajectory. A skilled computer using desk machines (Friden, Monroe, or Marchant) requires 15-45 minutes per point. A complete trajectory at half-second intervals would therefore require about 400 hours, or 10 weeks of working time."[43]

A Game with many players. As the 1950s came to a close, the number and variety of organizations involved in tracking launches and computing flight paths was considerable. The chief players were the armed forces, not surprisingly given their long-standing dominance of rocketry, with civilian scientists participating in military efforts since at least World War II. Three conferences held in 1959 and 1960 illustrate the gamut of agencies and scientists involved in tracking missiles and analyzing tracking data.

The first, hosted by the American Mathematical Society, was a conference on "Orbit Determination" held in New York City on April 4-6, 1959. The organizers, Garrett Birkhoff and Rudolph E. Langer, hoped to attract the attention of mathematicians to recent advances in the field highlighted by "the recent dramatic launching of man-made satellites and the currently accelerated developments of long-range ballistic missiles."[44] They interpreted the term "orbit" rather loosely, so that their meeting included papers on the motion of cosmic-ray particles and Störmer orbits. Virtually all of those discussing the determination of spacecraft orbits were astronomers, including such well-known figures as Fred Whipple and Dirk Brouwer. A notable exception was Krafft A. Ehricke, a rocket-propulsion engineer working for General Dynamics' Convair Division who formerly had been at Peenemünde from 1942 to 1945. He provided an analysis of lunar trajectories.[45]

NASA Headquarters, another civilian setting, hosted a second conference that took place on March 12-14, 1959. It dealt specifically with the problems of space trajectories and their computation. The agency invited some 238 guests, the largest portion of whom were NASA employees (at least 32 or 13.45%).[46] Of the rest, the largest contingent came from the Army Ballistic Missile Agency followed by Holloman Air Force Base. Located on the edge of the White Sands testing range and just outside the town of Alamogordo, Holloman was that service's main site for testing and developing guided missiles and other weaponry.[47] The other agency with a substantial

---

[42] Hoffleit, "DOVAP," 173.
[43] Hoffleit, "DOVAP," 174 & 175.
[44] Garrett Birkhoff and Rudolph E. Langer, "Introduction," v, in *Orbit Theory*.
[45] Ehricke, "Cislunar Orbits," 48-74 in *Orbit Theory*.
[46] "Conference on Orbit and Space Trajectory Determination 12-14 March 1959," [list of invitees], F269/B30, Vanguard.
[47] Holloman Air Force Base, "Holloman Air Force Base History,"

presence was the Smithsonian Astrophysical Observatory (SAO).[48] SAO astronomers were involved in devising optical tracking equipment and observing networks.

The list of attendees equally provides a snapshot of the broad range of military agencies collecting tracking data and calculating trajectories. They included the Naval Ordnance Test Station at China Lake; the Naval Proving Ground, Dahlgren, Virginia; the Army Signal Corps' Signal Engineering Laboratories, Ft. Monmouth, New Jersey; the Ballistic Research Laboratory at Aberdeen; the Army Map Service; the Air Force Cambridge Research Center, Bedford, Massachusetts; and the Wright Air Development Center, Dayton, Ohio. The attendees also represented several civilian laboratories that conducted a substantial amount of military research: MIT's Lincoln Laboratory; Johns Hopkins University's Applied Research Laboratory (APL); the University of California's Radiation Laboratory, Livermore, California; and the RAND Corporation.[49]

An analysis of those who read papers[50] at the NASA conference brings to light the collaborative role between the armed forces and industry. Employees of NASA, the host institution, contributed the largest number of papers followed by speakers from the Army Ballistic Missile Agency and Space Technology Laboratories (STL), a TRW subsidiary. A smaller number of papers were read by delegates from the SAO, Lincoln Laboratory, the Naval Ordnance Test Station, the National Bureau of Standards Central Radio Propagation Laboratory (Boulder, Colorado), and the Lockheed Aircraft Corporation (Palo Alto, California). The remaining speakers came from the armed forces, defense research laboratories (APL and the Livermore Radiation Laboratory); the U.S. Naval Observatory; Convair Astronautics (San Diego, California), the prime contractor for the Atlas missile; the Cincinnati Observatory; and the Jet Propulsion Laboratory, which recently had joined NASA.

JPL made up for its scant participation in the NASA Headquarters conference by holding its own "seminar" on February 23-26, 1960. It was an opportunity to bring in individuals and organizations that were on or near the West Coast. An analysis of the speakers' institutional homes shows a greater participation by West Coast defense firms and JPL.[51] The event took place in two parts: one on tracking, the other on orbit determination. The organizers credited the conference's success to the session chairs whose institutional bases were NASA, JPL, STL, UCLA, the Army Ballistic Missile Agency, and Lockheed.[52]

The majority of paper authors were associated with Ford's Aeronutronic Systems, JPL, the Army Ballistic Missile Agency, and STL. NASA's Goddard Space Flight Center and the Lockheed Missile Systems Division (Sunnyvale, CA) accounted for somewhat fewer talks. Additional presenters worked for the General Electric Company,

---

http://www.holloman.af.mil/library/factsheets/factsheet.asp?id=4361 (accessed January 31, 2008).
[48] "NASA Symposium," February 26, 1959, [list of attendees by institution] F269/B30, Vanguard.
[49] Ibid.
[50] This analysis used the list of paper authors from the program, found in F149/B15, Vanguard, and matched them with the stenographic transcription of the conference, "Conference on Orbit and Space Trajectory Determination," F150/B15, Vanguard.
[51] Information given in the Table of Contents, Appendix A, and Appendix D of *Tracking and OD*, v-vii, 205-206 & 215.
[52] Lorell and Yagi, "Forward," iii & iv in *Tracking and OD*.

North American Aviation, the SAO, IBM, RAND, and the Air Force Cambridge Research Center. The participants generally did not deal with lunar or planetary tracking or navigation. Two notable exceptions were Carl Tross, Aeronutronic Systems, who discussed guiding a spacecraft to the Moon, and the famed Victor G. Szebehely, then with General Electric's Missiles and Space Vehicle Department in Philadelphia. The latter talked about the firm's work on lunar and planetary trajectories.[53]

Aeronutronic, General Electric, Lockheed, Convair, and RAND, to name just a few, were interested in orbital mechanics because their clients, the armed forces, were paying for it. The goal no longer was solely to test missiles or to put a payload into a terrestrial orbit. The second half of the 1950s saw the mushrooming of military space programs that reached out to the Moon and beyond[54] even before the Sputniks. Moreover, despite the formation of NASA from a mélange of civilian and military facilities and projects, the armed forces, especially the Army and Air Force, continued to propose and launch a multitude of Earth satellites to undertake a myriad of tasks from weather to communications, from positioning to surveillance. As late as 1963, Sen. Barry M. Goldwater (R-AZ), a former Air Force pilot who soon would run as a Republican presidential candidate, loudly agreed with the military's astronaut program on grounds of national security.[55]

---

[53] Information from Table of Contents, v-vii; Addendum A, "Seminar Attendees," 205-206; Addendum C, "Seminar Papers not Published in Proceedings," 208, all in *Tracking and OD*.
[54] See, for example, Lee Bowen, *The Threshold of Space: The Air Force in the National Space Program, 1945-1959* (Washington: AF Historical Division, September 1960), esp 6-7 & 49.
[55] Hall, *Lunar Impact*, 217.

Poster for NASA Headquarters conference, March 12-14, 1959
(Siry Papers, B30/F269A)

Chapter 2

Applied Celestial Mechanics

The leap from test range to Earth-orbiting satellites and planetary spacecraft presented a number of new problems that called for a greater reliance on the expertise of astronomers. Long before rockets went to the Moon, astronomers had been computing the orbits and determining the location of celestial bodies, be they planet, moon, asteroid, or comet. The branch of astronomy known as celestial mechanics quickly became central to the leap from proving grounds to launching pad.

The adaptation of celestial mechanics to space navigation followed two independent, but not entirely unrelated, courses. One path, built on a long tradition of cometary research, led to the creation of a new branch of astronomy, known as astrodynamics, concerned with the study of the motion of rockets and vehicles in space. The other, founded on centuries of practical and theoretical adaptation of celestial mechanics to oceanic navigation, provided the mathematical and scientific basis for guiding probes in Earth orbit. This small community of astronomers also pioneered the automation of astronomical calculations. As a result, when the United States began sending craft into space, astronomers possessed the theoretical and mathematical knowledge and techniques requisite for designing and determining trajectories. They also had mastered the application of computing machinery to these tasks.

Modern cometary research. Astronomers and others believed that calculating the orbits of comets was very much akin to what navigators might do one day to determine the flight path of a spacecraft. Illustrative of this belief is a passage written by Arthur C. Clarke and published in 1951: "Once his position *and velocity* had been accurately determined—which would be done as soon as possible after leaving Earth—the navigator could calculate his future position at any time, since nothing could alter the predetermined path of the ship, save the firing of the motors or the approach to another planet. This type of calculation is, of course, exactly what an astronomer does to determine the orbit of a comet, but at present it requires a great many man-hours of work."[1]

Much like the trajectories of space probes, the orbits of comets frequently undergo changes that can be small or large making it difficult to establish their orbital parameters with any exactitude. Gravitational influences—such as from Jupiter or Saturn—commonly perturb cometary orbits just as they alter spacecraft flight paths. The analysis of these perturbations is at the heart of space navigation and vital to the determination of cometary orbits. A key difference between how astronomers calculate the orbits of comets and how navigators estimate those of spacecraft is the kind of data

---

[1] Clarke, *Interplanetary Flight*, 124 (italics in original). In 1958, the aerospace engineer Krafft Ehricke used the term "instrumented comets" to emphasize the "interplanetary and solar research mission" of spacecraft. Ehricke, "Instrumented Comets—Astronautics of Solar and Planetary Probes," 76, in Friedrich Hecht, *Proceedings of the VIIIth International Astronautical Congress* (Vienna: Springer Verlag, 1958).

used. Astronomers traditionally have relied on optical observations, while space navigators use radiometric readings (range and Doppler).

Between the two world wars, the study of comets (along with asteroids and meteors) received fresh attention. Astronomers in the United States studied cometary and planetary orbits to uncover clues as to their origin and the origin of the solar system itself. It was not clear, for example, whether comets were native to the solar system or came from beyond. In the face of the rise of astrophysics, these questions maintained the viability of celestial mechanics as a field of research. Smaller observatories lacking the capability to undertake work in the upcoming field of astrophysics, such as the one at the University of Cincinnati, looked to celestial mechanics as a way for investigators to engage in research being conducted by bigger observatories. Among the larger institutions studying cometary orbits was the Nautical Almanac Office of the U. S. Naval Observatory. Its *American Ephemeris and Nautical Almanac* rested on basic work regarding solar, lunar, and planetary motions.[2]

Another major U.S. cometary research center was the University of California at Berkeley, the home of astronomy professors Armin O. Leuschner (1868-1953) and Russell T. Crawford (1876-1958). Leuschner's doctorate at the University of Berlin had been on cometary orbits, and Crawford was a co-author with Leuschner of an important 1930 work on determining the orbits of comets and asteroids. Leuschner, who "single-handedly founded" the Berkeley astronomy department, insisted that all graduate students—even those preparing for careers in astrophysics—learn orbit computation methods and mathematical analysis. According to one former student, Paul Herget, "It was reputed that only one student in the department had ever succeeded in graduating without having computed at least one orbit. That student was said to be Frederick Leonard, and as a meteoriticist it would have stood him in good stead."[3]

Historian Ronald Doel characterizes Leuschner as a skilled scientific entrepreneur who succeeded in obtaining outside support for research programs that greatly exceeded what his school and its observatory (founded by Leuschner) could supply. His entrepreneurship was particularly impressive in light of his raising these sums in the midst of the Great Depression. The Berkeley astronomy department, aided by its close ties to the Lick Observatory (thanks largely to Leuschner), graduated an average of three to six new Ph.D.s per year, more than the combined departments of the University of Chicago and Princeton University.[4]

One of those graduates was Samuel Herrick. He turned the celestial mechanics of comets into a new field (astrodynamics), taught the first course on space navigation, and provided JPL with a number of navigators from the ranks of his graduate students. The calculation of cometary orbits was at the heart of his introduction to astronomy and the well on which he drew for rocket navigation techniques. The inspiration for his

---

[2] Doel, 16 & 123-125.
[3] Herget, "Leuschner," 131-132, quote 135; Doel, 19; Makemson, 503; Russell T. Crawford, Armin O. Leuschner, and Gerald Merton, *Determination of Orbits of Comets and Asteroids* (New York: McGraw-Hill 1930); Crawford, "Elements and Ephemeris of Comet a 1905 (Giacobini)," *Lick Observatory Bulletin no. 73* (1905). Several other cometary orbital elements and ephemerides also appeared in the Lick Observatory's *Bulletin*.
[4] Doel, 17, 19 & 21; Herget, "Leuschner," 131.

interest in space travel, however, was his correspondence with rocket pioneer Robert Goddard. Herrick began writing to Goddard in 1931,[5] when he was a year away from graduating from Williams College in Williamstown, Massachusetts.

As Herrick told Goddard, he already was thinking about "rocket navigation." He pondered "whether or not I am to devote my life, as I wish to, to the development of rockets and rocket navigation, and if so, to find how I may best accomplish that end." Herrick had been interested in space for some time, "because of its romantic appeal to a boyish imagination." "Now, however," he wrote Goddard, "my interest, having grown the stronger with increasing knowledge and years of study, is (I believe) an intelligent and practical one."[6]

His interest in "interplanetary flights" and space transportation led Herrick to ask Goddard repeatedly for a job working beside him in New Mexico. He gave the name of his astronomy professor as a reference who would "vouch that I am a serious and intelligent student." Goddard replied: "The scientific development of rockets is so new that there are, so far as I know, no courses given anywhere on the subject." He suggested mechanical engineering courses at MIT and Caltech as "the next best thing," but held little promise of work for Herrick. Goddard's rocket work was "of a very special sort" that required "special training." Throughout 1931 and 1932 Goddard encouraged Herrick, although their planned encounters never took place. Writing three decades later, Herrick reflected that the correspondence "did start me on a research career that developed in me a very great love for celestial mechanics for its own sake as well as for its applications to rocket vehicles in space."[7]

After graduating from Williams College, Herrick entered graduate school at Berkeley, where he did his doctoral thesis in astronomy under Leuschner and Crawford. Given the department's predilection for cometary research, it is not surprising that his dissertation analyzed various methods for determining their orbits. Primarily, he compared the so-called Laplacian and Gaussian methods and their usefulness in determining a comet's orbit using only three observations. Leuschner rejected the existing method of computing orbits, because he believed it to be too cumbersome and time consuming. He developed instead a statistical approach that used general tables of perturbations.[8] Earlier, in 1914, Leuschner had developed similar "short methods" for calculating an orbit from three observations upon which Crawford elaborated.[9] Herrick's dissertation showed the advantages of Leuschner's method, not surprisingly,

---

[5] This general statement appears in such Herrick biographical items as: Sarterios S. Dallas and Michael S. W. Keesey, "In Memoriam: Samuel Herrick (1911–1974)," *Journal of Celestial Mechanics and Dynamical Astronomy* 10 (August 1974): 2-3; J. L. Junkins, "Samuel Herrick, 1911-1974," *Acta Astronautica* 3 (November-December 1976): 53-54.

[6] Samuel Herrick to Robert Goddard, July 9, 1931, B2-1-4, Goddard.

[7] Herrick to Goddard, July 9, 1931; Goddard to Herrick, July 13, 1931; Goddard to Herrick, September 5, 1931; Goddard to Herrick, March 4, 1932; Goddard to Herrick, June 13, 1932; Herrick to Howard B. Jefferson, President, Clark University, November 28, 1966, all in B2-1-4, Goddard.

[8] Makemson, 503; Doel, 19.

[9] Leuschner, "A Short Method of Determining Orbits from Three Observations," *Publications of the Lick Observatory* 7 (1914): 1-20; 7 (1914): 217-376; 7 (1914): 455-483. See also Crawford, "Application of Leuschner's Method of Direct Solution of Orbits of Disturbed Bodies," *Publications of the Lick Observatory* 7 (1914): 487-503.

and as part of his thesis he performed a series of orbit calculations based on three observations of Minor Planet 1909 HC (683) made by Herbert C. Wilson at the Lick Observatory.[10]

"Rocket navigation." In 1937, Herrick became a professor of astronomy and engineering at UCLA. Instruction in astronomy at that school had begun in 1922 with the appointment of Frederick C. Leonard—also a graduate of Berkeley—as instructor of astronomy in the Department of Mathematics. A separate Department of Astronomy did not exist until 1931, and Herrick was the new department's first addition. He taught a course called "Determination of Orbits," and in 1942 established a new course—the first of its kind—called "Rocket Navigation."[11]

As taught in 1947, both courses owed much to Herrick's and his professors' research on cometary orbits. The "Determination of Orbits" course used Leuschner's method and in part recapitulated the work Herrick had carried out in his dissertation on alternative orbit-computation techniques. In fact, students received a mimeograph of equations taken from the dissertation that they applied to the determination of original orbital calculations.[12]

The 1947 "Rocket Navigation" course featured the same mimeographed equations. It began with such basic celestial mechanics topics as Kepler's three laws, the equations of motion, coordinate systems, and the solution of the two-body and three-body problems. Other subjects covered included geocentric orbits and the determination of trip trajectories to the Moon, Mars, and Venus. Herrick spent several classes on issues related specifically to space navigation. For example, one dealt with the "Rectification of orbits," that is, how to get a probe back to its desired flight path. Several lectures considered how to determine a spaceship's actual flight path from its position and velocity. Here were clear examples of Herrick applying celestial mechanics to the practical problems of guiding spaceships to the Moon and beyond well before any space program was underway to achieve those goals.[13]

---

[10] R. K. Young and Herbert C. Wilson, "Positions of Minor Planets Determined from Photographs Taken with the Crocker Telescope," *Lick Observatory Bulletin* 6 (1910): 89-90; Herrick, *The Laplacian and Gaussian Orbit Methods* (Berkeley: University of California Press, 1940), esp. 1-2 & 55-56, where Herrick acknowledged "his debt" to professor Leuschner, "especially for advice and criticism and for a final painstaking review of the text," and recognized both Leuschner and Crawford "for the inspiration of their teaching and counsel."

[11] In 1961, celestial mechanics and space navigation were transferred to the College of Engineering. The M.A. degree in astronomy was offered first in 1953; the Ph.D., in 1963. University of California Digital Archives, "Los Angeles: Departments and Programs," http://sunsite.berkeley.edu/UCHistory/general_history/campuses/ucla/departments_a.html (accessed October 1, 2007); Aller, Barnes, and Abell, 58-59.

[12] Letter, Gibson Reaves to Spencer R. Weart, June 18, 2002, in B1, Gibson Reaves; Reaves notes and materials from Herrick's course "Determination of Orbits" in F1, B1, Gibson Reaves. It appears that Herrick's students did not receive these mimeographs during the 1960s. Dallas, interview, 3.

[13] Reaves' materials from Herrick's course "Rocket Navigation" F2, B1, Gibson Reaves. In contrast, the "Orbit Course" that Herget taught between 1941 and 1948 at the University of Cincinnati contained no such problems; rather, it focused specifically on the orbits of solar-system objects. Course materials in "Orbit Course 1940," D2, Herget Papers.

Herrick found a number of means at his disposal to diffuse his techniques of "rocket navigation." For example, he explained how one might navigate a spaceship to Venus in a University of California radio program, "A Trip to the Moon," broadcast in 1946 over CBS stations in California. And TIME magazine in October 1946 reported that "red-haired Professor (of astronomy) Samuel Herrick, 35, of the University of California at Los Angeles" was "teaching an eight-man, two-girl class the delicate art of interplanetary gliding" to Venus and Mars.[14]

In 1945, and again in 1952, Herrick won a fellowship from the John Simon Guggenheim Memorial Foundation, an unusual occurrence. He used the time away from teaching to develop his theories about rocket navigation and cometary orbits. The result was a book of tables for calculating both rocket and cometary trajectories. With funding from the Office of Naval Research, the National Bureau of Standards' Institute for Numerical Analysis compiled the actual tables and published them at government expense for all to use. The tables allowed the user to determine the position and velocity of a rocket or comet at any given time. At the heart of the table was Herrick's theoretical work that sought to unify the calculation of rocket and cometary trajectories.[15]

Herrick also resorted to more institutional measures to spread astronomy-based orbit determination by founding the Institute of Navigation in 1946 with Captain Philip Van Horn Weems (1889-1979).[16] Weems, a graduate of the Naval Academy, devised his eponymous system of simplified celestial navigation in the 1920s and wrote a book, published in 1927, that became a standard text in navigation courses. He later taught navigation at Annapolis and the Navy's postgraduate school. In 1961, as a NASA consultant, he developed a simplified mathematical navigational formula to assist astronauts in determining their position in space within seconds by visual sighting.[17]

Automating astronomy. Alongside celestial mechanics, high-speed digital computers would become essential to space navigation. As early as 1951, Arthur C. Clarke explained how such machines would aid in guiding spacecraft. Navigational calculations, he wrote, would take place "in a few minutes or seconds by specialized calculating apparatus of the type now being developed on a very large scale; it is quite possible that the spaceship would merely take the observations while the 'fix' and resulting orbit would be calculated by ground stations with which it was in radio contact. In this way much more elaborate and powerful computing equipment could be used than could ever be carried

---

[14] University of California Public Information Radio Service, "A Trip to the Moon," November 24, 1946, in F2, B1, Gibson Reaves; "Gliding, Gliding," TIME, 48, October 21, 1946, 60-61.
[15] John Simon Guggenheim Foundation, "Fellows whose last names begin with H," http://www.gf.org/hfellow.html (accessed October 29, 2007); Herrick, Tables, esp. v and xxii.
[16] Aller, Barnes, and Abell, 58-59.
[17] His company, Weems Systems of Navigation, based in Annapolis, later operated as Weems and Platt, Inc., developed and distributed navigational literature and equipment. Various items in F2534, NHRC.

in a spaceship. (Even the observations might be done from Earth, as the ship could easily be followed in large telescopes during its initial few days of flight)."[18]

The marriage of computational automation and celestial mechanics was a necessity for future space navigation. The Naval Observatory astronomers charged with preparing tables (known as ephemerides) had embarked on a course of automating their calculations starting in the 1930s as part and parcel of their preparation of navigational tables. These astronomers combined celestial mechanics and automated computing with the mathematical technique of numerical integration. This technique later became the method of choice for determining spacecraft orbits, and JPL navigators adopted it for creating their own tables of planetary motions.

The preparation of Nautical Almanac Office products at the Naval Observatory underwent a radical transformation during the twentieth century first with the adaptation of punched-card business technology and second through the application of high-speed computers to astronomy. The British pioneer behind this transformation was the head of the British Nautical Almanac, Leslie J. Comrie (1893-1950). He was the leading exponent of punched-card machines in astronomy and demonstrated their value by employing them to overhaul the Almanac Office. In addition to introducing punched-card technology, Comrie hired clerical workers to operate standard business calculating machines to produce a scientific tool (the almanac), rather than a commercial product.[19]

The pioneer of punched-card astronomy in the United States was Wallace J. Eckert (1902-1971), who had a doctorate in astronomy from Yale. While teaching celestial mechanics at Columbia University, Eckert became familiar with Comrie's work and learned about IBM's recently announced new series of business accounting machines. He decided that the time was ripe for setting up a scientific laboratory based on IBM equipment adapted specifically to scientific needs. So, in 1933, he founded the Thomas J. Watson Astronomical Computing Bureau operated jointly by Columbia University, the American Astronomical Society, and IBM.[20]

This unique facility enabled faculty and students to solve the differential equations of planetary motion via numerical integration. The computations yielded a number of highly accurate orbits extending over long periods of time. The center undertook two initial large projects in cooperation with Yale University, specifically Dirk

---

[18] Clarke, *Interplanetary Flight*, 124-125.
[19] Comrie, "The Application of the Hollerith Tabulating Machine to Brown's Tables of the Moon," *Monthly Notices of the Royal Astronomical Society* 92 (1932): 694-707; H. S. W. Massey, "Leslie John Comrie, 1893-1950," *Obituary Notices of the Royal Society* 8 (November 1951): 97-105; Mary J. Croarken, "L. J. Comrie and the Origins of the Scientific Computing Service," *IEEE Annals of the History of Computing* 21 (1999): 70-71; Croarken, "L. J. Comrie: A Forgotten Figure in the History of Numerical Calculation," *Mathematics Today* 36 (August 2001): 114-118; Croarken, *Early Scientific Computing in Britain* (New York: Oxford University Press, 1990), especially Chapter 3, "The Mechanization of Computation at the Nautical Almanac Office," and Chapter 4, "The Nautical Almanac Office as a Computing Centre and the Founding of the Scientific Computing Service."
[20] Olley, esp. Chapters 3 and 4; *Astronomy at Yale*, 197; Brennan, 3-4 & 7-8; Eckert, "The Computation of Special Perturbations by the Punched Card Method," *The Astronomical Journal* 44 (1935): 177-182; Eckert and Brouwer, "The Use of Rectangular Coordinates in the Differential Correction of Orbits," *The Astronomical Journal* 46 (1937): 125-132; Dick, *Sky and Ocean*, 519-520; Eckert, *Punched Card Methods in Scientific Computation* (New York: Watson Astronomical Computing Bureau, 1940).

Brouwer, to compute the orbits of about 20 minor planets and to perform calculations on over 120,000 stars for the Yale star catalog.[21]

Eckert became head of the Almanac Office on February 1, 1940, and quickly began automating Naval Observatory astronomy. Upon his arrival, he found that the Almanac Office did "everything by hand," using slide rules, desk calculators, logarithms, and multiplication tables to produce the *American Ephemeris* and the *Nautical Almanac*. Even earlier, in 1928, the office had very few desk calculators, and some employees spent their entire careers never touching one. In Eckert's words, "Every digit was written by hand and read and written repeatedly."[22]

Eckert transformed the Almanac Office into a major center for mechanizing astronomical computations. He acquired several IBM punched-card machines, such as tabulators and sorters, and hired astronomer Paul Herget in 1942 to automate the *Nautical Almanac*. Herget deserves some attention because of the important role he played in pioneering space navigation. His long career in astronomy started when he landed a position at the University of Cincinnati's observatory in 1931 as a computer.[23]

Herget became interested in the problem of determining orbits and began to learn astronomy by "thumbing through" the *Astronomical Journal* and reading books, such as Russell Crawford's 1930 work on orbits. At the same time, he acquired a keen interest in mechanizing astronomical calculations being attracted especially to the work of Leslie Comrie. Thus inspired, Herget's 1935 dissertation developed a method for determining preliminary orbits and its adaptation to punched-card techniques.[24]

Thanks to the Alexander Morrison Fellowship, Herget spent a year at Berkeley with professors Crawford and Leuschner. He had his own approach to computing orbits, and Herget now saw firsthand how others did it. He recalled getting into trouble several times with the department over how to determine orbits and even how to organize the course on orbits. Leuschner's method, Herget recalled, "was the law and the gospel, and I didn't go along with that kind of stuff." As a result, "I wouldn't go so far as to say I was persona non grata, but at least I was under suspicion or under surveillance."[25]

Later, Herget opined: "Unfortunately [Leuschner's Method] was not without its disadvantages, as came to light on several occasions. One can perceive that on occasion

---

[21] Olley, 104-105 & 151; J. M. A. Danby, "Dirk Brouwer," *Quarterly Journal of the Royal Astronomical Society* 8 (1967): 84-88; Brennan, 8; Herget, "Eckert Memoir."

[22] Olley, 105-108; Duncombe, interview, 5; Herget, interview, 64; Brennan, 10; Dick, *Sky and Ocean*, 517-518.

[23] Olley, 108-110; Duncombe, interview, 5-6 & 7-9; Eckert, "Air Almanacs," *Sky and Telescope* 4 (1944): 4-8; Herget, interview, 17 & 64; Brennan, 10; Dick, *Sky and Ocean*, 517, 518 & 520.

[24] Osterbrock and Seidelmann, 62-63; Herget, interview, 15, 16 & 20; Herget to Prof. Jan Shilt, January 20, 1940, "National Research Fellowship," D2, Herget Papers. His dissertation was Herget, "The Determination of Orbits," Ph.D. Thesis, University of Cincinnati, 1935. In the same year, he published "The Determination of Orbits," *The Astronomical Journal* 44 (1935): 153-161, based on the dissertation and reprinted as "A Method for Determining Preliminary Orbits Adopted to Machine Computation" *Publications of the Cincinnati Observatory* 21 (Cincinnati: Cincinnati Observatory, 1936).

[25] Herget, interview, 35, 42, 44, 45; Osterbrock and Seidelmann, 63. Herget, "Leuschner," 132, states: "'Leuschner's Method' was the Gospel, the Law and the Prophets in the Students' Observatory." Leuschner's method is discussed in Herget, "Leuschner," 133.

the Lick astronomers sat up all night in order to complete the computations of a comet orbit as soon as the necessary observations became available because they were apprehensive of the reliability of the predictions that came from Berkeley when Leuschner was first developing his new method. But over the years the Harvard Announcement Cards bear testimony to the large amount of work done by Leuschner's students in computing the orbits of newly discovered comets."[26]

At the end of the war, both Eckert and Herget returned to their previous institutions and continued their efforts to expand the automation of astronomical computations. Unlike Eckert, Herget quickly found himself in need of even more computer time in 1947, when he took charge of the Minor Planet Center. The center, originally set up by the International Astronomical Union (IAU) at the Astronomisches Rechen-Institut (Computing Institute) in Berlin, collected observational data on asteroids, minor planets, and comets; calculated their orbits; and published this information.[27]

Meanwhile, Gerald M. Clemence had succeeded Eckert as supervisor of the Almanac Office. Clemence was intent on returning the observatory to analyzing and improving the theories of planetary motion and the associated refinement of the astronomical constants that underlay the almanacs. In this quest, Clemence collaborated closely with Eckert, Brouwer, and Herget. The "congenial and constructive relationships" that existed among the four astronomers ended only with the deaths of Brouwer in 1966 and Eckert in 1971. A 1947 Office of Naval Research contract supported the collaboration in celestial mechanics of Yale, the Almanac Office, and Eckert's laboratory and set the Almanac Office's research agenda for more than a decade. To aid in carrying out the laborious calculations necessitated by this research agenda, Clemence, Eckert, and Brouwer took advantage of the IBM Selective Sequence Electronic Calculator that Eckert had pioneered after leaving the Naval Observatory.[28]

Brouwer and Yale. As Clemence, Eckert, Brouwer, and Herget advanced the art of celestial mechanics through the application of computers and numerical integration, the field of celestial mechanics itself lingered as an underappreciated discipline not commonly taught in colleges at advanced levels. Astronomy was shifting its main focus to astrophysics. Allying celestial mechanics with the nation's space program seemed to hold out hope of revitalizing the field.

In 1947, for example, Clemence argued that "a significant demand" existed in both government and industry "for celestial mechanicians with the doctorate." "Strangely enough," he maintained, the demand was not for people to do research in celestial mechanics, "neither is the demand primarily for applications to rocket

---

[26] Herget, "Leuschner," 132-133.

[27] Brennan, 55; Herget, interview, 70-71. Following Herget's retirement in 1978, the center moved to the SAO under the direction of Brian G. Marsden. Marsden, "The Minor Planet Center," *Celestial Mechanics and Dynamical Astronomy* 22 (July 1980): 63-71.

[28] Olley, 172; Raynor L. Duncombe, "Gerald Maurice Clemence, 1908-1974," National Academy of Sciences, *Biographical Memoirs* 79 (2001): 7-8; Dick, *Sky and Ocean*, 524, 525 & 526-527; Clemence, "On the System," 169; Brennan, 21-23.

navigation, artificial satellites, or interplanetary travel." The real demand was for "experts in the art of computation," by which he meant "the art of formulating a problem so as to make it susceptible of numerical treatment, of doing this in such a way as to make it soluble by machines in the available time."[29]

Clemence later recalled that the launch of the first Sputniks: "led to an upsurge of interest in dynamical astronomy unprecedented in the history of the subject." Whereas in previous decades the number of active workers in the field in the United States was about six (and only slightly more around the world), "now within a single year scores of students were seized with the desire to study dynamical astronomy, and the demand for teachers of the subject far exceeded the supply."[30]

Dirk Brouwer, as head of the Yale Astronomy Department, exploited the dramatic upsurge of interest to raise celestial mechanics to a position of worldwide preeminence. He reportedly had more candidates for the graduate school than he could accommodate. In 1959, riding on the wave of demand for expertise in celestial mechanics, Brouwer obtained funding from NASA and the NSF to hold an experimental four-week Summer Institute in Dynamical Astronomy at Yale. The idea was to instruct college teachers who wanted to introduce a course in dynamical astronomy at their own institutions as well as to teach governmental and industrial researchers who lacked formal training in the celestial mechanics needed for orbit and trajectory studies. Of the more or less 100 initial participants, 20% were college teachers and 80% were divided nearly equally between government and industrial workers.[31]

Yale astronomers and invited specialists from various domestic and foreign institutions, including Clemence and Raynor L. Duncombe, Clemence's successor at the Almanac Office, gave the courses. With funding from a local aerospace firm, Brouwer had the proceedings of the first summer, covering fifteen sessions, printed in 1959 and reprinted in 1960. With NSF funding, the summer institutes ran for the next six years. In all, the program taught some 600 college teachers and government and industry employees.[32]

In 1962, Brouwer also created the Research Center for Celestial Mechanics within the Yale Astronomy Department thanks to joint underwriting from the Office of Naval Research and the Air Force Office of Scientific Research. The intent was for the center to play a role in solving problems related to Earth-circling satellites as well as lunar and interplanetary probes and in planning trajectories for future interplanetary spacecraft. Brouwer used the funding in part to double the number of celestial mechanics professors to six. His Research Center also sponsored annual seminars in celestial mechanics as well as the 1964 Conference on Ephemeris Calculations. It continued to thrive until 1968, when funding declined after Brouwer's death.[33] The

---

[29] Clemence, "The Need for Training Students in Celestial Mechanics," *Journal of the Royal Astronomical Society of Canada* 41 (1947): 291.
[30] *Astronomy at Yale*, 153; Dick, *Sky and Ocean*, 530.
[31] *Astronomy at Yale*, 147, 150 & 153-154; Dick, *Sky and Ocean*, 531.
[32] *Astronomy at Yale*, 154; George E. Bowden and Joanne Flis, eds., *Notes of the Summer Institute in Dynamical Astronomy at Yale University*, 2 Vol. (Hartford: United Aircraft Corporation, Missiles and Space Systems Division, July 1959).
[33] *Astronomy at Yale*, 154; Yale University Observatory, Celestial Mechanics Research Center, "5th Seminar

civilian (NSF) and military (Navy and Air Force) funding of the summer institute and research center reflected the dual civilian-military nature of the country's space effort that characterized the 1960s

As historian Steven J. Dick has argued, the collaboration of Brouwer, Clemence, Duncombe, Eckert, and Herget constituted a distinct era in celestial mechanics.[34] One can consider their research agenda in dynamical astronomy (a set of problems) combined with common research tools, such as punched-card and computer technologies and numerical integration, as forming what the philosopher of science Thomas S. Kuhn has termed "a paradigm."

Stated simply, Kuhn defined a paradigm as a core of consensus within a group of practitioners. The essence of the paradigm consensus is a set of problems and their solutions. For Kuhn, "normal science" was a specific phase of scientific development distinguished by universal consensus within a given scientific community over the problems to be solved and the ways of solving those problems. In other words, normal science was paradigm science. In this case, the paradigm consisted of a consensus on a particular set of problems in celestial mechanics and agreement on a particular way of solving those problems (computers, numerical analysis).[35]

The elements of this paradigm, particularly the marriage of celestial mechanics and computers, equally were at the heart of space navigation. As a result, this handful of astronomers from the Naval Observatory, Yale, and Cincinnati University formed the wellspring from which the nascent civilian space program drew for expertise, as we shall see in the next chapter. Also, IBM, which had provisioned and supported Eckert at Columbia University, would furnish the computers and programmers to calculate the flight paths of the country's pioneering civilian space effort known as Vanguard.

---

in Celestial Mechanics," January 1967, "Yale Seminar in Celestial Mechanics, 1967," D2, Herget Papers; "Informal Minutes," Yale University Observatory, "Conference on Ephemeris Calculations," November 19-20, 1964, "Yale Correspondence, 1954-1965," D3, Herget Papers.

[34] Dick, *Sky and Ocean*, 530.

[35] The works of Kuhn, which span over thirty years, have been summarized, explained, and analyzed in Paul Hoyningen-Huene, *Reconstructing Scientific Revolutions: Thomas S. Kuhn's Philosophy of Science*, trans. Alexander T. Levine (Chicago: University of Chicago Press, 1993). Especially relevant to the discussion presented here are 134-135, 143-154, 169, 188-190 & 193-194, which are summarized in Butrica, *See the Unseen*, 117-118.

Chapter Three

Vanguard

The Vanguard program heralded the beginning of civilian space navigation and paved the way for NASA's entry into the field. It brought together for the first time the sine qua non of space navigation: test-range expertise, celestial mechanics, automated computing, and numerical integration. Vanguard's goal, to put probes in Earth-circling orbits, provided the very first opportunities to determine the trajectory of an actual civilian satellite. For navigation expertise, Vanguard called on astronomers from the Naval Observatory, Yale, and Cincinnati University and the computer and software experts of IBM, the very same institutions that had pioneered the automation of celestial mechanics. However, as Samuel Herrick knew so well, navigating an actual satellite differed in certain significant ways from calculating the orbit of a solar-system object, even a comet. Astronomer-navigators, moreover, would have to deal with both optical and radio tracking observations gathered by two parallel networks set up specifically for the program.

Vanguard marked the beginning of navigation's many contributions to science. Heretofore, those guiding ships across deep waters—where mariners could no longer rely on keeping land within eyesight—depended on instruments and the application of science (celestial mechanics) in the form of tables and almanacs. At no point, though, did navigational observations contribute to the betterment of science or the almanacs. Space navigation transformed this unidirectional relationship between science and navigation. Observations and their analysis now served both to improve navigational accuracy and to add to the scientific knowledge on which navigation relied. In this and other ways, deep-space navigators would become scientists in their own right.

Vanguard is born. Vanguard was ostensibly a civilian endeavor. In the United States, rocketry had developed within an institutional framework that was almost entirely military thanks largely to the exigencies of World War II. The sole exception was the NACA's Wallops Island range, which, all the same, collaborated with armed forces' missile programs. Given the close blending of civilian and military elements, distinguishing one from the other was a daunting task.

Indeed, the integration of the armed forces, academia, and industry was a lesson learned during World War II. As General Dwight David Eisenhower wrote:[1]

> The lessons of the last war are clear. The armed forces could not have won the war alone. Scientists and business men contributed techniques and weapons which enabled us to outwit and overwhelm the enemy.

---

[1] Quoted in Stuart W. Leslie, *The Cold War and American Science: The Military-Industrial-Academic Complex at MIT and Stanford* (New York: Columbia University Press, 1993), 24-25.

Their understanding of the Army's needs made possible the highest degree of cooperation. This pattern of integration must be translated into a peacetime counterpart which will not merely familiarize the Army with the progress made in science and industry, but draw into our planning for national security all the civilian resources which can contribute to the defense of the country.

Later, of course, as President Eisenhower was leaving office, he cautioned against the influence on the country of the "immense military establishment": "The total influence—economic, political, even spiritual—is felt in every city, every statehouse, every office of the federal government." He further warned that: "In the councils of government, we must guard against the acquisition of unwarranted influence, whether sought or unsought, by the military-industrial complex."[2] The Vanguard project was an expression of Eisenhower's desire to establish a civilian space program—or at least one that had a civilian face—but whose realization hinged largely on military assets.

The initiation of the Vanguard project took place as part of the International Geophysical Year (July 1, 1957, to December 31, 1958). It allied the armed forces, which had the technical means to carry off the project, and civilian scientists with long track records of military collaboration. On the civilian side, Vanguard brought into play the National Academy of Sciences and the newly-founded National Science Foundation (NSF) as well as the older Smithsonian Astrophysical Observatory (SAO) whose creation dated to 1890. The panel that promoted the project had both civilian and military members, including Wernher von Braun (Army Ballistic Missile Agency) and William Pickering (JPL). The United States did not have a civilian space agency to oversee the project, nor did any unified military space agency exist either. The Naval Research Laboratory (NRL), after some debate, took over Vanguard management.[3]

Project Vanguard was born on July 29, 1955, when President Eisenhower announced that the United States would launch a small instrumented Earth-circling civilian satellite as part of the country's effort for the International Geophysical Year.[4] Vanguard had four distinct tracking networks. Consistent with practices at White Sands and other test ranges, both radio and optical tracking equipment and methods came into play. In addition, the optical and radio networks each had an analog composed of citizen volunteers.[5] The participation of both professionals and nonprofessionals contributed to Vanguard's civilian appearance and undoubtedly made for good Cold War propaganda, but it also bequeathed worldwide tracking organizations that outlasted Vanguard and the IGY.

---

[2] Eisenhower, "Farewell Address to the Nation," January 17, 1961, *Public Papers of the Presidents of the United States, Dwight D. Eisenhower, January 1, 1960 to January 20, 1961*, 8 (Washington: GPO, 1961), 1035-1040.

[3] Hagen, "Viking and Vanguard," 125; Newell, 47. One of the oldest tellings of the IGY satellite tale is: J. Tuzo Wilson, *IGY: The Year of the New Moons* (New York: Knopf, 1961).

[4] Hagen, "Viking and Vanguard," 125 & 126; Thomas, 31; Pickering, 411-413; Hall, "Satellite Proposals," 410-434; Koppes, 79; Green and Lomask, 43, 51 & 148.

[5] In place of "amateur," the term "citizen" is used following the arguments laid out by W. Patrick McCray, *Keep Watching the Skies! The Story of Operation Moonwatch and the Dawn of the Space Age* (Princeton: Princeton University Press, 2008).

For Vanguard organizers, the establishment of a system of professional observers equipped with proper optical scientific equipment was a given. They felt that not only would optical tracking be more accurate, but that it might be more dependable than an untried electronic system. Implementation of the optical networks began in the fall of 1955, as soon as the Vanguard project became official. Fred Whipple and J. Allen Hynek, SAO Director and SAO Associate Director, respectively, drew up plans for the professional optical network, which expanded on Whipple's groundbreaking camera observations of meteorites made during the 1940s.[6]

The web of professional observers, known as the Precision Photographic Program, consisted of twelve stations set up around the world under the supervision of SAO senior astronomer Karl G. Henize. The SAO was home to the central administrative, computing, and analysis facilities.[7] The superiority of the Vanguard optical net derived from its advanced high-precision telescopic camera, the so-called Baker-Nunn camera, created specifically for Vanguard under the direction of Whipple and Hynek.[8]

Citizen observers (Moonwatch[9]) relied on less risky, simpler technology. They acquired or "found" the satellite in the sky, while the Precision Photographic Program tracked it. Whipple and Hynek referred to Moonwatch as a "visual" rather than an "optical" program, because members did not use cameras. Citizens gathered preliminary orbital information and sent their data to Cambridge, where the SAO distributed it in the form of ephemerides to project technicians charged with gathering satellite orbital data.[10] If nothing else, Moonwatch was the much-desired civilian face of Vanguard.

Minitrack and Moonbeam. Vanguard also featured two radio tracking networks. Minitrack (Minimum-Weight Tracking System), the professional system, consisted eventually of fourteen stations arranged in three groups according to their role in tracking the three-stage Vanguard rocket. The so-called Prime Minitrack stations were strung in a north-

---

[6] Thomas, 121; Green and Lomask, 99; Whipple and Lyman Spitzer, Jr., "Tentative Optical Observing Program for Earth Satellites," nd, "SAO Vanguard," D3, Herget Papers.

[7] Green and Lomask, 99, 149 & 151-152; Whipple, 125-126; Hagen, "Viking and Vanguard," 134; Thomas, 38; Peebles, 39-40. For SAO orbit computation methods: George Veis, "Optical Tracking of Artificial Satellites," *Space Science Reviews* 2 (August 1963): 250-296; Veis and Charles H. Moore, "SAO Differential Orbit Improvement Program," 165-184 in *Tracking and OD*; John P. Rossoni, "Technical Aspects of Satellite Tracking on IBM Computers at Smithsonian Astrophysical Observatory in Cambridge, Massachusetts," 185-191 in *Tracking and OD*.

[8] Green and Lomask, 152, 153 & 240; Thomas, 34-36 & 39-42; Whipple and Hynek, "A Research Program Based on the Optical Tracking of Artificial Earth Satellites," *Proceedings of the IRE* 44 (June 1956): 762-763; Whipple, 126. For the Baker-Nunn camera, see Henize, "The Baker-Nunn Satellite-Tracking Camera," *Sky and Telescope* 16 (January 1957): 108-111.

[9] Various names were used, such as Project Moonwatch, Volunteer Moonwatch Program, and Operation Project Moonwatch.

[10] Green and Lomask, 149 & 154; Whipple, 125 & 128; Herbert O. Johansen, "You Can Be First to See a Satellite," *Popular Science*, May 1956, 110-115. Frederick I. Ordway, III, "Project Vanguard Earth Satellite Vehicle Program: Characteristics, Testing, Guidance, Control, and Tracking," *Astronautica Acta* 3 (1957): 77-86, provides a contemporary account.

south line along the east coast of North America and the west coast of South America in order to catch the satellite in any of the numerous expected flight paths.[11]

Minitrack creators John Mengel and Roger Easton of the NRL chose to detect satellite signals using a radio interferometer. The latter's use in missile-tracking dated at least to 1948, when Convair engineers created the AZUSA for the Army. Named for the California town where it was devised, the AZUSA interferometer compared the phases of signals received at two antennas and determined position by triangulating from a baseline. It also could obtain range (distance) measurements.[12]

Each Minitrack station had eight separate antennas connected to form three pairs in the north-south direction and two pairs in the east-west direction. The Minitrack antenna array lay flat and parallel to the ground. By means of triangulation with the distance between the antenna pairs (the baseline of the interferometer), the Minitrack determined the direction of the signals, that is, the angle of the signal wave relative to the baseline. In this way, the system calculated the position of the satellite, but not its distance.[13]

The first JPL navigators did not rely on interferometer readings, but the importance of station location and time information for Vanguard held true for deep-space navigation. Minitrack stations employed a coordinate system whose central point was the geometrical center of the Earth. Calculating a satellite's position required determining the distance from the center of the Earth to the satellite, the distance from the Earth's center to the station, and the distance from the station to the satellite. Critical to establishing these distances was an exact knowledge of a Minitrack station's location within the coordinate system.[14]

The Army Map Service played a major role in determining station locations. A 1956 Site Selection Team set out the criteria needed for Minitrack locations and selected candidate domestic and foreign sites. The Army Map Service, U. S. Coast and Geodetic Survey, and the Inter-American Geodetic Survey conducted surveys and collaborated with agencies in several countries in order to tie the Minitrack facilities into existing geodetic grids within an Earth-centered coordinate system. In actuality, the stations did not align themselves perfectly to this coordinate system, so a series of tests using airplanes and telescopes determined the amount of correction required for each Minitrack location.[15]

Minitrack satellite observations included a notation of the time of day (time tag) when it collected the data. To achieve the requisite temporal exactitude, Minitrack relied on its own digital chronometers that technicians checked daily against radio station WWV. The National Bureau of Standards' short-wave radio station in

---

[11] Thomas, 19 & 117-118; Green and Lomask, 155 & 163; Hagen, "Radio Tracking," 62; Whipple, 126; Smitherman, 2-3 and Appendix D.

[12] Thomas, 12, 15, 16 & 120; Green and Lomask, 10, 45 & 145; Hagen, "Radio Tracking," 63; Hagen, "Viking and Vanguard," 133.

[13] Smitherman, Appendix A, np; Mengel and Herget, 24; Hagen, "Radio Tracking," 64; Thomas 15; Green and Lomask, 146 & 147.

[14] Mengel and Herget, 25; Thomas, 19-20.

[15] Smitherman, 2-3; Mengel and Herget, 25; Thomas, 19-20.

Greenbelt, Maryland, beeped once every second and provided the time to an accuracy of about 1 millisecond.[16]

Using WWV time signals was a convenient, but not trouble-free, solution. The Bureau of Standards tended to make corrections on Wednesday evenings, but not every Wednesday and not even at regular intervals. Several weeks often elapsed between corrections. The time signals sufficed for short-term predictions, astronomer Gerald Clemence concluded, but detailed orbital analyses for scientific and other purposes would have to refer to the log of time signal corrections maintained by the Naval Observatory. He believed that correcting the WWV time signals during Minitrack satellite observations would introduce confusion. Accordingly, the Vanguard project corrected WWV time signals only during post-flight analysis of tracking data.[17]

The NRL also set up a collaborative net of citizen participants, called Project Moonbeam, to encourage radio amateurs ("hams") individually and in groups—universities, professional groups, and advanced amateur radio clubs—to build their own stations and to participate in tracking satellites in orbit. Here again was another go at covering Vanguard with civilian paint. Ham operators built the Mark II (or "Poor Man's") Minitrack rather than full-blown Minitrack stations. These were substantially different. The electronics were less sophisticated; some had as few as two antennas. The Mark II included a receiver to pick up WWV time signals. Builders in places too far away to receive WWV or WWVH (based in Hawaii) or any other standard time signals reliably had to acquire an auxiliary precision crystal clock unit.[18]

Working Group on Orbits. One of the fundamental challenges facing the Vanguard project was that of computing an orbit for an actual satellite in Earth orbit. In July 1956, the NRL created the five-member Working Group on Orbits to establish procedures for computing orbits, preparing ephemerides, and extracting geodetic and geophysical information from orbital data, among other navigation-related duties. The NRL furnished two members. One was James J. Fleming, a data reduction specialist who brought extensive test-range experience.[19] The other was Joseph W. Siry. From 1946, Siry had worked under Homer E. Newell as a member of the team researching the upper atmosphere with rockets. From 1949 to 1953, he led that group's "theoretical analysis branch," which was involved in generating missile trajectories and determining flight paths.[20]

---

[16] Mengel and Herget, 25; Hagen, "Radio Tracking," 64; Working Group on Orbits, "Minutes of 6 December 1956 Meeting," December 6, 1956, "Working Group on Orbits, 1956-1957," D3, Herget Papers; Dick, *Sky and Ocean*, 500.

[17] Working Group on Orbits, "Minutes of 25 October 1956 Meeting," October 25, 1956, "Working Group on Orbits, 1956-1957," D3, Herget Papers; Memorandum, Director, NRL, to Director, NBS, "The Use of WWV Time Standards in the Vanguard Earth Satellite Program," January 18, 1957, F148/B15, Vanguard.

[18] Green and Lomask, 148; Hagen, "Radio Tracking," 64 & 65.

[19] Hagen, Vanguard Project Directive, July 13, 1956, "Vanguard," D8, Herget Papers; Green and Lomask, 159.

[20] "Joseph William Siry," 1582-1583; Thomas, 22; Siry, 77-144.

The Working Group on Orbits also included three specialists in celestial mechanics. Gerald Clemence and Raynor Duncombe were astronomers from the Naval Observatory. Duncombe took over the Almanac Office after Clemence in 1963. Both moved the Naval Observatory farther and farther into the computer age.[21] Paul Herget, an alumnus of the Almanac Office, rounded out the Working Group and contributed his considerable knowledge of orbit determination methods and computational equipment.

The Working Group addressed not a small number of questions related to the computation of Vanguard orbits. For his part, Siry concerned himself with the effects of the ionosphere on Minitrack observations and the Vanguard's flight path from launch pad to orbital insertion. Herget tackled the values for certain constants that figured in determining satellite orbits and their perturbation by the Earth's gravitational force, a problem with which he was familiar because of his earlier work for Convair on the Atlas long-range missile.[22]

Shooting missiles over such unprecedented distances revealed that the Atlas test units simply were not tracing the trajectories predicted by mathematics and classical mechanics. Accordingly, in March 1951, Convair sought Herget's help. He characterized the challenge as being "how to hit Moscow from Kansas." Herget realized that: "one of the problems that they had not touched yet was the effect on a ballistic trajectory of the oblateness of the [E]arth." The planet's equatorial bulge, he observed, might throw off a missile by several miles and cause it to miss its target.[23] As a result, Vanguard navigation relied on a thorough knowledge of the effect of the Earth's irregular gravitational field on spacecraft orbits.

The Working Group on Orbits also oversaw the computerization of the equations for determining Vanguard orbits and took an active part in planning the project's computing center. They had detailed questions about the nature of the radio and optical input data, the computer output data for both radio and optical networks, and data storage, that is, the data to be used for "Post-Mortem Calculations."[24] Such post-flight analyses, de rigueur among JPL navigators, provided critical insights that helped to improve navigation on future missions, but also yielded valuable scientific insights, as we shall see.

The Vanguard Computing Center. Vanguard orbit determination took place at the computing center. IBM, which won the contract over Univac, brought to bear its computer hardware and programming expertise. The firm viewed Vanguard as a

---

[21] Memorandum, Director, NRL, to Superintendent, USNO, "Vanguard Orbit Calculations," July 17, 1956, "Vanguard," D8, Herget Papers; Duncombe, "Gerald Maurice Clemence: August 16, 1908–November 22, 1974," National Academy of Sciences, *Biographical Memoirs* 79 (2001): 51-65; Green and Lomask, 159; Thomas, 22; Hagen, "Viking and Vanguard," 134; Dick, *Sky and Ocean*, 524-532.

[22] Working Group on Orbits, "Minutes," January 31, 1957, F278/B30, Vanguard; Siry, "Satellite Launching Vehicle Trajectories," nd, F78/B6, Vanguard; Memorandum, Herget to distribution, "Basis of Numerical Constants for the Orbit Computations and Equations for the 'Bulge' perturbations," September 1, 1956, F278/B30, Vanguard.

[23] Herget, interview, 76-77.

[24] D. H. Gridley to Herget, May 11, 1956, and attachment, "Vanguard Computing Center," D3, Herget Papers.

business opportunity to expand its sales of scientific computers, and IBM staff involved in computing satellite orbits viewed Vanguard as a research effort.[25] IBM supplied its high-speed, large-capacity magnetic-tape-memory IBM 704 computer (introduced in 1955)[26] plus a number of free services, such as orbit computations during the lifetime of the satellite for the first three successful satellites. They also offered at no charge the services of their mathematicians for coding, programming, numerical analysis, and related tasks as well as for any rehearsals needed to work out the routine of calculating flight paths from Minitrack data. Given this generous offer—and the company's low bid—it is not surprising that IBM won the contract.[27]

The facility that housed the IBM computer, the Vanguard Computing Center, was in a leased building in downtown Washington, DC,[28] a bit distant from the NRL, which maintained its own computing center. The NRL digital computer—the Electronic Digital Computer (Narec)—and staff calculated Vanguard satellite positions and compared the results with those of the civilian center. In addition to verifying the equations used to determine the Vanguard's orbit, the goal of the exercise was to determine the speed and accuracy attainable with Narec. As the NRL report explained: "The potential military and scientific applications of artificial [E]arth satellites make desirable the ability to perform real-time computations of the positions of all known satellites."[29] Here was another aspect of Vanguard's dual military-civilian persona: the NRL center hidden from public view; the IBM center on public display.

Preparation of the IBM 704 for launches started at the NRL, where the Working Group on Orbits—Herget for the most part—determined the required mathematical formulas. As he explained: "I became the nominal head of the Vanguard Computing Center, from the fall of 1955 until the spring of 1958." Assisting him was his colleague from the Cincinnati Observatory, Peter Musen, a native of the former Yugoslavia who had worked as a civilian at the Astronomisches Rechen-Institut.[30]

IBM mathematicians and programmers also participated in translating the Working Group on Orbits' mathematics into IBM punched cards and programming language. Thomas R. Horton was IBM's manager at the Vanguard Computing Center, while A. Robin Mowlem was Manager of Vanguard Operations. Herget and the Working Group frequently interacted with Donald A. Quarles, Jr., IBM's Chief Mathematician for Vanguard. Starting in December 1957, the project added an IBM 709 computer, already

---

[25] Mowlem, 119-120.
[26] Fact sheet, IBM, "704 News Sheet," July 1957, "Vanguard Box," D1, Herget Papers; Press release, IBM, "IBM to Provide Computing Facility for Project Vanguard," December 21, 1956, "Vanguard Box," D1, Herget Papers.
[27] Green and Lomask, 160 & 161; Thomas, 22; Mowlem, 119; Hagen, "Radio Tracking," 65; Working Group on Orbits, "Minutes of 25 October 1956 Meeting," October 25, 1956, "Working Group on Orbits, 1956-1957," D3, Herget Papers.
[28] The address was 615 Pennsylvania Avenue. Green and Lomask, 160; Thomas, 21. An alternative name was the IBM Space Computing Center. See, for example, "IBM 709: IBM 709 Data Processing System," http://ed-thelen.org/comp-hist/BRL61-ibm0709.html (accessed September 18, 2007).
[29] W. D. Dahl, *Satellite Coordinates and Real-Time Position Computation*, Report 5659 (Washington: NRL, September 12, 1961), 1, 6-7 & 10-11.
[30] Herget, "Of Computing and Astronauts," *Sky and Telescope* 60 (November 1980): 373; Herget, interview, 71; Osterbrock and Seidelmann, 64.

in operation at Patrick Air Force Base, to follow the early stages of Vanguard launches as the rocket was still climbing out of Earth's gravity well.[31]

"Dry runs" of the orbit determination program, initially scheduled for May 1957,[32] suffered delays. The purpose of the exercise was to evaluate equipment, processes, and methods as well as the ability of individuals belonging to multiple organizations to function as a coordinated team.[33] In January 1957, Herget already had complained to Quarles that the company was taking too long on the Fourier series work. "One has to look at the interval between the dates when results were sent to us to realize that at that rate we will never make the schedule. . . . we can no longer stand the long delays between when I send you something and when we get the results back."[34] They were near victory in March 1957, when Herget wrote to Quarles: "We have nearly reached the end of the trail on the Fourier series work" thanks to the "elegant" solutions devised by Peter Musen.[35]

Deciding on the best method for calculating Vanguard orbits was a challenge. A test using three observations—standard in astronomy—gave the wrong distances, so Herget proposed a new method based on four observations. Further analysis of preliminary tests of the orbit-determination software led Herget to "suspect some kind of a systematic programming error, or a programming error which would produce a systematic effect," the cause of which eluded him. Herget was "[b]affled." He proposed reexamining the computations and submitting everything to Duncombe and Clemence for their opinion. The next day, however, he found the trouble. "It was in the simulated observations I computed," he explained to Quarles. "I made only a small mistake, but it was enough to ruin the whole thing: I had an error (due to an oversight) in the rate at which the missile was traveling around in a circular orbit."[36]

Orbit determination. All Vanguard data from launch until the failure of the satellite's battery went to the Vanguard Computing Center for use in orbit determination. In case of a failure at that facility, IBM maintained a similar backup computer in Poughkeepsie, New York. Orbit determination owed a great debt to astronomy. The method for computing flight trajectories came from astronomy, specifically from cometary studies, and numerical integration was a key mathematical tool for calculating orbits, as it had been for the astronomers of the Almanac Office.

In computing satellite flight paths, Vanguard navigators had recourse to a technique known simply as Cowell's method. Astronomers considered it to be superior to other numerical integration approaches.[37] The method took its name from Philip H.

---

[31] "IBM Personnel," nd, "Vanguard," D8, Herget Papers; Green and Lomask, 161; Mowlem, 120; Hagen, "Radio Tracking," 65; Thomas, 22.
[32] Herget to Quarles, January 1, 1957, "D. A. Quarles Correspondence," D3, Herget Papers.
[33] D. H. Gridley to Herget, May 11, 1956, and attachment, "Vanguard Computing Center," D3, Herget Papers.
[34] Herget to Quarles, January 1, 1957.
[35] Herget to Quarles, March 24, 1957, "D. A. Quarles Correspondence," D3, Herget Papers.
[36] Herget to Quarles, January 1, 1957; Herget to Quarles, January 16, 1957; Herget to Quarles, January 17, 1957, all in "D. A. Quarles Correspondence," D3, Herget Papers.
[37] Mowlem, 122.

Cowell (1870-1949), a British astronomer who was Superintendent of the British Nautical Almanac Office from 1910 to 1930. His field of research was celestial mechanics, especially the orbits of comets and asteroids. His most notable studies probably were those of Halley's Comet.[38] Here again was the influence of cometary astronomy on space navigation. The main disadvantage of the Cowell method, as Robin Mowlem and the IBM Computing Center staff saw it, was the time it took to set up a so-called converged function table at the start of integration.[39]

One of the major disadvantages of numerical integration was that, in order to obtain predicted quantities corresponding to the time of a specific observation, one had to perform all the previous steps leading up to that point in the flight. The process additionally was subject to cumulative errors caused by rounding off the data, which set a limit on how far in advance one could predict the orbit with any accuracy, usually anywhere from four to seven days. Once the satellite's orbit stabilized, and changes from one revolution to the next became minor, the Computing Center substituted oblateness perturbation techniques for numerical integration.[40] "Oblateness" referred to the fact that the Earth's polar axis was shorter than its diameter at the equator. In other words, the planet had an equatorial bulge.

Navigation and science. To make the oblateness techniques work, Vanguard navigators had to possess detailed information about the Earth's shape. Analysis of tracking observations revealed even more information about the shape of the Earth, because localized gravity changes appeared as deviations in the flight path. Vanguard orbit calculations took into account the known perturbations caused by the Earth's equatorial bulge. These perturbations shifted the perigee of the orbit and the orientation of the orbital plane. The General Oblateness Perturbation (GOP) program, as it was known, addressed this problem. It was easier to use than integration methods, resulted in greater accuracy, and yielded predictions farther in advance for any particular time in the orbit. The process also avoided cumulative errors and did not require any of the intermediate steps necessitated by integration.[41]

Ultimately, the Computing Center had few opportunities to determine Vanguard orbits: only three of the eleven attempted launches succeeded in putting a payload in orbit.[42] Nonetheless, the orbit determination effort resulted in not only flight paths computed with the required accuracy, but also some key scientific discoveries. For example, the Vanguard flight paths showed the degree to which the gravitational fields of the Moon and Sun influenced the orbits of Earth satellites as well as how the pressure of

---

[38] Edmund T. Whittaker, "Philip Herbert Cowell, 1870-1949," *Obituary Notices of Fellows of the Royal Society* 6 (November 1949), 375-384; Cowell and Andrew Claude de la Cherois Crommelin, *Investigation of the Motion of Halley's Comet from 1759 to 1910* (Edinburgh: Neill & Co., 1910).
[39] Mowlem, 122.
[40] Mowlem, 124.
[41] Mowlem, 124; Herget, "Vanguard Computational Problems Status Report," May 15, 1956, "Vanguard Computing Center," D3, Herget Papers; Herget, "General Theory of Oblateness Perturbations," 29-35 in *Orbit Theory*; Herget and Musen, "Modified Hansen," 430-433; Herget and Musen, "Erratum," 73.
[42] The successful launches were: Vanguard 1 on March 17, 1958; Vanguard 2 on February 17, 1959; Vanguard 3 on September 18, 1959.

radiation from the Sun affected the movement of satellites.[43] This knowledge in particular would be vital for interplanetary navigation.

Furthermore, a retrospective analysis of trajectory calculations made for the first Vanguard satellite over a period of 90 days using the General Oblateness Perturbation program yielded results that revealed that the Earth had a "pear" shape. Ann Eckels, while measuring the satellite orbit perturbations caused by the Moon and Sun, noticed that whenever the perigee of the probe's orbit was in the southern hemisphere, its altitude was always lower than when its perigee was in the northern hemisphere. Eckels concluded that the southern hemisphere must have an overall greater land mass exerting a greater gravitational attraction on the craft than the northern hemisphere. This discovery, of course, also bespoke the accuracy of the Computing Center's calculations.[44]

Joining Eckels in this discovery was John A. O'Keefe, who usually receives the main credit. O'Keefe, an astronomer by training, had worked as a civilian for what became in 1945 the Army Map Service, the branch of the Army Corps of Engineers responsible for creating topographic maps and the geodetic data required for intelligence and artillery targeting.[45] In 1954, then focused on geodetic problems, O'Keefe argued for the utility of Earth-orbiting satellites—albeit in conjunction with ground-based searchlights—for conducting a range of geodetic research, including more precise measurements of intercontinental distances, terrestrial gravity, and the size of the Earth's semi-major axis.[46] These suggestions became part of a general proposal made by the American Rocket Society's Space Flight Committee to encourage the NSF to underwrite the launch of a satellite, although without reference to plans for the International Geophysical Year.[47]

Subsequently, in 1957, O'Keefe showed mathematically how one could use satellite tracking data to study the shape of the Earth. By applying O'Keefe's mathematics, he and Eckels made a major contribution to geodesy. They showed that the Earth is shaped like a pear and has a slight bulge around the North Pole. The "pear-shaped Earth" became front-page news and was even the subject of a "Peanuts" cartoon.[48]

---

[43] Musen, Robert W. Bryant, and Ann [Eckels] Bailie, "Perturbation in Perigee Height of Vanguard I," Science 131 (March 25, 1960): 935-936; Bryant, "The Effect of Solar Radiation Pressure on the Motion of an Artificial Satellite," The Astronomical Journal 66 (1961): 430-432.

[44] Green and Lomask, 244; Hagen, "Radio Tracking," 63; Hagen, "Viking and Vanguard," 139; Mowlem, 120 & 125; E. Upton, [Eckels] Bailie, and Musen, "Lunar and Solar Perturbations on Satellite Orbits," Science 130 (December 18, 1959): 1710-1711; Myron Lecar, John Sorenson, and Ann Eckels, "A Determination of the Coefficient J of the Second Harmonic in the Earth's Gravitational Potential from the Orbit of Satellite 1958 β2," Journal of Geophysical Research 64 (1959): 209-216. "1958 β2" was COSPAR's label for the first Vanguard.

[45] "O'Keefe," BAAS, 1683; "O'Keefe," EOS, 55.

[46] O'Keefe, "The Geodetic Significance of an Artificial Satellite," Jet Propulsion 35 (February 1955): 775-776.

[47] "On the Utility of an Artificial Unmanned Earth Satellite," Jet Propulsion 25 (February 1955): 71-78.

[48] O'Keefe and Charles D. Batchlor, "Perturbations of a Close Satellite by the Equatorial Ellipticity of the Earth," The Astronomical Journal 62 (August 1957): 183-185; O'Keefe, Eckels and R. Kenneth Squires, "The Gravitational Field of the Earth," The Astronomical Journal 64 (September 1959): 245-253; O'Keefe, "IGY Results on the Shape of the Earth," ARS [American Rocket Society] Journal 29 (December 1959): 902-904; "O'Keefe," EOS, 55; "O'Keefe," BAAS, 1684.

NASA's Vanguard inheritance. The Vanguard project was a largely military endeavor with a civilian "face" provided by civilian astronomers and civilian observers. The addition of civilian scientists in consulting roles, such as on the Working Group on Orbits, was typical of the usual practice of blending civilian experts into military enterprises within the military-industrial complex. This same blending of civilian and military elements characterized the new space agency, NASA, that inherited richly from Vanguard and the world of missile test ranges.

The new organization subsumed the totality of the NACA, including its Wallops Island proving ground and the Langley facility that managed it. One of the major range experts was Edmond C. Buckley, who had been the head of the NACA Langley Instrument Research Division. That division had selected, developed, built, and installed all tracking instruments at Wallops. Buckley joined NASA, and in 1961, when NASA Administrator James Webb set up Headquarters tracking and operations as a separate program, Buckley became the first Director of the Office of Tracking and Data Acquisition.[49]

The Vanguard project found a new home at NASA's Goddard Space Flight Center. O'Keefe left the Army Map Service and became the Assistant Chief of Goddard's Theoretical Division. Eckels, along with the rest of the NRL contingent working on Vanguard, also transferred to Goddard. The transferred assets included the Computing Center—renamed the Space Computing Center—and its staff. Joseph Siry (NRL) and Peter Musen (Cincinnati Observatory) also went to Goddard from the Working Group on Orbits. Siry became Chief of the Theory and Analysis Staff; Musen continued to develop satellite navigational methods.[50]

Vanguard thus became the foundation for Goddard's tracking and navigation expertise and facilities. Vanguard made multiple contributions to the history of deep-space navigation. It brought into being the first worldwide tracking system, Minitrack, and established a major civilian orbit computation center. It produced the first practical orbit-determination software. In the process, the project's astronomer-navigators borrowed techniques from astronomy, such as the Cowell method, and mathematics, such as numerical integration and the least-squares method, all of which found their way into deep-space navigation at JPL. Moreover, like Vanguard, interplanetary navigation would rely at least initially on collaboration with astronomers and the Naval Observatory's Almanac Office.

---

[49] Hall, *Lunar Impact*, 87.
[50] "O'Keefe," *EOS*, 55; "O'Keefe," *BAAS*, 1684; Smitherman, Appendix B through Appendix G; Mowlem, 120; "Joseph William Siry," 1582-1583; Siry, 77-144; Thomas, 22. For Musen's work at Goddard, see, among others, Musen, *A Modified Hansen's Theory as Applied to the Motion of Artificial Satellites*, D-492 (Greenbelt: Goddard, November 1960); Musen, *On the Long Period Luni-Solar Effect in the Motion of an Artificial Satellite*, Technical Note D-1041 (Greenbelt: Goddard, July 1961); Musen, "On Stromgren's Method of Special Perturbations," *Journal of Astronautical Science* 8 (Summer 1961): 48-51.

Part Two

Pioneering Space Navigation

Chapter 4

Pioneers of Space Navigation

Before its merger into NASA as a facility managed by Caltech, JPL remained largely a developer of missiles for the Army. During the late 1950s, its defense role mushroomed in step with the armed forces' space programs. Despite the attempt to create a civilian space program with Vanguard, space remained an essentially military enterprise. The armed forces initiated and carried out the nation's first attempts to reach the Moon within the framework of a unified military space agency. Because of its ongoing Army affiliation, JPL was admirably positioned to profit from the escalation and new direction of the space program. The Army already had introduced JPL to satellite tracking (the Microlock system), and now the services' lunar endeavors would extend JPL's tracking capabilities into space and initiate the lab in space navigation. This experience laid the foundation for JPL to become one of the leaders in both tracking and navigating probes and triggered a rapid transformation of the laboratory as a whole.

The military spending that reoriented JPL from the testing range to outer space also established a competing tracking and navigation system built and operated by the TRW subsidiary STL. The two were not truly in competition, each serving a different service, but their parallel operation was a reminder that, by the end of the 1950s, the number and variety of organizations involved in tracking launches and computing flight paths was considerable. The organizations were either military agencies or civilian institutions paid by the armed services to develop navigational expertise. JPL was not alone.

The main driver of the nation's space efforts during this period and into the next decade was the Cold War. The successful launch of the world's first satellite by the U.S.S.R. provided a concrete rationale for going to the Moon. The greatest attention has been paid to the United States' dash to put an astronaut on the Moon by the end of the 1960s, a race that in retrospect appears to have been one-sided. Quite the opposite was true of the multiple lunar and planetary probes launched by both sides. They were the frontline of the projection of the Cold War into space. The Sputniks dramatically impacted not just the space program—such as it was—but also U.S. society. In many ways, Soviet successes sparked and inspired what the country did in space, because in no small part they revealed the country's lack of preparedness for rivalry with the Soviet Union.

Network competition. The Soviet Union followed its initial Sputnik successes with more and more space triumphs. Luna I, launched on January 2, 1959, was the first probe to travel to the Moon and the first to enter a heliocentric orbit. Luna II followed on September 12, 1959, and demonstrated very precise guidance inasmuch as the probe, like its predecessor, set out on a lunar trajectory propelled solely by the launch vehicle and without the need for a midcourse maneuver. Luna III, launched October 4, 1959, returned the first images of the Moon's far side. The next year saw the first successful launch of a spacecraft toward Mars, the Korabi 5 (known as Marsnik 2 in the United States), on October 14, 1960, and on February 12, 1961, Venera 1 left Earth for Venus.

The United States lacked a centralized agency to manage its response to the U.S.S.R., until Congress established the Advanced Research Projects Agency (ARPA) on February 12, 1958 (Public Law 85-325). A few months later, on July 29, 1958, President Eisenhower signed the National Aeronautics and Space Act. The agency did not begin operations officially until October 1, 1958, when Eisenhower directed the transfer of a number of ARPA-directed projects to the new organization, including Vanguard and associated NRL staff plus the military lunar programs.[1]

Two months later, on December 3, 1958, Eisenhower ordered the transfer of JPL to NASA. The space agency promptly defined JPL's roles. Its responsibilities included designing, developing, engineering, installing, and operating NASA's world-wide tracking network and performing the research and development required to maintain the network as a state-of-the-art system.[2] Before the creation of NASA, though, ARPA already had initiated JPL into its role as a tracker of spacecraft. JPL was not alone. ARPA had divided the military space program between the Air Force and Army, and each service begot its own tracking system. JPL built for the Army; STL built for the Air Force. As a result, STL developed tracking facilities and navigational expertise alongside JPL well into the first Mariner launches. Furthermore, as we shall see, JPL called on STL expertise for assistance on the Ranger Moon shots on more than one occasion.

The ARPA-directed Lunar Program, as it was known, originated on March 27, 1958, just five months after the first Sputnik's flight. It consisted of five launches: three undertaken by the Air Force and two by the Army. Their basic scientific goals were the measurement of cosmic radiation, a more accurate determination of the Moon's mass, and the verification of tracking and communication systems.[3]

The three ARPA Air Force launches eventually flew under the aegis of NASA, although "flew" is perhaps not an accurate term. All three Pioneer launches failed. Failure or not, STL built the tracking system, called the Air Force Deep Space Net (alternatively known as the SPAN network) and authorized by ARPA in March 1958. STL had to assemble the system quickly and relied on several existing facilities, including the Jodrell Bank radio telescope in Britain and Lincoln Laboratory's Millstone Hill radar station in Massachusetts. Prior to the first launch, the company also designed, built, and tested three steerable 60-foot (20-meter) parabolic dishes that operated at the Vanguard frequency of 108 MHz.[4]

A central computing office at STL's Operations Center (southwest of Los Angeles in Inglewood) performed orbit calculations based on flight path data received by teletype from these tracking sites. The data came in a variety of types: 400-MHz radar echoes from MIT's Millstone site, phase-difference readings from the Minitrack and Hawaii antennas, and azimuth and elevation observations from Millstone, Jodrell Bank, and Hawaii. Hawaii also acquired two-way Doppler. In addition, the Operations Center

---

[1] Hall, *Lunar Impact*, 47, 53 & 69.
[2] Hall, *Lunar Impact*, 75 & 78.
[3] Hall, *Lunar Impact*, 57.
[4] Hall, *Lunar Impact*, 58 & 64; STL, *1958 NASA/USAF Space Probes*, 2, 48-49, 52, 64, 69 & 73; STL, *Capabilities Presentation*.

furnished these facilities with predicts, that is, information on where to direct their antennas as the probe flew overhead.[5]

The STL Operations Center was in contact with NRL's Control Center in Washington, DC, which relayed Minitrack data to Inglewood via the STL communication center at Cape Canaveral. STL and NRL also were in telephone contact with each other, so that NRL interpretation of Minitrack data was available rapidly to STL staff. The NRL also stationed technical personnel at the STL Operations Center to assist further in interpreting Minitrack data.[6]

Acquiring optical tracking information was not part of the STL system. Nonetheless, STL did attempt to track and photograph the probes with the large Palomar Observatory telescope, but without success. The hope had been to demonstrate the value and capabilities of optical instruments for measuring angular positions by fixing the precise trajectory of the vehicle with several moderately-spaced observations. The other motivation was the "obvious prestige value" of a "photograph of a missile on its way to the [M]oon," according to STL.[7] The experiment was an omen that ground-based optical tracking data had no future in deep-space navigation, but it was not the last try to have an observatory follow a spacecraft.

The idea, for instance, persisted into NASA's Ranger program and its first probe sent to the Moon, Ranger 3. In preparation for the launch, JPL considered ten observatories as potential eyewitnesses to its lunar impact. Two facilities were in Chile, while the others were in Texas, Arizona, and California. Using the Chilean observatories, however, raised a "security problem." JPL would have to connect all participating observatories with a telephone link to its communication center and notify them of the time and location of impact. Allan R. Sandage, who managed the Mount Wilson and Palomar Observatories, agreed to partake in the venture.[8] The experiment, though, never took place, because Ranger 3 missed the Moon.

The STL network met its greatest test with Pioneer V. That mission differed from its predecessors (aside from actually achieving its mission) in that its primary purpose was to measure cosmic rays, magnetic fields, and micrometeorites while in a heliocentric trajectory, that is, while orbiting the Sun. The challenge arose from the great distances measured in tens of millions of miles that Pioneer V traversed. Human-generated radio waves had never before traveled over such an expanse. Of all the stations that STL called on to observe Pioneer V, only the Jodrell Bank dish succeeded in following the probe beyond 10 million miles (16,100,000 km). The craft was entirely out of its range, though, when it attained its maximum distance from Earth, about 168 million miles (270,370,000 km). If nothing else, STL demonstrated that it was possible to communicate with a spacecraft that was 22.5 million miles (36 million kilometers) away.[9]

---

[5] STL, *Capabilities Presentation*; STL, *1958 NASA/USAF Space Probes*, 64, 65 & 75.
[6] STL, *1958 NASA/USAF Space Probes*, 64 & 75.
[7] Ibid., 65-66.
[8] Memorandum, Carl B. Guderian to Justin Rennilson, "List of Local Observatories Interesting to Ranger 3," January 17, 1962; Memorandum, Guderian and Stan Rubinstein to Ted Pounder, "Systems Analysis of Several Lunar Impacts," January 15, 1962, both F401/B19, JPL421.
[9] Paul F. Glaser, "Interplanetary Spacecraft Communications Systems," Seventh Annual Meeting of the American Astronautical Society, January 16-18, 1961, Dallas, TX, F3/B64, Willy Ley; "Observing the

Explorer. JPL's introduction to tracking a civilian satellite came rather suddenly. The laboratory succeeded thanks to its earlier work for the Army. The event, the launch of Explorer I, marked a watershed moment in the history of JPL navigation and the laboratory as a whole. The hastily readied satellite beat the Vanguards and Pioneers into space. It launched on March 17, 1958, just months after the second Sputnik, and became the first successful U.S. satellite. Explorer I was a critical victory for the United States vis á vis its Cold War space rivalry.[10]

The satellite was ready to launch in record time thanks to salvaging from previous projects. The Microlock tracking system was part of that heritage. Setting up and running that system gave JPL its first taste of network development and operation. Microlock electronics originated in classified missile work carried out for the Army during the early 1950s. The architecture reflected the philosophy of minimizing the power and weight of the satellite's transmitter, while adding complexity (as well as weight and power) to ground equipment. This philosophy also compensated for the lower boost power of U.S. launchers compared to those available in the Soviet Union. For Explorer I, a portion of JPL's Microlock tracking system was available at locations in Earthquake Valley (Borrego Springs) near San Diego, California, and Cape Canaveral (the Air Force Missile Test Center). JPL added new ones in Singapore and Nigeria in cooperation with the British International Geophysical Year committee, which facilitated the diplomatic formalities and negotiations to obtain permission to operate in those lands.[11]

JPL found the Explorer I tracking Doppler to be rather surprising. The strength of the signal received from the satellite's transmitter was not what the laboratory had expected, Eb Rechtin reported to a meeting of the American Rocket Society. Rechtin, the future architect and head of JPL's Deep Space Network, explained that JPL expected to receive a relatively constant signal with only small variations resulting from the satellite spinning on its axis, a normal motion for spin-stabilized probes. Instead, the signal faded roughly every 7 seconds, and the degree of fading increased over time until the signal stabilized some days later. The pattern of the radio signals, according to Rechtin, suggested that the satellite's spin rate rapidly decelerated, and that it had begun to spin through space like a propeller, rather than like a bullet.[12]

Satellites," *Sky & Telescope* 19 (May 1960): 409-412; Evert Clark, "Pioneer V Transmits Deep Space Data," *Aviation Week*, March 21, 1960, 28-30.

[10] Pickering, 417 & 418; Rechtin, 343; Koppes, 90.

[11] Koppes, 87-88; Pickering, 414, 416-417; Rechtin, 326 & 327; Corliss, *History of the DSN*, 2; Hall, *Lunar Impact*, 13; Hall, *Ranger Chronology*, 12, 13 & 45; Hall, "Origins," 110; Henry L. Richter, William F. Sampson, and Robertson Stevens, "Microlock: A Minimum Weight Instrumentation System for a Satellite," *Jet Propulsion* 28 (August 1958): 532-534; Mudgway, *Big Dish*, 10-13 & 162-163; Mudgway, *Uplink-Downlink*, 472 & 521-523.

[12] Rechtin, "Satellite Tracking," American Rocket Society Meeting, June 9-12, 1958, Los Angeles, CA, F76/B7, Willy Ley.

Despite its role in observing Explorer I, JPL was not involved in determining the probe's flight path. Other agencies shared that task. The Vanguard Computing Center, for instance, focused on computing the orbit from readings derived from both Minitrack and Microlock stations. In the end, though, JPL did have the opportunity to analyze retrospectively the telemetry it received via telegraphy and tape recordings shipped to Pasadena from Earthquake Valley, Singapore, and Nigeria.[13]

Data for space navigation. Although the Microlock system handled the Earth-circling Explorer I, the Army's ARPA Moon flights would be beyond their reach. For those launches, known subsequently as Pioneer III and Pioneer IV, JPL built a tracking system that became the core of today's Deep Space Network. One of the ARPA mission's stated goals was to determine the probe's actual flight trajectory from tracking data, thereby making it a test of JPL's orbit-determination capabilities. This was the beginning of space navigation at JPL. Pioneer III failed to achieve escape velocity because of a premature booster stage cutoff, but Pioneer IV, launched on March 3, 1959, headed for the Moon. It became the first U.S. spacecraft to escape Earth's gravity, but missed its target by some 37,000 miles (59,546 km) and went inadvertently into a heliocentric orbit. The feat of orbiting the Sun, however, already had been achieved by Luna 1.[14]

The first ARPA dish—based on a radio telescope design—went up at the JPL Goldstone Dry Lake facility on the Army's Fort Irwin Mojave Desert military reservation. The 85-foot (26-meter) diameter dish, dubbed the Pioneer station, is now a National Historic Landmark as the first antenna built for what became the Deep Space Network.[15] The ARPA net featured communication links with JPL, Goldstone, the Air Force Cape Canaveral site, and the Mayagüez, Puerto Rico, facility. The system was not worldwide, but confined to the Western Hemisphere (Goldstone, Cape Canaveral, and Mayagüez); and it could not provide 24-hour tracking. Aware of this deficiency, ARPA purchased two more 26-meter antennas to fill in the coverage gap, but they were not ready in time for Pioneer III and Pioneer IV.[16]

The overseas 26-meter ARPA dishes were outside Woomera, Austalia, and Johannesburg, South Africa. Goldstone inherited an additional antenna from the Echo communication satellite experiment. It played an essential role in tracking lunar probes because, unlike the other JPL facilities, it had a transmitter that enabled it to send commands at least as far as the Moon unlike the weaker Pioneer 25-watt transmitter. The two overseas dishes plus Goldstone formed the Deep Space Instrumentation

---

[13] Rechtin, 349, 350 & 353; Fletcher Kurtz and Fridtjof Speer, "Early Orbit Determination Scheme for the Juno Space Vehicle," 62-63 in *Tracking and OD*; Memorandum, Herget to Hagen, "Definitive Orbit of 1958 Alpha: Explorer I," April 8, 1958, "Doppler Observations," D4, Herget Papers.

[14] Hall, *Ranger Chronology*, 54, 55, 75, 76 & 88; Wolverton, 21 & 22.

[15] Hall, *Ranger Chronology*, 76; Wolverton, 20; Corliss, *History of the DSN*, 16-17.

[16] Henry Curtis and Dan Schneiderman, *Pioneer III and IV Space Probes*, TR 34-11 (Pasadena: JPL, January 29, 1960), 3; Hall, *Ranger Chronology*, 60, 65 & 74; Corliss, *History of the DSN*, 14 & 19; Mudgway, *Uplink-Downlink*, 9.

Facility (DSIF), which, on December 24, 1963, became the Deep Space Network (DSN).[17]

Because JPL was in the business of both collecting and analyzing tracking data, a central question was: what kind of data? A significant difference between the Pioneer and later missions was the type of data processed to compute flight paths. For the Pioneers, JPL relied more on angular measurements than on Doppler. This was the opposite of future navigational practice. Navigators abandoned their preference for angles as spacecraft crossed greater and greater distances. When probes were relatively close to Earth, angle determinations were enough to initiate orbit computations, but as their distance from Earth grew, the angular rate of motion across the celestial sphere became increasingly smaller. The craft scarcely appeared to move against the background of fixed stars.

Navigators also hoped to stop relying on one-way Doppler. The Goldstone and Mayagüez dishes collected one-way Doppler along with position (azimuth and elevation) angles and telemetry. One-way Doppler meant that the antennas did not transmit to the probe, but simply received the signals coming from the Pioneer transponder. The reliance on one-way Doppler rested on two assumptions. One was that the frequency transmitted by the probe did not change during the flight. The other was that the variations in the signals received by the ground antennas represented actual variations in flight speed. JPL correctly judged that these assumptions were open to question. One-way Doppler inherently suffered from frequency drifts in the spacecraft's oscillator, and ground electronics could not separate the frequency drift from the Doppler frequency shift caused by the spacecraft's motion.[18]

On future flights, therefore, JPL wanted to put a transponder on the probe with which they could communicate and receive "an unambiguous measurement of range rate [Doppler]." JPL argued that NASA needed two-way Doppler for all future missions as a critical prerequisite for obtaining precise measures of flight paths.[19] Meeting that need, though, meant installing additional hardware on probes, which already were weight-challenged by contemporary standards. Luckily for NASA navigation, JPL's arguments carried the day.

NASA switched to two-way Doppler beginning with the Rangers. This data type required that a transmitter and receiver on the ground work in tandem with the spacecraft's transponder. The transponder received an 890-MHz signal from Earth, modulated the signal by multiplying it by 96/89, and transmitted it at this new frequency (960 MHz). Technicians were able to make more accurate Doppler measurements by comparing the a priori frequency (890 MHz) with the frequencies of the received signals. The local frequency standard allowed trackers to discount oscillator drift and to obtain more accurate measures of Doppler shift.[20]

---

[17] Mudgway, *Uplink-Downlink*, 9-15 & 61; Hall, *Ranger Chronology*, 60 & 74; Hall, *Lunar Impact*, 149; Corliss, *History of the DSN*, 14, 46, 49 & 54; Renzetti, "DSIF," 36-37 & 70; *Techniques and Performance*, 1-3; Renzetti, *DSN History*, 39; JPLSPS 37-10:1, 64.

[18] *Tracking Moon Probes*, 3-5; *Pioneer IV OD Program*, 3-4; *Tracking Pioneer IV*, 4-5; Corliss, *History of the DSN*, 9; Renzetti, "DSIF," 70; "Ranger Program," 64-65.

[19] *Tracking Moon Probes*, 12.

[20] Corliss, *History of the DSN*, 9 & 54; Renzetti, "DSIF," 36, 37 & 70; "Ranger Program," 64 & 65;

Another, but experimental, data type was three-way Doppler. DSIF engineers tested it during the Ranger 4 flight in hopes that it might become a primary data type in the future. Three-way Doppler differed from one-way and two-way in that it called upon observations made at two different antennas. One station collected two-way Doppler, while a second "listened" to the transponder signal, acquiring one-way Doppler. The goal of the study—carried out over all subsequent Block III Ranger flights—was to determine the accuracy of three-way Doppler. The main factor limiting its utility arose from variations in the ground stations' reference frequency.[21]

Subsequently, JPL addressed this issue by initiating the use of crystal-controlled oscillators within the DSIF. These oscillators were relatively stable, yet frequency drift still happened. Therefore, in preparation for Mariner 2, NASA installed special low-noise, extremely-sensitive equipment, namely a combination of parametric amplifier and synthetic-ruby maser, at the Goldstone Pioneer facility. The significant improvement wrought by the crystal oscillators dwarfed in comparison with the performance of the new hardware. The higher stability of the experimental Goldstone frequency standard led William R. Corliss, a historian of NASA tracking networks, to hail its adoption as the "most significant improvement from the standpoint of navigation" that resulted in a dramatic leap in the precision of two-way Doppler.[22]

Data preparation. The JPL Computing Center processed the readings flowing from Goldstone via teletype and telephone. The facility benefited from automatic IBM teletype-to-punched card converters that put the observations into a format usable by the IBM 704 machine. Personnel at the Pasadena Computing Center also called on smaller electronic computers, desk calculators, and pre-computed charts. The RAND Corporation's IBM 704 in Santa Monica served as a backup. For Ranger 3 through Ranger 5, STL provided the backup computer.[23]

Computing Center staff analyzed incoming data to provide quick and accurate pointing angles (the predicts) for the tracking antennas as well as to estimate the spacecraft's trajectory. This information, encoded on punched cards, went through a card-to-tape converter before going out to the tracking stations via teletype. In this way, the primary tracking network (Goldstone and Mayagüez) was able to count, encode, transmit, and convert tracking data, and then compute, convert, transmit, and display acquisition data nearly "automatically." The only step that was not automatic, JPL

Techniques and Performance, 6 & 23; Hall, *Lunar Impact*, 149; *Tracking Mariner Venus 1962*, 19.
[21] *Ranger 4 OD*, 10; *Ranger 6 OD*, 3 & 4; *Tracking Mariner Venus 1962*, 31 & 74; *Ranger 7 OD*, 3, 8 & 82; *Ranger 9 OD*, 9, 13, 103 & 104.
[22] *Ranger 9 OD*, 9; *Ranger 6 OD*, 57; Corliss, *History of the DSN*, 39, 61, 64 & 90; Wheelock, 70; Hamilton, interview, 1992, 2.
[23] *Tracking Pioneer III and Pioneer IV*, 3; Hall, *Lunar Impact*, 149; Renzetti, "DSIF," 70; Corliss, *History of the DSN*, 54; *Tracking Moon Probes*, 5; *Pioneer IV OD Program*, 4-5 & 11; *Tracking Pioneer IV*, 5.

boasted, "was the carrying of data cards the 25 ft. [7.6 meters] between converters and machine input and output."[24]

One of the critical operations on which navigation relied was the processing of tracking data in preparation for its use by navigators. During the Pioneer days, tracking data underwent a number of procedures by a vital computer program known originally as the Tracking Data Editing Program (TDEP). Its main purpose was to winnow out bad data points from the batches of tracking data arriving by teletype. Appropriately, it was a joint product of the navigation and programming departments, having been developed by Marshall S. Johnson, one of the lab's "computer wizards," in navigation and J. H. Brown in the Computer Applications and Data Systems Branch.[25]

Navigators also visually scrutinized the data. For example, they looked at a display of incoming teletype data printed in real time to detect systematic errors. They studied printed and verbal reports from DSIF facilities to detect any regularly occurring flaws. The data also was available on punched cards. One of the navigators who visually scrutinized Ranger data was Bill Sjogren. He recalled: "We had IBM punch cards—all our data came in on these IBM punch cards—editing the data by throwing out punch cards, 'Oh, that looks like a bad data point.' We then refitted the orbits with these edited punched cards. . . . Sometimes the data were coming in on teletype machines, paper tapes. From the paper tapes we'd punch IBM cards, and then read the IBM cards in for our data."[26]

Human visual inspection of the data had its limits, so the Tracking Data Editing Program took over. It checked the format, data condition code, data range, station identification, and time sequence against a deck of master format and control cards. It next printed out a list of all data along with the reason for rejecting any particular data point, so that JPL staff could review it. Technicians then entered the data that the program did not reject onto a master data tape that contained a record of all good data points. This master data tape fed into the navigation software. The editing program also assigned weights to the data in accordance with information provided by the navigators.[27] Data weighting recognized that not all data were equal; they varied in accuracy and degree of reliability.

The JPL approach to weighting data was the conception of A. R. Maxwell Noton. Originally from England and armed with a doctorate in mathematics from Cambridge University, Noton started at JPL in 1958. His task was to develop the mathematical analysis of radio-commanded midcourse guidance for lunar missions. Elliott Cutting and Fred Barnes were assigned to work with him. Noton laid out his method for weighting data in an internal memorandum of August 1959 and further elaborated it in a report dated September 1960. The latter, Cutting noted, was "the first comprehensive analysis of midcourse guidance for space missions." Noton derived equations for calculating the midcourse velocity correction based on orbit determination done from DSN data. He also determined guidance errors resulting from deficient maneuver executions or orbit

---

[24] *Tracking Moon Probes*, 5-6; *Pioneer IV OD Program*, 5 & 6; *Tracking Pioneer IV*, 6.
[25] *Ranger 4 OD*, 82; Hall, *Lunar Impact*, 89.
[26] *Ranger 4 OD*, 82; Sjogren, interview, 2000, 1-5.
[27] *Ranger 4 OD*, 36; *Ranger 5 OD*, 37.

determinations. At the time, JPL navigation (orbit determination) was directed mainly towards spacecraft guidance.[28]

JPL referred to Noton's method as a modified least squares method, because it weighted data in a somewhat different way than the usual least squares method. One result of its application was a reduced sensitivity to "blunder points" or small "hidden" errors whose effect might be rather substantial. "Blunder points" were data points that deviated excessively from limits established in the weighting process. JPL validated its weighting method by commissioning a study by T. A. Magness and James B. McGuire of STL that compared the performance of the Noton method versus the usual least squares methods.[29]

The data editing program underwent an upgrade in time for the Mariner 2 mission that took advantage of the new IBM 7090 in the Central Computing Facility. The software rewrite, the work of Robert E. Holzman and C. Coltharp in the Computer Applications and Data Systems Section, sported a new feature aimed at decreasing the mass of data that the navigation software had to process by applying data-compression techniques.[30] Data compression was a significant departure from the Ranger Block II (Rangers 3 through 5) approach. The quantity of data gathered from the Mariner Venus voyage over a period of months was dramatically greater than the amount acquired during the Ranger flights which lasted about 70 hours. Without compression, the IBM 7090 would have needed three hours to process each iteration of computations. When combined with the fact that the Mariner navigation program processed data three times faster than its Ranger predecessor, compression kept iteration times under 30 minutes.[31]

The data editing program added yet one more feature: it constructed new data types through a process called differencing. Tests of differencing took place during the Mariner 2 mission. The purpose of the technique was to decrease the effect of errors common to one-way and two-way Doppler. In differenced one-way Doppler, two stations collected one-way Doppler simultaneously, and the program differenced the two one-way signals. Only the Goldstone complex was capable of deriving differenced two-way Doppler. It differenced the two-way Doppler received at the Echo antenna and that received simultaneously at the Pioneer station.[32] Differenced Doppler eventually became a common data type for JPL navigators.

---

[28] Noton, JPL internal memorandum, "Effect of Correlated Data in Orbit Determination from Radio Tracking Data" August 1959, cited in *Ranger 5 OD*, 37; Noton, Cutting, and Barnes, *Analysis of Radio-Command Mid-Course Guidance*, TR 32-28 (Pasadena: JPL, September 8, 1960); Personal communication, Elliott Cutting, June 14, 2012.

[29] *Ranger 5 OD*, 36-37; Carleton B. Solloway, *Elements of the Theory of Orbit Determination*, EPD-255 (Pasadena: JPL, December 9, 1964), 6-6, 7-1, 7-2 & 7-3; Magness and McGuire, "Comparison of Least Squares and Minimum Variance Estimates of Regression Parameters," *The Annals of Mathematical Statistics* 33 (June 1962): 462-470; Magness and McGuire, *Statistics of Orbit Determination: Weighted Least Squares* (Redondo Beach, CA: STL, January 15, 1962).

[30] Holzman and Coltharp, *JPL Tracking Data Editing for the IBM 7094*, TM 33-170 (Pasadena: JPL, August 1, 1964), 13-15; *Mariner-Venus 1962*, 60; *Mariner R Project*, 300 & 302.

[31] *Mariner R Project*, 301.

[32] *Mariner R Project*, 302. The Echo two-way Doppler was coherent, because the Echo station had transmitted and received. The Pioneer Doppler was not coherent, because that antenna was receiving

In 1964, Mariner 4 made available a new navigation computer, an IBM 7094. The transition to the IBM 7094 necessarily meant rewriting all programs used by navigators and mission designers. The emphasis was on adapting the programs to be compatible with the various display media in the flight operations center. The Tracking Data Editing Program now became the Tracking Data Processor and Orbit Data Generator (TDP-ODG) program. It still cast out bad data points and stored the good ones on master tapes that went into the Master Data Library, a compilation of all the good telemetry and tracking data received and recorded during the Mariner mission plus important commentary information that explained the peculiarities and anomalies in data quality. The hope was that the history of the mission embodied in the library would serve as the basis for post-flight analyses of spacecraft subsystems, the scientific payload, the spacecraft's trajectory, and the performance of the Deep Space Network. The library maintained raw tracking data as well as digital magnetic tapes of the data in formats compatible with the tracking data processor and orbit data generator programs.[33]

---

signals sent by the Echo station. They were pseudo-coherent, because the DSIF attempted to match up the sent and received signals as if a single station had been involved.

[33] J. N. James to Donald P. Hearth, October 28, 1966, F50/B3 JPL421; Corliss, *History of the DSN*, 131; Mudgway, *Uplink-Downlink*, 62; Null, "Mariner IV Flight Path," 23; *MM64 Final Report*, 226; *MM64 Operations Report*, 24.

Chapter 5

Adapting Astronomy to Navigation

Navigation requires not just tracking data, but also science, especially astronomy, and software. At the outset, like Vanguard, JPL navigation borrowed heavily from Naval Observatory astronomers, especially the practices, values, and formulas embedded in the ephemerides distributed by the Almanac Office. Adapting conventional celestial mechanics to space navigation was not always straightforward. Navigators and astronomers measured differently. Navigators, for example, used kilometers instead of Earth radii and Greenwich Mean Time instead of astronomers' Ephemeris Time. Increasingly, too, navigators wanted better physical constants for their calculations.

Navigators' software reflected its indebtedness to astronomy and astronomers' practices, but also requirements specific to determining spacecraft trajectories. These needs, for example, led to the development of specialized ephemerides tapes suited to JPL navigation software. The distribution of the navigators' ephemerides on computer tapes to the growing number of research and governmental users was the first step in JPL navigation becoming the source of the constants, models, and ephemerides for almanac offices. The navigation software also owed a debt to the computer and other expertise that JPL had acquired as an Army contractor and missile developer, not to mention its debt to the world of subatomic physics.

The SPACE Program. The program that calculated and forecast flight paths both before and during missions was known as the Space Trajectories Program or simply SPACE for short. It served two distinct purposes. During missions, it operated as an auxiliary to the main navigation program, which itself was a collection of programs, routines, and subroutines designed to work interactively both during and after flights. Between launches, mission designers ran the program independently in order to generate candidate trajectories for future space projects.

The creation of the trajectory program was the work virtually of a single person, Douglas B. Holdridge,[1] who worked in JPL's Computer Applications and Data Systems Section. He wrote and rewrote the program in FORTRAN Assembly Program language, as navigation migrated from the IBM 704 to the IBM 7090.[2] SPACE was a bit simplistic in a number of ways, yet upgraded versions served as JPL's trajectory software for many years.

The original program calculated for the Moon, Venus, and Mars as target destinations. Much like the Vanguard software, when the probe was close to Earth, it took into account perturbations arising from the Earth's oblateness as well as flight path disturbances caused by lunar gravitation. The mathematical formulas for handling these

---

[1] Ekelund, "History of ODP at JPL," 2; *Ranger 4 OD*, 82.
[2] *Tracking Moon Probes*, 5; *Pioneer IV OD Program*, 4-5 & 11; *Tracking Pioneer IV*, 5.

perturbations were due to subroutines developed by Bernard E. Kalensher, who previously had worked at the laboratory on missile path perturbations for the Army Ordinance Corps (the ORDCIT Project).[3]

One of the principal characteristics of trajectory programs is the ability to map flight paths in any one of four different frames of reference commonly used by astronomers. These reference frames were the mean equator and mean equinox of 1950.0, the mean equinox and ecliptic of 1950.0, the true equator and equinox of the current date, and the true equinox and ecliptic of the current date.[4] The 1950.0 reference frame was a convention similar to the one adopted in 1960 by astronomers and which related directly to the keeping of time.

In 1950, during a meeting held in Paris, the International Astronomical Union (IAU) recommended adopting Ephemeris Time. As a result, the basis of timekeeping was no longer the diurnal spinning of the Earth on its axis (such as Greenwich Mean Time), but the period of its orbit around the Sun (known as Ephemeris Time). The adoption of Ephemeris Time still left many questions unanswered, such as the value of the basic unit of time, namely the ephemeris second. In 1957, the IAU, and in 1960, the Bureau International des Poids et Mesures, addressed this particular issue. They based their resolution on the length of the tropical year defined as the interval between consecutive crossings of the equator by the Sun as it journeyed from the southern to the northern hemisphere. The length of this interval was not constant for two reasons. In the first place, the period of the Earth's orbit varied because of the gravitational influences of other planets. Secondly, the point of the Sun's crossing—the vernal equinox—was not fixed with respect to the stars, because the Earth undergoes certain motions known as precession and nutation. Astronomers decided upon an arbitrary tropical year that began at 12h0 on January 0, 1900.[5]

In lieu of Ephemeris Time, JPL utilized Universal Time, specifically Greenwich Mean Time. The choice was consistent with earlier space practice, such as the Vanguard project. Holdridge adopted the astronomers' 1950.0 reference frame, but his began at 0h0 on January 0, 1950. The Ranger lunar shots called for a further alteration. Kalensher developed a subroutine that performed the rotation necessary to transform data in the 1950.0 reference system to lunar Cartesian coordinates. Yet another deviation from astronomical practice in SPACE was the use of kilometers instead of classical units of celestial mechanics (the astronomical unit for example).[6]

Holdridge designed SPACE so that it could use either of two methods favored by both astronomers and navigators for computing orbits. Both approaches bore the names of astronomers: Johann Franz Encke (1791-1865) and Philip H. Cowell (1870-

---

[3] Kalensher, *Equations of Motion of a Missile and a Satellite for an Oblate-Spheroidal Rotating Earth*, No. 20-142 (Pasadena: JPL, April 12, 1957).
[4] Holdridge, 1 & 3.
[5] *Constantes fondamentales de l'astronomie, 27 mars-1 avril 1950* (Paris: Centre National de la Recherche Scientifique, 1950), 128-135; Dick, *Sky and Ocean*, 528; Anderson, 11.
[6] Kalensher, *Selenographic Coordinates*, TR 32-41 (Pasadena: JPL, February 24, 1961); Holdridge, 2 & 3; *Tracking Moon Probes*, 5; *Pioneer IV OD Program*, 4-5 & 11; *Tracking Pioneer IV*, 5.

1949). Although the two methods dated from an earlier era, they found a new purpose with the advent of space exploration.[7]

The Cowell method, as discussed in Chapter 3, is a straightforward integration of spacecraft accelerations and deviations from its predicted orbit. One could adapt it easily to large-scale mainframe digital computers, and it did not consume as much scarce computer space as other approaches, both of which were navigators' desiderata. But, the Cowell method was about twelve times slower than the Encke. For lunar and longer missions, JPL avoided the Cowell method, because errors resulting from rounding off soon took their toll.

The Encke method involved integrating only the deviations from the predicted orbit rather than the probe's total accelerations. It assumed that the spacecraft's motion deviated only slightly from that of an ellipse, and it integrated only those amounts representing differences from the elliptic motion thereby reducing the effects of rounding errors. The Encke's main drawback was that it required more work than the Cowell to compute a single step. Encke had many proponents among those involved in computing Earth-satellite orbits, including Samuel Herrick and his former pupil, Robert M. L. Baker, Jr., who was an Air Force consultant. Her Majesty's Nautical Almanac Office also preferred the Encke approach, but for determining the paths of comets as they approached the Sun.

STL, on the other hand, favored the Cowell method for its Pioneer navigation software. Samuel D. Conte, the head of STL's orbit-determination effort, had managed the STL research group involved in ballistic missile and satellite studies from 1956 to 1962.[8] Under his direction, STL developed three computer programs for determining spacecraft flight paths. One, derived from the work of Samuel Herrick, served strictly for Earth satellite calculations. In those instances, they discovered that Herrick's approach was more accurate than the Cowell. Nonetheless, STL called on the latter for the Pioneer launches and planned to use it on future interplanetary missions.[9]

These adaptations of astronomical practices to the needs of space navigation persisted into the SPACE upgrades introduced for Mariner 2. JPL referred to this iteration as simply DBH07, which stood for Douglas B. Holdridge (DBH) version 7 (07). The advantages inherent in the latest rewrite significantly influenced the organization of the Orbit Determination Program, which itself had undergone improvements for the first Mariner flight. The subsequent Block III Rangers took advantage of further improvements in SPACE that Holdridge added.[10]

---

[7] For the following discussion of the Cowell and Encke methods, I relied on: Holdridge, 2; Lorell, Carr, and Hudson, 4; *Techniques and Performance*, 13; Robert M. L. Baker, Jr., "Astrodynamics," 38, 39 & 57 in Helvey; James A. Ward, "Trajectory Computation and Optimization," 96-97 in Helvey; Paul Herget, *The Computation of Orbits* (Ann Arbor: Edwards Brothers, 1948), 98; Pedro Ramon Escobal, *Methods of Astrodynamics* (New York: John Wiley & Sons, 1968), 224-227; Edward V. B. Stearns, *Navigation and Guidance in Space* (Englewood Cliffs, NJ: Prentice-Hall, 1963), 59.
[8] Purdue University, Department of Computer Sciences, "Computer Science Pioneer Samuel D. Conte Dies at 85," July 1, 2002, http://www.cs.purdue.edu/feature/conte.html (accessed January 30, 2008).
[9] Conte and Lem Wong, "Computing Methods for STL Space Probe," 349-351, stenographic transcription, NASA, "Conference on Orbit and Space Trajectory Determination," F150/B15, Siry Papers.
[10] Ekelund, "History of ODP at JPL," 2; *Mariner R Project*, 302; *Ranger 7 OD*, 2 & 188; *Ranger 6 OD*, 193; *Ranger 9 OD*, iv.

The Orbit Determination Program. The Orbit Determination Program (ODP) was the principal navigation program for estimating spacecraft trajectories. The first ODP dated back to the Pioneer III and Pioneer IV launches, and JPL subsequently applied it to analyze tracking data from the Tiros I weather satellite, the Navy's Transit Ib navigation probe, the first Vanguard (1958β), and the NASA-Bell Telephone Echo experimental communications balloon as well as Pioneer V.[11]

The crafting of the earliest iterations of the ODP for the Pioneer and Ranger probes was the work of a rather modest number of individuals. R. Henry Hudson was the chief computer programmer. Jack Lorell was another key contributor. Hired into Homer Joe Stewart's Research Analysis Section in 1946, Lorell later was in the Systems Analysis Section from 1960 and for a while taught at Caltech.[12] Russell E. Carr, in JPL's Space Sciences Division, also contributed to the software. During the mid-1950s, he and Lorell had worked together on a GALCIT propulsion project,[13] and in 1959 Carr wrote a short report for public consumption on the Pioneer tracking and telemetry acquisition method, which borrowed somewhat from radio telescope practices.[14]

The ODP computer code underwent upgrades rather quickly and under trying circumstances. For starters, JPL did not receive the IBM 704 navigation computer until two months before the launch of Pioneer III. Then, during the period of less than three months between the launches of Pioneer III and Pioneer IV, JPL altered its approach to combining data from various sources and its methods for weighting data. The resulting software was the product of a major reorganization of computing procedures that enabled it to handle greater quantities and types of data.[15]

The first Ranger launch saw a similar rush to undertake a major overhaul of the navigation software. Tom Hamilton, the navigation chief, recalled:[16]

> Our orbit program was disassembled the week before the first Ranger launch to add in some capability that it never had before. It got put back together the day before launch with these new capabilities in and we went out and ran it and found out it took three times as long to run now. But it had accuracy that was required to accomplish the mission. We didn't have very much time. Afterwards we fixed the thing so it would run faster and still have the capabilities but we couldn't do that in a day.

---

[11] *Tracking Moon Probes*, 10 & 12.

[12] Nead, "Reminiscences," 4; "Deaths of AMS [American Mathematical Society] Members," *Notices of the AMS* 56,6 (June-July 2009): 748; "Around the Jet Lab," *Lab-Oratory* 5,12 (July 1956): 9; "Lorell, Jack," *Pasadena Star-News*, March 21, 2008, http://www.legacy.com/pasadenastarnews/Obituaries.asp?Page=SEARCHRESULTS (accessed February 28, 2009).

[13] Lorell and Henry Wise, "Steady-State Burning of a Liquid Droplet: I. Monopropellant Flame," *Journal of Chemical Physics* 23 (1955): 1928-1932; Lorell, Wise, and Carr "Steady-State Burning of a Liquid Droplet: II. Bipropellant Flame," *Journal of Chemical Physics* 25 (1956): 325-331.

[14] Carr, *The Jet Propulsion Laboratory Method of Tracking Lunar Probes*, EP-793 (Pasadena: JPL, June 4, 1959), esp. 2-3 for radio telescope practices.

[15] *Pioneer IV OD Program*, 4-5, 9 & 11; *Tracking Moon Probes*, 5; Lorell, Carr, and Hudson, 3; *Tracking Pioneer IV*, 5.

[16] Hamilton, interview, 1992, 18.

Such a major software revision so close to launch was neither normal nor to be repeated.

The program's original name—Lunar-Probe Tracking and Orbit-Determination Program[17]—made clear that its main purposes included providing predicts for aiming the DSIF antennas.[18] As rewritten for the first Rangers, the ODP integrated two kinds of angular measurements (azimuth and elevation; local hour angle and declination) and two types of Doppler (one-way and two-way).[19] Like all future navigation programs, it determined the differences between the calculated and observed values (the residuals), and endeavored to reduce the sum of the squares of the residuals to a minimum.[20]

A significant ODP feature was the determination of the extent to which a spacecraft might miss its target. This calculation was critical for devising course correcting maneuvers and evaluating the maneuver's success or failure. In this sense, the target was not a celestial body, but a place along the trajectory. It was not an aiming point, but rather a point in a plane—a two-dimensional area called the B-plane—along the flight path that JPL navigators termed the miss parameter. The miss parameter was notionally similar to the use of range and azimuth on Earth to specify impact points. The B-plane concept and terminology were the invention of Bill Kizner, a navigator with a background in physics who first described them in 1959. JPL lore attributes the origin of the B-plane to the world of nuclear physics. In particle physics, "B" is the "miss distance" between something in motion and the thing it was supposed to hit.[21] The B-plane was exceptional, an original JPL concept that did not come from astronomy, but from nuclear physics.

The ODP factored in several parameters—aside from the Earth's oblateness—that influenced flight paths as well as the accuracy of navigators' computations. For instance, it accounted for the gravitational pull of the planets out to Jupiter, the long-term and short-term nutations of the Earth, and solar radiation, a phenomenon recently measured during the Echo communication balloon experiments conducted by JPL and Bell Telephone.[22] However, the first ODP iterations had no corrections for the impact

---

[17] Lorell, Carr, and Hudson; Duane Muhleman, *Satellite Orbit Determination and Prediction Utilizing JPL Goldstone 85-Foot Antenna and the JPL Tracking Program*, TR 34-27 (Pasadena: JPL, February 23, 1960), 1. By the launch of Ranger 3, the software bore the name JPL Orbit Determination Program. Ranger 3 OD, 2.

[18] Carr and Hudson, 1.

[19] *Techniques and Performance*, 10-11 & 12-13; Carr and Hudson, 26 & 43; Lorell, Carr, and Hudson, 4 & 5-6.

[20] *Tracking Pioneer III and Pioneer IV*, 13 & 14.

[21] Ranger 3 OD, 24; Kizner, *A Method of Describing Miss Distances for Lunar and Interplanetary Trajectories*, EP-674 (Pasadena: JPL, August 1, 1959), 1. On the attribution of the B-plane to Kizner and its derivation from particle physics: Efron, interview, 27; Mottinger, interview, March 9, 2009, 61; Wood, interview, 76; William M. Owen, Jr., "How We Hit That Sucker: The Story of Deep Impact," *Engineering and Science* 4 (2006): 10-19. Kizner's publications do not provide any written clues as to its derivation.

[22] Harrison M. Jones, R. W. Parkinson, and Irwin I. Shapiro, "Effects of Solar Radiation Pressure on Earth-Satellite Orbits," *Science* 131 (March 25, 1960): 920-921; Shapiro and Jones, "Perturbations of the Orbit of the Echo Balloon," *Science* 131 (November 18, 1960): 1484-1486; Duane O. Muhleman, R. Henry Hudson, Douglas B. Holdridge, R. L. Carpenter, and K. C. Oslund, *Observed Solar Pressure Perturbations of Echo I*, TR 34-114 (Pasadena: JPL, August 22, 1960); R. W. Bryant, "The Effect of Solar Radiation Pressure on the Motion of an Artificial Satellite," *The Astronomical Journal* 66 (October 1961): 430-432; Paul Musen, Bryant, and Ann [Eckels] Bailie, "Perturbations in Perigee Height of Vanguard I," *Science* 131 (March 25, 1960):

of the ionosphere on tracking signals. A preliminary model was available for the Ranger flights, but the relatively short distance to the Moon seemed to justify not correcting for those effects.[23] By the 1964 Mariner 4 mission, the ODP had a model of the ionosphere, but it still did not address the pole-wandering motion of the Earth.[24]

The second generation ODP. Mariner 2 and Ranger 5 were the first to benefit from a significant improvement in navigational capabilities thanks to the replacement of the IBM 7090 with an IBM 7094 mainframe. They equally profited from a greatly revised navigation program referred to as the second generation ODP. The software and mainframe combination yielded a more flexible and more capable navigation process. At the same time, though, the improvements introduced a greater degree of complexity.

Undertaken in January 1962, the writing of the second generation software was the work of Michael R. Warner, Melba W. Nead, and Henry Hudson, all in the Computer Applications and Data Systems Section. Melba Nead was a veteran of the GALCIT project, having joined JPL in 1942 as one of the first human computers. At the time, she was only one of three such computers. Nead computed trajectories on a Marchant calculator for the Corporal E tests conducted for the Army. It was slow, laborious work. As Nead recalled, "Each of us could compute a complete trajectory in only one week!"[25]

Nead learned machine language programming for the IBM 701 and IBM 704 at North American Aviation, where she worked from 1950 until 1956, when she transferred to the inertial guidance section of the firm's Autonetics Division and programmed the Minuteman onboard computer. Upon returning to JPL in October 1961, Nead brought with her computing expertise funded by the armed forces for Cold War purposes.[26] Additionally, she, like Jack Lorell, Russell Carr, and others, reflected the indebtedness of JPL navigation to its past as an Army contractor and developer and tester of missiles.

With Hudson in charge, Nead and Warner completed the first operational version of the new ODP in August 1962.[27] A young John D. Anderson, a graduate student of Samuel Herrick at UCLA, devised the equations that went into the software. In 1960, while still a graduate student, Anderson started work in the small JPL navigation section and received help on his dissertation (a topic related to navigation) from fellow navigators. He was one of the first of Herrick's students to enjoy a navigation career at JPL.[28] The second-generation equations and code were capable of handling an unprecedented amount of Doppler, more than for any previous NASA flight. This

---

935-936; Musen "The Influence of the Solar Radiation Pressure on the Motion of an Artificial Satellite," *Journal of Geophysical Research* 65,5 (1960): 1391-1396.

[23] *Techniques and Performance*, 11 & 12.

[24] *MM64 Final Report*, 25 & 226; Null, "Mariner IV Flight Path," 6 & 23; *MM64 Operations Report*, 588; Donald W. Trask and Leonard Efron, "DSIF Two-Way Doppler Inherent Accuracy Limitations: III. Charged Particles," 3-11 in JPLSPS 37-41:III.

[25] Nead, "Reminiscences," 1.

[26] Ibid., 1, 4 & 6.

[27] *The JPL ODP*, 1-3, 7 & 54; Nead, "Reminiscences," 6.

[28] Anderson dissertation materials, B4, Herrick Papers.

capacity was critical, because the Deep Space Network provided Mars Mariner coverage "equivalent to 24-station-hr/day" for the duration of the mission.[29]

During space flights, the need to run the software without human involvement was imperative. Most options and parameters were preset; human intervention was constrained. The limitations of the mainframe were paramount in constraining practice. When the program operated in the "utility mode" between missions, the user could alter all options and parameters. The ability to treat certain physical constants and station locations as variables, rather than as unvarying parameters, distinguished the new software from its predecessors. Thanks to the greater capability of the IBM 7094, the ODP could solve for 12 physical constants plus the latitude, longitude, and distance from the geometric center of the Earth at 15 tracking stations (even though the network had far fewer antennas). In all, the program potentially could have worked with 63 parameters, but not simultaneously. Instead, the user had to select up to 20 of the 63 parameters to form a subset on which the program performed computations.[30]

Ephemeris tapes and constants. Neither the ODP nor SPACE could function without access to ephemerides indicating the location of the Earth, Moon, Sun, and planets at any given time. The ephemerides, moreover, had to be written on computer tapes, so that the software could access the information. The ephemeris tapes largely were the product of JPL programmers who worked in conjunction with subcontracted personnel from STL. The latter's expertise, as we saw, derived from work carried out for the military. The creation of the tapes also depended on existing ephemerides generated by Naval Observatory astronomers.

The author of the first set of ephemeris tapes was the same Henry Hudson who had written the original ODP. Hudson created two separate ephemeris tapes. One was geocentric (with the Earth at the center) and contained the coordinates of the planets, the Sun, and the Moon tabulated at one-day intervals for a period of time expressed as spanning from 2437116.5 (June 30, 1960) to 2444132.5 (June 30, 1971) in Julian days, a common astronomical practice. The second tape was heliocentric (with the Sun was in the center) and held the coordinates of the planets and the Earth-Moon barycenter[31] tabulated at 4-day intervals for the period 2437116.5 (July 1, 1960) to 2451716.5 (June 21, 2000).[32]

Consistent with SPACE, both tapes used the mean equator and equinox of 1950.0 as their frame of reference, but they expressed all coordinates in astronomical units, not kilometers, and they gave the distance from the Earth to the Moon in Earth radii.[33] Computing in astronomical units and Earth radii was a common astronomical

---

[29] *SPODP; Tracking Mariner Venus 1962*, 17 & 32; *MM64 Operations Report*, 587, 588 & 595; Null, "Mariner IV Flight Path," 1; *MM64 Final Report*, 26 & 27.
[30] *The JPL ODP*, 1 & 3; *Ranger 5 OD*, 1 & 19.
[31] The Earth-Moon barycenter center is a point in space about which the Earth and Moon appear to orbit as they travel around the Sun. It is located exactly along the line that connects the center of the Earth with the center of the Moon.
[32] Ekelund, "History of ODP at JPL," 3; Hudson, 1; Holdridge, 2.
[33] Hudson, 1.

practice, but it made the ephemeris tapes inconsistent with the trajectory program, which quantified distances in kilometers. The use of astronomers' units reflected the origins of the data written on the tapes.

The data came virtually entirely from the Nautical Almanac Office, including astronomical tables for the Sun, the Moon, Venus, and the outer planets. Because the Naval Observatory had been automating astronomical computations for some time, some data already was available on punched cards, and the Observatory supplied JPL with punched-card data. In addition, as part of the STL effort, the Lawrence Livermore Radiation Laboratory encoded some ephemerides information on magnetic tape, the so-called Themis Tapes, which JPL retained for future missions. Although the JPL tapes used certain astronomical units, Hudson had to write a program to rotate the coordinate system of the Naval Observatory data to make it consistent with JPL norms. Holdridge also created a subroutine that converted SPACE's Universal Time into Ephemeris Time to harmonize with the Naval Observatory information on the ephemeris tapes.[34]

Underlying all of the ephemerides, whether developed by JPL navigators or Naval Observatory astronomers, was a set of physical constants. The astronomical unit is perhaps the best known example. NASA had a wide-ranging interest in these constants that reflected the variety of space projects that it underwrote, including human exploration, Earth satellites, and deep-space probes. NASA dispersed responsibility for these projects among its field centers, each of which had its own history and approaches to tracking missions and calculating flight paths. NASA contractors brought their own practices to those projects, as well. In contrast, astronomers had a single set of constants whose values were accepted universally through international agreements. NASA navigators similarly wanted a standardized set of constants to help to guarantee the accuracy and consistency of orbit computations performed by NASA centers and contractors.

In late 1960, the concept of a single set of constants for use by all of NASA brought together engineers from a number of centers. Their numbers included R. Kenneth Squires (Goddard), Paul G. Brumberg (Manned Spacecraft Center, now the Johnson Space Center), and JPL's Victor C. Clarke, Jr. The Marshall Space Flight Center and Lewis Research Center also sent representatives. In 1961, this ad hoc group adopted a NASA-wide Standard Set of Constants, although the agency itself never sanctioned or published them. JPL, nonetheless, adopted them, and contractors on the Ranger, Mariner, and Surveyor programs implemented them. When Goddard and the Manned Spacecraft Center adopted them, the constants fell into general, if unofficial, use on both Earth-orbiting and human spaceflight missions.[35] Because the constants were of value to NASA engineers as well as astronomers, their continual incremental improvement became one of the great scientific contributions of JPL navigation, as we shall see in Chapter 9.

---

[34] Hudson, 2; Holdridge, 2; Lorell, Carr, and Hudson, 5; *JPL Ephemeris Tapes*, 9 & 10; *Ad Hoc NASA Standard Constants*, 10.

[35] *Ad Hoc NASA Standard Constants*, 1-2.

Chapter 6

Modeling Earth

Just as the drive to develop ephemerides and software involved navigators in celestial mechanics, the need for better determinations of tracking station locations led navigators into geodesy and geodynamics. The application of both celestial mechanics and geodesy typified the multidisciplinary nature of navigation science. The problem of determining the exact position of each station was complex, and the precision of the solution had a direct impact on computing spacecraft positions. Calculating a probe's flight path demanded knowing, among several other variables, the distance from the antenna to the Earth's center. It also required situating the antenna within the six datum sets in the World Geodetic System. In this case, just as JPL and other NASA navigators created their own ad hoc physical constants system, NASA adopted its own geodetic model for all spaceflight missions. Also, just as JPL's ephemerides effort led to improved official ephemerides, so the navigators' enhanced station locations enabled them to show the need to correct Army and Air Force topographic maps of the Moon.

When JPL navigators instituted a concentrated effort to raise the exactness of stations locations, their effort highlighted the critical role of timing errors. Timing accuracy has two elements: 1) the correctness of each station's clock compared to an accepted time standard, and 2) the synchronization of all overseas clocks with the master clock at Goldstone. The fix for the latter was a new synchronization method that relied on the accuracy of the lunar ephemeris, thereby linking celestial mechanics and geodynamics.

The kind of timing precision required for space navigation also entailed knowledge of polar motion. The JPL ephemerides effort involved navigators in an international organization (the IAU), while the search for good timing and polar motion information engaged them with the Bureau International de l'Heure and other international scientific institutions. Again, as with the ephemerides, here was a case of JPL navigators joining and contributing to an international scientific endeavor.

In order to model the impacts of media effects on Doppler and range, navigators similarly engaged with science and scientists. The media effects resulted from the passage of electromagnetic (tracking) signals through the troposphere and ionosphere as well as through interplanetary space. Addressing media effects required acting locally, by collecting data at each dish location, and internationally, by collaborating with scientists overseas.

Station locations. Precise knowledge of station locations in three-dimensional space is vital to navigational accuracy. Although seemingly on terra firma, the DSN antennas sit on a planet that moves and wobbles on its axis while traveling through space and whose crust also is in motion. For each mission, navigators establish a location set for each station and designate them with the letters "LS" followed by a number. The early navigation programs treated these location sets as known static quantities. Because station locations were incorporated in a flight's ephemeris, errors impacted course computations both directly and, through the ephemeris, indirectly. During post-flight analysis, when navigators reflew missions with their software, they solved for station locations as variables and discovered errors in station longitudes caused by a variety of factors, such as the Earth's polar motion and inaccuracies in Universal Time, which is a function of the Earth's rotation.[1] Their efforts to deal with these error sources drew navigators toward the sciences known as geodesy and geodynamics.

Surveys of DSN antennas situated them within a standard three-dimensional geodetic framework. The basis for the original survey was the Clarke spheroid of 1866, a standard reference for geodesy in the United States and elsewhere.[2] A new survey of DSN locations responded to a change in the geodetic reference from Clarke's spheroid to the "Kaula" or "185" spheroid. This new reference was the recent creation of Sydney-born William Kaula[3] at Goddard, where he led an ad hoc NASA committee formed to study the determination and standardization of station locations. In 1961, using gravity estimates and observations from two Vanguard satellites, Kaula set forth specific correctional amounts to the Earth's six datum sets in the World Geodetic System (the foundation for all surveys). The World Geodetic System was a geometric model of the Earth that JPL and other NASA centers had adopted unofficially for all spaceflight missions. Kaula's committee also laid out how NASA would document and describe its tracking station locations.[4] This was the geodetic equivalent of the Standard Set of Constants adopted by NASA navigators in 1961 for celestial mechanics.

JPL re-examined DSN locations using just Ranger 6 tracking data and compared the results with the locations designated in the old survey. Next, adding Ranger 8 and Ranger 9 data to the mix, they compared their new location estimates with those of both the old and new surveys.[5] The exercise led to more precise definitions of station locations. It also motivated navigators to question the accuracy of the lunar maps drawn up by the Air Force Aeronautical Chart and Information Center (ACIC) in St. Louis and the Army Corps of Engineers' Map Service.

The Army Map Service began issuing lunar topographic maps in late 1958. The first ACIC Moon maps produced under the direction of Robert W. Carder using photographs and airbrushing techniques followed in 1959. The Air Force subsequently

---

[1] Corliss, *History of the DSN*, 127.
[2] Col. Sir Charles Close, K.B.E., C.B., C.M.G., F.R.S., "The Life and Work of Colonel Clarke," *The Royal Engineers Journal* 39 (December 1925): 658-665.
[3] Gerald Schubert, "In Memoriam: William M. Kaula, Professor Emeritus of Geophysics, Los Angeles, 1926-2000," http://www.universityofcalifornia.edu/senate/inmemoriam/WilliamM.Kaula_000.htm (accessed December 16, 2009); James Ganz, "William Kaula, 73, Who Drew Maps of Earth Using Satellites," *New York Times*, April 13, 2000, http://query.nytimes.com/gst/fullpage.html?res= 9A05EFDD103EF930A25757C0A9669C8B63 (accessed March 17, 2008).
[4] *Ad Hoc NASA Standard Constants*, 2 & 13 & 17-18; Kaula, "A Geoid and World Geodetic System Based on a Combination of Gravimetric, Astrogeodetic, and Satellite Data," *Journal of Geophysical Research* 66 (1961): 1799-1811. Kaula subsequently revised his results with the addition of more observations. See, for example, Kaula, "Tesseral Harmonics of the Gravitational Field and Geodetic Datum Shifts Derived from Camera Observations of Satellites," *Journal of Geophysical Research* 68 (1963): 473-484; Kaula and John A. O'Keefe, "Stress Differences and the Reference Ellipsoid," *Science* 142 (October 18, 1963): 382.
[5] *Ranger 7 OD*, 46; *Ranger 9 OD*, 48.

collaborated with astronomer Gerard P. Kuiper, director of the Yerkes and McDonald Observatories, who already had initiated his own lunar atlas project with funding from the NSF and the Air Force Cambridge Research Laboratories.[6]

Navigators began to question the accuracy of the ACIC Lunar Astronautical Charts—many of which described regions of interest to space exploration—as a result of a discrepancy between the Ranger 6 impact time predicted by their software and the time observed on the ground by the DSN. The same recorders installed at tracking stations to measure time differences between antennas also measured the time of lunar impact with precision. The difference between the software-computed and the station-observed impact time was 1.5 seconds, a large amount when expressed in navigational time units of milliseconds. The discrepancy stimulated an extensive reexamination of the JPL navigation programs, their mathematical models, and tracking station hardware. But, navigators found no error sources that could account for the time gap.[7]

They next turned to the lunar maps. They theorized that the actual lunar elevation at the impact point differed from that shown on the ACIC charts issued in 1962, 1963, and 1964, as well as the Army Map Service's 1964 map. The Ranger 6 1.5-second difference required that the elevation at the point of impact be 3 km lower than the 1,738.3 km elevation shown on Lunar Chart LAC 60, that is, 1,735.3 ± 0.2 km. Ranger 7 also experienced a time difference, but of 1.14 seconds. Explaining the Ranger 7 time disparity required an elevation decrease of 2.7 km from the lunar elevation of 1,737.9 km given in the latest (April 1964) Lunar Chart, LAC 76, meaning the elevation of the Ranger 7 impact point was 1,735.2 ± 0.4 km.[8]

JPL navigators believed that their elevation corrections were warranted based on the consistency of the Ranger 6 and Ranger 7 results as well as new, more precise values for the radius of the Moon derived from radar observations.[9] Navigators took the revised lunar radius value into account during the Ranger 7 flight. Their predicted impact time based on tracking data acquired up to 2 hours before impact was within 0.1 seconds of the observed impact time. The improved accuracy also extended to computations of impact distance from the target point. Accordingly, Ranger 6 crashed only 27 km (17 miles) from its target point, and Ranger 7 impacted the Moon within 13 km (8 miles) of its target point. Ranger 7 television pictures provided another check of

---

[6] Raymond M. Batson, Ewen A. Whitaker, and Don E. Wilhelms, "History of Planetary Cartography," 24 & 40 in Ronald Greeley and Raymond M. Batson, eds., *Planetary Mapping* (New York: Cambridge University Press, 1990). Zdeněk Kopal and Robert W. Carder, *Mapping of the Moon: Past and Present* (Boston: D. Reidel, 1974), provides a detailed history of the military mapping endeavor.

[7] *MM64 TDA Report*, 41; *Ranger 7 OD*, 47; *Ranger 9 OD*, 49.

[8] Air Force, Aeronautical Chart and Information Center (ACIC), *Lunar Chart LAC 60* (St. Louis: ACIC, September 1962); ACIC, *Lunar Chart LAC 76* (St. Louis: ACIC, April 1964); ACIC, *Lunar Chart LAC 77* (St. Louis: ACIC, May 1963); Army Map Service, *Topographic Lunar Map 1:5,000,000*, Edition 2-AMS (Washington: Army Map Service, June 1964 [compiled in 1963]).

[9] Benjamin S. Yaplee, Stephen H. Knowles, Allan Shapiro, K. J. Craig, and Dirk Brouwer, "The Mean Distance to the Moon as Determined by Radar," 82 in Jean Kovalevsky, ed., *The System of Astronomical Constants* (Paris: Gauthier-Villars, 1965); Victor C. Clarke, Jr., "Earth Radius/Kilometer Conversion Factor for the Lunar Ephemeris," *AIAA Journal* 2 (February 1964): 363; *Ranger 6 OD*, 58; *Ranger 7 OD*, 47.

the navigators' calculations. They, too, indicated an impact 13 km from the target point.[10]

After the Ranger 9 flight, JPL navigators looked at the time differences for all Block III flights. They composed a table that compared Ranger lunar radius values with those of the Army and Air Force. They calculated a Ranger lunar radius value of 1,735.2 ±0.5 km, in contrast to the Army and Air Force values of 1740.1 ±3.1 km and 1737.8 ±0.6 km, respectively. The differences between the JPL and military elevation values were greater on Ranger 9 than on the preceding missions. The reason for this larger difference, navigators reasoned, arose from the impact point being in a crater as opposed to a lower elevation in a flat maria area.[11]

Timing is everything. The lunar map discrepancies called attention yet again to the importance of exact station locations. For that reason, Don Trask, the head of the Orbit Determination Group, hired Neil Mottinger to take on the daunting task of raising station location accuracy.[12] In Trask's words, timing was a most vexing issue, "a nontrivial problem in Orbit Determination."[13] The ultimate goal of the quest was to reduce uncertainties in the longitude and the distance off the Earth's spin axis of station locations to about 1 meter. This goal was far above current accuracy levels. Mottinger quickly realized that one of the chief impediments to achieving that goal was the navigation software itself, not to mention the limited computational capability of the IBM mainframe.[14] The software, for example, lacked models for the atmosphere, ionosphere, and polar motion. Pole changes were insignificant over the relatively short periods of lunar flights, but Mottinger's calculations indicated that planetary voyages would see station locations vary by as much as 12 meters.[15]

Mottinger began with a paucity of decent tracking data with which to undertake his task. Soon enough the amount of data from missions to the Moon—the Block III Rangers, the Surveyors, and the Lunar Orbiters—the planets—Mariner 2 and Mariner 4—and the Sun—Pioneer 6—began to pile up.[16] Still, not all data were alike in quality. Luckily, though, a few were better than the rest because the DSN had collected the Doppler at zero declination. Zero declination is the celestial equator, with declination

---

[10] William L. Sjogren and Donald W. Trask, "Radio Tracking of the Mariner Venus and Ranger Lunar Missions and its Implications," *Electronics in Transition* 4 (1965): IIIB-6.

[11] Ranger 9 OD, 50 & 51.

[12] Mottinger, interview, March 9, 2009, 2-8; Mottinger and Sjogren, "Consistency of Lunar Orbiter II Ranging and Doppler Data," 19-23 in JPLSPS 37-46:III.

[13] Trask and Muller, "Timing: DSIF Two-Way Doppler Inherent Accuracy Limitations," 15 in JPLSPS 37-39:II.

[14] Ranger 4 OD, 33.

[15] Mottinger, "Status of DSS Location Solutions for Deep Space Probe Missions: II. Mariner V Real-Time Solutions," 10, 11 & 22 in JPLSPS 37-49:II; Mottinger, "DSS Location Solutions for Deep Space Probe Missions, Using Mariner IV, Mariner V, and Pioneer VII Data," 46 in JPLSPS 37-56:II; Mottinger and Trask, 12, 14 & 18; Mottinger, "Status of DSS Location Solutions for Deep Space Probe Missions: Third-Generation Orbit Determination Program Solutions for Mariner Mars 1969 Mission," 82 & 87 in JPLSPS 37-60:II.

[16] Mottinger and Trask, 12 & 14; Moyer, "DPODP Basis."

being the degree to which an object is above that equator. At that viewing angle, the uncertainty in the spacecraft's position did not degrade his ability to compute the station's distance off the Earth's spin axis. These zero-declination observations did not take place during encounter, but rather during the cruise phase, when the Sun was the main body exerting a gravitational pull on the probe. Mottinger found no zero-declination observations among the lunar missions and only two from Mariner 5, one occurring before and the other after encounter.[17]

He further realized that station errors were arising from timing inaccuracies caused by the Earth's polar motion, Universal Time (a function of the Earth's rotation), and the lack of synchronization among station clocks.[18] Focus now shifted to raising the synchronicity of DSN clocks to prepare for upcoming missions. Timing accuracy depended on the agreement of the local station clock with an accepted time standard as well as the synchronization of all DSN clocks with the Goldstone master clock. NASA missions were demanding ever more accurate clock harmonization. For instance, the Lunar Orbiter program wanted DSN clocks to be synchronized within 50 microseconds of Goldstone. The upcoming Mariner Mars 1969 required a synchronization of 20 microseconds. In contrast, previous missions had needed a synchronization of only 10 *milliseconds*, a considerable difference.[19]

From the beginning, the DSN relied on radio signals from the WWV. The errors in synchronizing station clocks to WWV time typically were 1 millisecond at Goldstone and 2 to 5 milliseconds overseas. JPL navigators had to cut that lag to 20 microseconds, but they hoped to reach 5 microseconds as a target.[20] How would they accomplish this quantum leap of accuracy?

The DSN conducted a series of tests with four experimental methods that included using range readings from Lunar Orbiter 1, transporting a mobile Naval Observatory time standard from one station to another, and relying on the Very Low Frequency signals (15 to 60 KHz) transmitted by the National Bureau of Standards. The one that worked had an astronomical dimension: bouncing radio waves off the Moon. The system became operational in March 1967. The method's only Achilles' heel was the lunar ephemeris. JPL's best available ephemeris was accurate enough to predict the position of the Moon's center of mass to within 150 meters, a discrepancy that would put time synchronization off by 1 microsecond. Still, it was exactly what the DSN wanted: The cost was low, and operation was semiautomatic. One could make daily calibrations, and everything was completely under DSN control.[21]

---

[17] Mottinger and Sjogren, "Station Locations," 19-20 in *TSAC for MM69*.
[18] Corliss, *History of the DSN*, 127.
[19] Renzetti, *DSN History*, 60; Mudgway, *Uplink-Downlink*, 71; "Selenodesy Experiment," 24; Baugh, 97.
[20] Moyer, "DPODP Basis: Time," 36; Warren L. Martin, "Digital Communication Tracking: Resolving the DSN Clock Synchronization Error," 49-52 in JPLSPS 37-39:III; "Time-Synchronization System," 92.
[21] "Time-Synchronization System," 73, 75, 92-94 & 98; R. C. Coffin, Richard F. Emerson, and John R. Smith, "Time-Synchronization System," 72-73 in JPLSPS 37-45:III; Baugh, 97; Martin, "Digital Communication Tracking: Resolving the DSN Clock Synchronization Error," 49-52 in JPLSPS 37-39:III; Martin, "Digital Communication and Tracking: Time Synchronization Experiment," 61-67 in JPLSPS 37-42:III; Renzetti, *DSN History*, 69; Corliss, *History of the DSN*, 134.

Timing and polar motion. In addition to clock synchronization issues, timing errors also arose from the Earth's polar motion. Navigator Paul Muller recalled overhearing his colleagues talking about station location errors of 40 meters. He noticed that these "glitches" occurred around the first of the month and suggested that the cause was a timing error in the ODP. The next day his supervisor, Don Trask, summoned Muller to his office and asked him how certain he was that the fault lay with the software. Muller was quite certain. "Then you had better get set up, and fix it," Trask said with a smile, "don't you think?"[22]

Muller began working on the problem of polar motion and its impact on station locations in early 1967. The goal still was to establish station locations to within 1 meter, but the reality was that shrinking them below 10 meters was a major challenge. Muller believed that polar motion was a major error source, but he also knew that predicting polar motion "to high accuracy is quite difficult."[23]

Admitting that he knew nothing about polar motion, Muller came to believe that the key was to acquire "up-to-date time and polar motion," because the two were interrelated. Both station locations and time literally and figuratively revolved around the Earth and its rotational motion. Muller learned more about the motion of the Earth and he explored obtaining polar motion information from the Bureau International de l'Heure (BIH), the International Latitude Service (ILS), and the International Polar Motion Service (IPMS).[24]

Muller concluded that none of these agencies was equipped to supply the information he needed and certainly not in a timely fashion. Indeed, they at times located the position of the Earth's pole as much as 3 meters apart from each other. Navigators wanted to reduce the uncertainty of the pole location to around 0.5 meters. Moreover, the agencies' polar motion information was not especially up to date. Extrapolating from the time of the last measurement to the current date during a space flight could mean extrapolating over a period as long as 3 months. "Astronomers are happy with month-old and year-old quality fits," Muller opined, but not navigators.[25] The search for greater polar motion accuracy was not unlike the navigators' need for greater precision in celestial mechanics.

Muller carried out his own study of earthquakes that had occurred between 1904 and 1966. Earthquakes, especially those of sizeable magnitude—such as the February 27, 2010, earthquake near Chile and that of April 11, 2011, off the coast of Japan—redistribute the Earth's mass and consequently alter the length of the day. If correlations existed between earthquakes and polar motion, Muller reasoned, it might be possible to construct a model that would allow navigators to predict polar motion

---

[22] "Ten Years Service," *Lab-Oratory* 6 (1974): 15; Muller, "Time," 217-218.
[23] Muller, "Polar Motion," 10 & 14.
[24] Muller, "Time," 218 & 219; Muller, "Polar Motion," 10. On the BIH, see Bernard Guinot, "History of the Bureau International de l'Heure," 175-184 in *Polar Motion*. For IPMS and ILS, see K. Yokoyama, S. Manabe, and S. Sakai, "History of the International Polar Motion Service/International Latitude Service," 147-162 in *Polar Motion*; Joachim Höpfner, "The International Latitude Service: A Historical Review from the Beginning to its Foundation in 1899 and the Period until 1922," *Surveys in Geophysics* 21 (September 2000): 521-566.
[25] Chao and Muller, 69; Muller, "Time," 219.

and thereby achieve the accuracies requisite for future missions. He came to realize, however, that it was impossible to predict polar motion from earthquakes with the desired exactness.[26]

Muller and Chia-Chun Chao now took on the task of obtaining better polar and timing information for the upcoming Mariner Mars 1969 flights. They acquired the information from the Naval Observatory's Richmond station near Miami. That station made more observations and provided better results than their other choice, the Naval Observatory itself. Muller and Chao resolved that it was "wise to stay with a single station that produces consistent results."[27] They made a special arrangement to receive up-to-date time data from the Naval Observatory between 30 days before and 6 days after Mars encounter, the data arriving by teletype and/or telephone, in order to improve navigational accuracy.[28]

The BIH Rapid Service. Muller left JPL for England to work on a doctoral dissertation under the celebrated geophysicist Stanley Keith Runcorn who taught at Newcastle upon Tyne University.[29] The challenges of timing and polar motion now passed to Henry F. Fliegel in Bill Melbourne's Tracking and Orbit Determination Section, and fellow navigator Ray Wimberly in the celestial mechanics group. Fliegel brought to the task the expertise that he had acquired at the Naval Observatory's Time Service Department before joining JPL.[30] He and Wimberly oversaw the institution of a new source for timing and polar motion information: the world's first rapid time and polar motion service. Here was an example of JPL navigators joining an international scientific endeavor and helping to improve it.

The BIH collected, analyzed, and published polar motion and timing data from cooperating national time services throughout the world. Its published values for UT1 (Universal Time or Greenwich Mean Time) appeared in a publication known as BIH Circular D about one month later. The time-lag between observation and publication was a vexing error source for navigators and many other users. Fliegel and Wimberly, for instance, noted that the most recently published data was 40 to 70 days old. Its use would have led to unacceptable errors of 1 to 3 meters after extrapolation.[31]

Starting in 1971, with the help of BIH director Bernard R. Guinot, JPL contracted with the Bureau to obtain time and latitude data from 14 cooperating observatories by teletype, as soon as obtained, thus creating the world's first rapid time and polar motion service. The Rapid Service was somewhat comparable to the 48-hour, 24-hour, and 12-hour forecasts provided by the National Weather Service in which the later forecasts

---

[26] Chao and Muller, 69 & 73-74. A later relevant study of polar motion and earthquakes: Benjamin Fong Chao, Richard S. Gross, and Yan-Ben Han, "Seismic Excitation of the Polar Motion, 1977–1993," *Pure and Applied Geophysics* 146 (September, 1996): 407-419.
[27] Muller, "Timing Data and the Orbit Determination Process at JPL," 19-20 in JPLSPS 37-41:III; Muller, "Time," 218.
[28] Muller and Chao, "Timing Errors and Polar Motion," 37-39 in *TSAC for MM69*; Dick, *Sky and Ocean*, 460.
[29] Sjogren, interview, 2008, 22.
[30] O'Handley, Standish, and Fliegel, interview, np.
[31] Fliegel, 66; Fliegel and Wimberly, 78; Carlos Jaschek, *Data in Astronomy* (New York: Cambridge University Press, 1989), 190.

always superseded the earlier ones. JPL actually supported a second Bureau service, the so-called JPL-BIH Operation service, that furnished machine-readable decks of punched cards with the timing and polar motion information for space navigation.[32]

Thanks to the Rapid Service, the standard deviations of the timing and polar motion information supplied to the Mariner Mars 1971 mission were slashed to about half of those of previous missions. Random timing and polar motion errors also were significantly less. The decks of punched cards containing timing information prepared weekly also found their way into navigating the Pioneer 10 probe and the Mariner 9 extended missions, synchronizing DSN station time via Moon bounces, and running lunar laser measurements.[33] The creation of the Rapid Service brought JPL navigation into contact with several international organizations and led navigators into a direct relationship with the BIH in Paris. Navigation was becoming an intercontinental and global scientific enterprise.

Global networks. Navigation was inherently international because of its dependence on the DSN. The need to model the impacts of the stratosphere and ionosphere, like the quest for better station locations, focused navigators' attention on local conditions and local measurements. Obtaining local measurements led to arrangements with local organizations. This web of international measuring, institutions, and tracking stations came together within the models that navigators created to represent the influences of the stratosphere and ionosphere on tracking signals at all DSN antennas. The local measurements told navigators how to make corrections to their models to compensate for media effects. The models, like the DSN, had become global in reach.

The refraction of electromagnetic signals in the troposphere created distinctive error signatures at each tracking station. Initially, navigators corrected Doppler readings by applying a single tropospheric model for all stations. As computing power grew, navigators could tailor their model to the local conditions at each facility.[34] They relied on tropospheric information collected from radiosonde weather balloons that measured various atmospheric conditions and transmitted the data to a ground receiver. The tropospheric model developed by navigators John Ondrasik and Kathryn Thuleen, unlike earlier versions, reflected changes over time. They discovered a certain annual stability in the troposphere's effect on range, but the troposphere produced other refractive effects that had an impact on Doppler, and these remained to be understood.[35]

For Mariner Mars 1969, Tony Liu developed a refraction model of the stratosphere, but it was not perfect. Just a few weeks before Mars encounter, Dan Cain noticed that the model did not work well at elevation angles above 15 degrees, the very place in the sky where tracking stations acquired most of the data. He recommended a

---

[32] Muller, "Time," 219; Fliegel, 66-67 & 69 70 71; Fliegel and Wimberly, 77-78; Dick, *Sky and Ocean*, 482.
[33] Fliegel and Wimberly, 80.
[34] Corliss, *History of the DSN*, 126; *Ranger 6 OD*, 59 & 60; *Ranger 9 OD*, 52.
[35] Ondrasik and Thuleen, "Variations in the Zenith Tropospheric Range Effect Computed from Radiosonde Balloon Data," 25 & 35 in JPLSPS 37-65:II.

new model that solved the problem, and, as a result, the mission adopted Cain's model.[36]

The Mariner Mars 1971 tropospheric model incorporated data collected around the world. Tracking stations in the Northern Hemisphere took advantage of readings from Edwards Air Force Base, California; Yucca Flats, Nevada; and Madrid, Spain, while those in the Southern Hemisphere received observations from Pretoria, South Africa, and Woomera and Wagga in Australia. Thuleen and Ondrasik continued to study observations from six different weather stations.[37] By 1975, the DSN had installed a new Meteorological Monitor Assembly at each complex to provide local tropospheric and ionospheric conditions to allow tracking personnel to calibrate range and Doppler.[38]

The drive to model the ionosphere began in 1966.[39] Different approaches were available to correct for ionospheric effects. One way was to estimate the number of charged particles by measuring the magnitude of Faraday rotation, which is a measure of the rotation of a signal's plane of polarization. JPL obtained Faraday rotation data from the Stanford University Radioscience Laboratory, which, in turn, collaborated with other institutions conducting ionospheric research. For example, the Stanford Radioscience Laboratory and the University of Hawaii's Electrical Engineering Department measured the Faraday rotation of signals transmitted from the Syncom III geosynchronous communication satellite. Stanford also carried out a similar study using one of NASA's Applications Technology Satellites.[40]

Tom Hamilton and Don Trask, concerned over navigational errors arising from ignorance of ionospheric effects during the Mariner Mars 1969 encounters, requested ionosonde data collected by the Environmental Science Service Administration (the National Oceanic and Atmospheric Administration after 1970) headquartered in Boulder, Colorado.[41] That agency enabled navigators to benefit from measurements made around the world.

Collecting data directly at DSN facilities was a different route to take, but one that promised to yield local information. Thanks to the Pioneer 6 solar occultation experiment, Goldstone had its own polarimeter for measuring Faraday rotation. Overseas antennas lacked these instruments, so JPL put together a network of

---

[36] Liu, "Recent Changes to the Tropospheric Refraction Model used in the Reduction of Radio Tracking Data from Deep Space Probes," 93-97 in JPLSPS 37-50:II; Ondrasik, "The Troposphere," 69-70 in *TSAC for MM69*.

[37] Chia-Chun Chao, "The Tropospheric Calibration Model for Mariner Mars 1971," 64, 67-68, 69 & 73 in *TSAC for MM71*.

[38] Mudgway, *Uplink-Downlink*, 126 & 150; Henry E. Burnell, Horace P. Phillips, and Richard Zanteson, "Meteorological Monitoring Assembly," 152-159 in DSNPR 42-29.

[39] Donald W. Trask and Leonard Efron, "DSIF Two-Way Doppler Inherent Accuracy Limitations: III. Charged Particles," 3-11 in JPLSPS 37-41:III; Ondrasik and Brendan D. Mulhall, "Estimation of the Ionospheric Effect on the Apparent Location of a Tracking Station," 29-42 in JPLSPS 37-57:II.

[40] Corliss, *History of the DSN*, 185; Owen K. Garriott, Fred L. Smith, III, and P. C. Yuen, "Observations of Ionospheric Electron Content using a Geostationary Satellite," *Planetary and Space Science* 13 (August 1965): 829-838; Liu, "Results, Phase II," 33. On Syncom III and the NASA ATS Program, see Daniel R. Glover, "NASA Experimental Communications Satellites, 1958-1995," 51-64, esp. 55-60, in Butrica, *Beyond Ionosphere*.

[41] Hamilton and Trask, "Introduction," SPS 37-55; Lois M. Webb and Mulhall, "Evaluation of Errors in h_{max} on the DPODP Ionosphere Model," 13-15 in JPLSPS 37-55:II.

institutions to serve their needs. In South Africa, for instance, the National Institute for Telecommunications in Johannesburg furnished ionosonde readings in nearly real-time to the South Africa antenna, while stations of the Australian Bureau of Meteorology delivered data to Woomera and Tidbinbilla. They collaborated with a Southern Hemisphere expert on the ionosphere, Professor Francis H. Hibberd, who taught at the Department of Physics of the University of New England in Armidale, Australia.[42] La Institucion de Ebro, in Tortosa, Spain, provided Faraday rotation data for the Robledo and Cebreros facilities.[43]

The main problem with using Faraday measurements from a stationary satellite, such as Syncom III, was that the geometry between the satellite and the tracking station did not change. In contrast, spacecraft tracking involved a constantly changing signal path between the spacecraft and ground antenna. JPL navigators wrote software to convert these geostationary readings into data that more realistically reflected their needs. The program handled both Faraday rotation readings made by the Goldstone polarimeter and the ionosonde data collected by the Environmental Science Service Administration from around the world.[44] At the same time, navigators were working on alternative methods that would obviate the need to collect local tropospheric and ionospheric data and would deal with the effects of interplanetary plasma as well. Instead of making measurements to calibrate their models, they changed the arrangement for making spacecraft observations.

---

[42] Matthew Jordan, *A Sprit of True Learning: The Jubilee History of the University of New England* (Sydney: University of New South Wales, 2004), 142-143.

[43] Gerald S. Levy, Charles T. Stelzried, and Boris Seidel, "Pioneer VI Faraday Rotation Solar Occultation Experiment," 69-71 in JPLSPS 37-53:II; A. Robert Cannon, Stelzried, and John E. Ohlson, "Faraday Rotation Observations During the 1970 Pioneer 9 Solar Occultation," 87-93 in DSNPR:XVI; Mulhall, Ondrasik, and Thuleen, "The Ionosphere," 48, 50 & 53 in *TSAC for MM69*; Hamilton and Trask, "Introduction," *SPS 37-55*, 13; Hamilton and Trask, "Introduction," *SPS 37-58*, 65-66; Mulhall, "Calibration," 68 & 69.

[44] Mulhall, "Calibration," 68; Hamilton and Trask, "Introduction," *SPS 37-55*, 13; Mulhall and Thuleen, "Conversion of Faraday Rotation Data to Ionospheric Measurements," 15-19 in JPLSPS 37-55:II.

Chapter 7

Navigator Scientists

Navigators took on various roles as scientists. The improvement of ephemerides and the physical constants underlying them certainly involved navigators in celestial mechanics. During NASA mission, they also acted in various roles as practitioners of their own multidisciplinary science. For instance, they made scientific discoveries just by pursuing their customary occupation. The discovery of mascons is one such example. This original scientific discovery had wide-reaching consequences for theories about the Moon's history and constitution. The multidisciplinary nature of navigation also led them to propose performing radio occultations from spacecraft and to collaborate with Stanford University researchers in that undertaking. Perhaps somewhat surprising is navigators' role as project scientists initially performing experiments in celestial mechanics. NASA had no applicable policy, so it developed special protocols for navigators to be project scientists. That same policy also enabled scientists to use the DSN for research purposes. When the DSN opened its doorways to the galaxy to scientists, notably radio astronomers, from outside the space agency, it became more than a tracking network. The DSN was now an instrument of scientific research.

Finding anomalies. The Lunar Orbiter program did not rely on JPL navigation, yet it became a springboard for navigators to make a major scientific discovery. One of the Lunar Orbiter science experiments was an analysis of the Moon's gravitational field using tracking data. Current knowledge was rather limited, because it relied solely on Earth-based measurements. Detailed gravitational information was of vital importance to Apollo, because the astronauts would be flying at a low altitude above the lunar surface and, therefore, far more subject to gravitational variations. Ignorance of these variations potentially could cause astronauts to crash into, rather than alight on, the lunar surface. Thus, Apollo program needs—and secondarily scientific curiosity—led to the inclusion of the lunar gravity experiment (known officially as the Selenodesy Experiment) on the Lunar Orbiter project.[1] The design of the Selenodesy Experiment was the work of a group at NASA Langley led by William H. Michael, Jr., who also was the Principal Investigator. Michael had worked at Langley since 1948, and he later was the Principal Investigator on the Viking Orbiter and Lander Radio Science Teams.[2]

As Principal Investigator, Michael oversaw the extraction of gravimetric information from the tracking data for the Selenodesy Experiment. His team comprised

---

[1] Byers, 133 & 138-145.
[2] James R. Hansen, *Enchanted Rendezvous: John C. Houbolt and the Genesis of the Lunar-Orbit Rendezvous Concept* (Washington: NASA, December 1995), 3-5; LaRC, "William H. Michael, Jr." biographical sheet, August 1966, F19,510, NHRC; William H. Michael, Jr., "Viking Lander Tracking Contributions to Mars Mapping," *Earth, Moon and Planets* 20 (April 1979): 149-152; JPL, NASA Facts, "Viking Mission to Mars," 8, http://www.jpl.nasa.gov/news/fact_sheets/viking.pdf (accessed April 9, 2009).

representatives from both Langley and JPL, including Jack Lorell. The main data types were two-way and three-way Doppler collected by the DSN. Some analysis relied on the Lunar Harmonic Determination Program (LHD-L), the standard JPL orbit determination program modified by Boeing.[3] As the first data came back from Lunar Orbiter 1, the results were not quite what Michael and his team had anticipated. Jack Lorell explained: "The first data received were both illuminating and enigmatic. It contained an unmistakably strong signature of the Moon's gravity anomalies, and at the same time it presaged the difficulties inherent in extracting the information."[4]

Many lunar experts expressed surprise over the discovery that the Moon was "very rough." John O'Keefe, for example, had believed that the data would indicate a relatively "smooth" Moon, while UCLA's Gordon MacDonald correctly predicted that its gravitational field would be "rough."[5] The accuracy of the Doppler used to estimate the Lunar Orbiter 1 flight path was impressive. However, the difference between the predicted and observed values (the residuals) was surprisingly large during the perilune of its orbit (the orbital point closest to the Moon). The residuals seemed to have systematic errors that were three orders of magnitude larger than those observed during any previous lunar mission. Because of their recurring nature, these errors sparked speculation about their origin. Lorell and his JPL colleagues considered and tested several hypotheses, including DSN station anomalies. None held water. Careful analysis of the orbit determination program by JPL and Langley absolved the software.[6]

Then, beginning in October 1967, Boeing carried out its own study funded by NASA. They looked into DSN methods of acquiring data, the JPL software, the mathematics and assumptions of the software, the spacecraft hardware, and the motions of both the Moon and the tracking stations, among several other factors. Boeing vindicated the DSN, the spacecraft, and the navigation software. They reported "a very distinct correlation between the Doppler residual patterns (of orbit determination data arcs involving single orbits) and the ground track of the spacecraft. In some cases the residuals can be directly related to terrain changes on the lunar surface but this was not true in all instances." The "terrain changes" the study considered were "irregular surface features in the form of highlands and lowlands" that might "cause gravitational perturbations." The study noted too the inadequacy of the lunar gravitational model

---

[3] Mayo, 395 & 396; "Selenodesy Experiment," 22.

[4] Lorell, "LO Gravity Analysis," 194.

[5] Lorell, "LO Gravity Analysis," 194; Muller and Sjogren, "Mascons," 10; Michael, Robert Tolson, and John Gapcynski, "Lunar Orbiter: Tracking Data Indicate Properties of Moon's Gravitational Field," *Science* 153 (September 2, 1966): 1102-1103; Lorell, "Lunar Gravity from Orbiter Tracking Data," 356-365 in Zdeněk Kopal and Constance L. Goudas, ed., *Measure of the Moon* (Dordrecht, Holland: D. Reidel Publishing, 1967); Tolson and Gapcynski, "An Analysis of the Lunar Gravitational Field as Obtained from Lunar Orbiter Tracking Data," in A. P. Mitra, L. G. Jacchia, and W. S. Newman, ed., *Space Research: 8th Proceedings of Open Meetings of Working Groups* (Amsterdam: North-Holland Publishing Company, 1968); Lorell and Sjogren, "Lunar Gravity: Preliminary Estimates from Lunar Orbiter," *Science* 159 (February 9, 1968): 625-627; Gordon J. F. MacDonald, "The Moon's Gravity Field," 12-15 in [Proceedings] *NASA Conference on Celestial Mechanics*, USNO, Washington, DC, January 10-11, 1963, http://ntrs.nasa.gov/archive/nasa/casi.ntrs.nasa.gov/19660025727_1966025727.pdf (accessed March 11, 2008).

[6] Mayo, 397-398; Muller and Sjogren, "Consistency," 28 & 34; "Selenodesy Experiment," 30; Lorell and Sjogren, *LO Data Analysis*, 2-6.

used in the orbit determination program: "It is obvious from this study that there is a serious need for more analysis on the lunar gravitational model."[7]

The Boeing study conclusions reflected work already undertaken by Lorell and JPL navigator Bill Sjogren, who had been a consultant to Boeing on its adaptation of the JPL navigation software. Sjogren recalled how he, Lorell, and their colleagues discovered the gravity anomalies. Sjogren and Paul Muller, who was working on a different project, were talking about the anomalous Doppler residuals. Muller believed that gravity had to be the root cause, specifically "something to do with the surface." Was there some way to separate the gravity effects from the Doppler? Both were changes in velocity: Doppler measured velocity; gravity mathematically was an acceleration. Along with Jack Lorell, they set about developing the mathematics for differentiating between the Doppler and gravity in the tracking data. They hoped to achieve two main goals. One was to describe the Moon's gravity field for scientific purposes; the other was to create a representation of the Moon's gravity field sufficient for the Apollo program.[8]

Their attention, though, soon turned to a determination of whether the changes in gravity correlated with specific features in the lunar landscape. Was there a direct connection between the Doppler residuals and the lunar surface immediately beneath the Lunar Orbiter's path? The task was mathematically more complex than teasing the gravitational accelerations out of the Doppler. They needed a new model of the Moon. They evolved seven different gravity models each purporting to represent the Lunar Orbiter data, or at least portions of it.[9]

Complications arose because the Moon is not a perfect sphere, and their model of its gravitational field had to reflect its irregularities. They applied a mathematical tool known as gravitational harmonics. They tested their hypotheses by analyzing tracking data from two different spacecraft flying similar trajectories weeks apart. Lunar Orbiters 4 and 5 had similar polar orbits and perilunes. Their flight paths differed by about 120 kilometers, so the two orbiters traversed roughly the same stretch of lunar landscape. Their Doppler residuals were in agreement.[10] The navigators now were on the road to linking residuals and features in the lunar landscape.

Mascons. Sjogren started by matching many of the largest residuals with visible features such as rough highlands and maria, the large dark plains on the Moon surface. He found that the largest residual variations seemed to correlate with the largest changes in lunar elevation. He knew that this correlation did not imply that *all* residuals correlated in this way. In fact, he failed to find consistent relationships between the more typical residuals and lunar elevations.[11]

Given the working assumption that the Doppler residuals were a direct measure of the local gravitational field, Sjogren's immediate plan was to estimate a small set of

---

[7] Gayle D. Barrow, *Lunar Orbiter Doppler Residual Study* (Seattle: The Boeing Company, October 1968), esp. 8, 29, 69, 80, 82 & 85.
[8] Sjogren, interview, 2008, 1-2, 5 & 8-9; Muller and Sjogren, "Consistency," 28.
[9] Muller and Sjogren, "Consistency," 30; Lorell, "LO Gravity Analysis," 197.
[10] Wood, interview, 23-24; Muller and Sjogren, "Consistency," 30, 34 & 36.
[11] Muller and Sjogren, "Consistency," 36.

mass points along the trajectory of a single orbit. Next, he planned to analyze several sequential orbits, building up a "mass-grid" to demonstrate a consistent correlation of Doppler residuals and lunar landscape. The residuals were no longer errors but useful scientific information. By the summer of 1968, Lorell, Sjogren, and Muller had demonstrated a correlation between Lunar Orbiter 3 Doppler variations and the landscape that it traversed. The correspondence between residuals and major visible features pointed to essentially two enormous areas covering 600 to 1,000 kilometers, namely, the Mare Nubium and its adjacent maria and the highlands lying between 0° and 30° east longitude and 0° and 30° south latitude. When they looked at polar orbits, they again found a good correspondence between the residuals and the lunar landscape (the regions of Mare Imbrium and Mare Serenitatis).[12] Soon they found more of these concentrations of underground mass, and their distribution correlated with known surface features. They gave these mass concentrations the shorthand name of mascons.

Looking at Lunar Orbiter 5 Doppler, they successfully identified large mass concentrations beneath the centers of five ringed maria (Imbrium, Serenitatis, Crisium, Nectaris, and Humorum). They also found a mascon in the area between Sinus Aestuum and Sinus Medii that probably represented an ancient ringed maria that had become obliterated partially over time by debris from Serenitatis and Imbrium. Initially, their contour map showed two nearby mascons of intermediate magnitude between Sinus Aestuum and Sinus Medii, but they lacked tracking data to determine whether these mascons were truly distinct. "However," they concluded, "this location certainly contains at least one definite mascon." They found an additional mascon in the area of Mare Orientale, a ringed mare discovered by Lunar Orbiter photography.[13] By sifting through Apollo Doppler, Sjogren, Muller, and Lorell found even more mascons.[14]

In August 1968, Muller and Sjogren published their findings in the journal *Science*. The scientific community soon found—or at least claimed to have found—additional mascons. Malcolm J. Campbell, Brian T. O'Leary, and Carl Sagan at Cornell University speculated that a basin about a thousand kilometers in diameter on the far side of the Moon was a mascon (but no data was available to confirm their speculation). They also interpreted the Mare Marginis to be a flooded portion of a mascon basin about 900 km in diameter. Goddard navigators James P. Murphy and Joseph W. Siry reported finding a mascon at a depth of about 100 kilometers in the area of Sinus Aestuum using Apollo 8 Doppler.[15] These discoveries raised some fundamental questions that scientists already

---

[12] Sjogren and Muller, "Correlations Between Major Visible Lunar Features and Systematic Variations in Radio-Tracking Data Obtained From Lunar Orbiter III," 20, 21, 23 & 36-37 in JPLSPS 37-52:II.

[13] Muller and Sjogren, "Detailed lunar Gravimetric Map of the Mare Humorum Area," 38-40 in JPLSPS 37-60:II; Sjogren and Wollenhaupt, 25-32; Muller and Sjogren, "Mascons," 10, 12 & 15; Muller and Sjogren, "High-Resolution Gravimetric Map of the Mare Imbrium Region," 14-16 in JPLSPS 37-54:II.

[14] Peter Gottlieb and Sjogren, "The Paucity of Gravimetric Information Due to Spurious Anomalies in Apollo 10 and 11 LM Tracking Data," 111 in JPLSPS 37-60:II; Gottlieb, Muller, Sjogren, and Wilber R. Wollenhaupt, "Lunar Gravity Over Large Craters from Apollo 12 Tracking Data," 40 in JPLSPS 37-60:II; Sjogren, Gottlieb, Muller, and Wollenhaupt, "Lunar Gravity via Apollo 14 Doppler Radio Tracking" *Science* 175 (January 14, 1972): 165-168.

[15] Muller and Sjogren, "Mascons," *Science* 161 (August 16, 1968): 680-684; Muller and Sjogren, "Lunar Gravimetry and Mascons," *Applied Mechanics Reviews* 22 (September 1, 1969): 955-959; Campbell, O'Leary, and Sagan, "Moon - Two New Mascon Basins," *Science* 164 (June 13, 1969): 1273-1275; Murphy

were debating. What were the origins of the mascons? What did they say about the Moon's structure and history?

Lunar science. Muller and Sjogren hoped to answer, or at the very least shed new light on, these and other questions. They were finding large mascons embedded under nearly every ringed mare and their relative absence elsewhere. Did each mascon represent an asteroid-sized body that had caused its associated mare by impact? If the mascons were not the original impact bodies, by what processes were they formed in the lunar interior? Was the presence of mascons consistent with a molten lunar interior?[16]

Sjogren believed that the mascons were objects, such as asteroids or meteors, that had struck the Moon long ago. His belief was consistent with the long-standing impact theory of how the lunar landscape came to have its current appearance. By 1960, as NASA set out to explore the Moon, planetary astronomers and geophysicists generally accepted the impact thesis. Among those scientists was Harold C. Urey, a Nobel Laureate chemist at the University of California, San Diego, who had played a key role in initiating the NASA lunar program.[17]

Urey contended that the craters, mountains, and maria mostly were the result of collisions of bodies in the Earth-Moon system with its surface. He quickly embraced the discovery of mascons. Urey maintained that they were objects similar to meteorites in composition and density that had struck the Moon's surface, flattened out, and left a high-density deposit that remained following the formation of the maria.[18]

Urey was, in fact, one of the first scientists to see the study of Doppler residuals done by Sjogren, Muller, and Lorell. They used a program created by Muller to print out the residuals and to plot them over the Moon's surface. Muller "ran down and saw Harold Urey," Sjogren recalled. "Harold Urey found all this stuff about the meteorites sitting in there. It was really an explosion after that."[19]

Three competing interpretations of mascons soon emerged. Urey and other scientists suggested that they were composed of heavy metallic material derived from the slow-moving meteorites that had gouged out the basins where the mascons were found, while others emphasized the role of volcanism in the Moon's geologic past. The latter scientists argued that mascons were remnants of lavas of different densities that

---

and Siry, "Lunar Mascon Evidence from Apollo Orbits," *Planetary and Space Science* 18 (August 1970): 1137-1138.
[16] Muller and Sjogren, "Mascons," 16.
[17] Hall, *Lunar Impact*, 11-12, 15, 52, 79, 181, 226, 319 & 341; MacDonald, "Origin of the Moon: Dynamical Considerations," 165-209 in Brian G. Marsden and A. G. W. Cameron, ed., *The Earth-Moon System* (New York: Plenum Press, 1966); MacDonald, "Origin of the Moon: Dynamical Considerations," *Annals of the New York Academy of Sciences* 118 (September 1965): 742-782; Hall, *Ranger Chronology*, 79, 264 & 395.
[18] Urey, "The Origin and Significance of the Moon's Surface," *Vistas in Astronomy* 2 (1956): 1667-1680; Urey, "The Origin of the Moon's Surface Features," Pt I & II, *Sky and Telescope* 15 (January 1956): 108 and (February 1956): 161; Urey, "Meteorites and the Moon," *Science* 147 (March 12, 1965): 1262-1265; Urey, "Mascons and the History of the Moon," *Science* 162 (December 20, 1968): 1408-1410; Urey, "The Contending Moons," *Astronautics and Aeronautics* 7 (January 1969): 37-41; M. von Reinhardt, "Maria, Mascons, and the History of the Moon," *Astrophysical Letters* 4 (1969): 225.
[19] Sjogren, interview, 2008, 9-10; Sjogren and Wollenhaupt, 25.

had filled the basins with the material observed in the maria. Yet other scientists posited that the mascons were plugs of denser material from a lunar mantle that had been shoved upward after the basins had been formed by meteor impacts. The geophysicist William Kaula, who thought that the mascons were too large to be the result mainly of impacts, believed that they probably resulted from the rapid cooling of the outer crust and other factors.[20]

Sjogren was of the opinion that the presence of the mascons under some of the maria suggested that the lunar crust was thicker and more rigid than was previously supposed. Had the crust been more plastic, the mascons would have settled deeper into it over geologic time, eliminating the gravitational anomalies. The outer layer of the Moon had been cold and rigid for a very long period, yet lava flows large enough to fill the basins had occurred as well. The implication was that the Moon's structure was not uniform and that it probably had had a complex evolutionary history. In 1977, though, Sjogren had to concede: "How the mascons formed is still an issue."[21]

He later searched unsuccessfully for mascons on Mars using data from Mariner 9 and techniques similar to those that he, Lorell, and Muller had applied to the Lunar Orbiter and Apollo data. Lorell also undertook a more ambitious effort to map the entire planet with the help of the Aerospace Corporation and JPL funding. The gravity map became a reality, but mascons did not turn up.[22] Eventually, Sjogren did find Martian mascons, several of them, using Viking Orbiter 2 readings. Data from the Mars Global Surveyor mission verified their existence.[23]

The discovery of mascons and their contribution to our knowledge of the Moon's long evolution represented an important scientific discovery by navigators. This scientific breakthrough also received the first recognition of navigators' work outside of NASA. For their "discovery of the lunar mascons (mass concentrations) leading to the first detailed gravimetric map of the Moon," Sjogren and Muller received the American Philosophical Society's Magellanic Premium in 1971. The society has presented the prize, an engraved solid gold oval plate, to only 36 individuals since it was first awarded in 1790. Funded by a donation from the Portuguese natural philosopher Jean-Hyacinthe Magellan, the society confers the Magellanic Premium for the best discovery or useful improvement in the areas of navigation, "natural philosophy," and astronomy.[24]

---

[20] Newell, 333; William M. Kaula, "Interpretation of Lunar Mass Concentrations," *Physics of the Earth and Planetary Interiors* 2 (October 1969): 123-137.

[21] Sjogren, "Lunar Gravity Determinations and their Implications," *Philosophical Transactions of the Royal Society of London, Part A* 285 (March 31, 1977): 220. On this point, see Urey and MacDonald, "Origin and History of the Moon," 213-289 in Kopal, ed., *Physics and Astronomy of the Moon*, 2d ed. (New York: Academic Press, 1971).

[22] Sjogren, Lorell, Stephen J. Reinbold, and Ravenel N. Wimberly, "Mars Gravity Field Via the Short Data Arcs," 393-400 in *MM71 Final Report, Vol. IV*; Lorell, John D. Anderson, J. Frank Jordan, Robert D. Reasenberg, and Irwin I. Shapiro, "Celestial Mechanics Experiment," 15-16 in *MM71 Final Report, Vol. V*.

[23] Sjogren, "Mars Gravity: High Resolution Results from Viking Orbiter 2," *Science* 203 (March 9, 1979): 1006-1010; Dah-Ning Yuan, Sjogren, Alexander Konopliv, and Algis B. Kucinskas, "Gravity Field of Mars: A 75th Degree and Order Model," *Journal of Geophysical Research* 106 (October 25, 2001): 23377-2340.

[24] APS, "The Magellanic Premium," http://www.amphilsoc.org/library/exhibits/magellan/magprem.htm (accessed March 27, 2008).

The science of the occult. The JPL navigators who discovered mascons were not official Lunar Orbiter project scientists. Mariner 4, on the other hand, featured navigators for the first time as project scientists. Their participation was exceptional, but even more remarkable was their admission to the project science teams *after* NASA had selected all of the mission science experiments and scientists.[25] The experiment, a radio occultation of Mars, also was unlike other mission science projects in that it required no additional hardware or changes to the spacecraft. On the other hand, it did require installing special hardware at Goldstone to obtain the requisite Doppler accuracy as the probe passed behind Mars as viewed from Earth.

At that point in the probe's trajectory, Mariner radio signals traveled through the Martian atmosphere and were refracted by it. As the waves traveled from one medium (space) into another (the atmosphere), the refraction altered the signals' signature (frequency, phase, and amplitude). After taking into account all other influences on the signals—such as the motion of the spacecraft, the movement of the DSN stations on the rotating Earth, the lengthening of the signals' transit times, and the refractivity of the Earth's lower atmosphere—navigators could attribute the remaining signal changes to the effects of the Martian atmosphere.[26]

Radio occultation would yield valuable knowledge unobtainable by direct measurements from Earth. Mariner project managers deemed the occultation to be of the highest importance. For example, at a press conference held on June 22, 1965, while the spacecraft was approaching Mars, Dan Schneiderman, the JPL Project Manager, told an audience: "we rate the occultation experiment on virtually an equal value with the pictures."[27] The value of the scientific results stood on their own as a contribution to planetary science; however, a better understanding of the Martian atmosphere was a necessity for future Mars exploration.

NASA already was considering a mission (known as Voyager) to put a lander on the planet's surface. The agency needed good knowledge of the Martian atmosphere to decide how to enter the atmosphere and land on the surface. For example, if the atmospheric pressure were low, say around 10 millibars, slowing the probe's descent with parachutes was out of the question. On the other hand, if the pressure were much higher, say 40 millibars, the landers could use parachutes. The pressure also had a bearing on the shape of the landers and other mission aspects.[28]

The idea for the experiment arose independently at Stanford University and JPL, but earlier at Stanford (1962) than at JPL (1963).[29] At Stanford, the idea originated with Von R. Eshleman and Gunnar Fjeldbo (later Lindal), who subsequently joined JPL. Eshleman was a professor of electrical engineering as well as co-director and founder of the Center for Radar Astronomy. Fjeldbo was a research associate at the radar center

---

[25] For the selection of the scientists and experiments, see: *MM64 Final Report*, 13; NASA, News Release, "NASA Selects Mariner Experimenters," April 11, 1963, F5194, NHRC.
[26] *MM64 Final Report*, 316, 317 & 318; *M4 Occultation Instrumentation*, 2; "Preliminary Results," 258.
[27] NASA, "Mariner IV Pre-Encounter Press Conference," June 22, 1965, 11 & 29, F5193, NHRC.
[28] Ibid., 11 & 29; Eshleman, interview, 27; *MM64 Final Report*, 317.
[29] "M4 Occultation Experiment," 73; Asmar and Renzetti, 13; Mudgway, *Uplink-Downlink*, 515; Eshleman, interview, 23-25.

and a former professor at the Norwegian Institute of Technology.[30] The occultation experiment was an offshoot of Fjeldbo's theoretical work (and the subject of his 1964 Stanford dissertation) on a general method for exploring the Moon and planets with a technique called bistatic radar.[31]

Eshleman failed in his attempt to include an occultation on Mariner 2. He suggested the experiment during a meeting at JPL. He recalled: "We were almost thrown out of the room!" After all, he explained, "The idea of going behind a planet was studied in great detail explicitly as something to avoid. The concept of losing contact with this spacecraft that they had nursed all the way to the planet was unacceptable. They have a string tied to the spacecraft, and you don't want anybody clipping that slender thread connecting you to the spacecraft. So flying behind Venus was completely out of the question."[32]

Interest in the Martian atmosphere "rose to a high level while Mariner 4 was in flight," Eshleman continued. He and his colleagues proposed the same experiment for Mariner 4. "Some people, especially Dan Cain, Tom Hamilton, and Arvydas Kliore at JPL, raised the question of using the communication system at the same time. I explained that we had already proposed that for the other mission, and a lot of argument had taken place. Yet NASA wound up doing a bistatic occultation experiment on Mariner 4 using the communications system; this was the only way it could be done at that late date, and opposition apparently had dissipated."[33] Mariner 4 thus became the first of many radio occultation experiments carried out by Stanford.

The occultation experiment had rather different roots at JPL in navigation, not radar. According to Arv Kliore, NASA approved the experiment "based on the demonstrated precision of the radio-tracking technique and the accuracy of the equipment of the NASA/JPL" DSN. Kliore compared the exercise to bowling in an alley that stretched from Pasadena to San Francisco. The occultation, he explained, "would be analogous to rolling a strike."[34]

Tom Hamilton, in discussing the origins of the occultation experiment, called attention to the role of Dan Cain, one of JPL's more inventive navigators. Hamilton characterized him as "a funny guy" who liked "to think out of the box." Cain came upon the idea of the occultation while "playing around with atmospheric refraction corrections to make the tracking a bit more accurate." Cain became "really intrigued" with the idea of an occultation experiment and proposed it to Bill Ekelund, a group supervisor, who thought it sounded like a good idea. He wondered whether they could do the experiment on Mariner 4. Ekelund, who had just hired "this young Ph.D." Arv Kliore, suggested to Cain that they work together. When Hamilton told Jack James, the

---

[30] "M4 Occultation Experiment," 72; Butrica, *See the Unseen*, 19-20; Eshleman, interview, 14 & 17-18.
[31] Asmar and Renzetti, 13; Fjeldbo and Eshleman, "The Bistatic Radar-Occultation Method for the Study of Planetary Atmospheres," *Journal of Geophysical Research* 70 (July 1, 1965): 3217-3225; Fjeldbo, Eshleman, O. K. Garriott, and F. L. Smith, III, "The Two-Frequency, Bistatic Radar-Occultation Method for the Study of Planetary Ionospheres," *Journal of Geophysical Research* 70 (August 1, 1965): 3701-3710; Butrica, *See the Unseen*, 57 & 155.
[32] Eshleman, interview, 28.
[33] Ibid., 25 & 28.
[34] "M4 Occultation Experiment," 73; NASA, "Mariner IV Pre-Encounter Press Conference," 30-31.

Mariner 4 Project Manager, about the experiment and explained that it involved flying behind the planet, James told him: "Drop dead."[35] The moment seemed to echo JPL's rejection of Eshleman's Mariner 2 occultation proposal.

In fact, preliminary calculations suggested that the data obtained during the occultation would not be good. Nonetheless, because "we had looked at the [spacecraft's] atomic oscillator performance on Venus," Hamilton explained, "we knew how good it was. It was probably two orders of magnitude better than the specification where we needed it. So we said no, you really can see this. So we ended up getting the project to change the aiming point for Mariner 4, getting Jack James to do this which was an amazing accomplishment." The experiment also required revising the probe's trajectory and calculating a new aiming point that would make Mariner appear to pass behind Mars (as seen from Earth) and emerge on the other side.[36]

The occultation science team comprised both Stanford and JPL personnel. The three JPL members were Principal Investigator Kliore, Dan Cain, and Gerry Levy, who designed and built the critical ground equipment. The Lithuania-born Kliore became the key figure behind JPL occultation research on subsequent missions to Venus, Jupiter, Saturn, and their moons. The Stanford team included Eshleman, Fjeldbo, and Frank Drake, a Cornell astronomy professor perhaps remembered best for the search for extraterrestrial life and the Drake equation (also called the Green Bank equation or formula).[37] Stanford scientists and engineers and JPL navigators thus joined together to become the world leaders in studying planetary atmospheres and rings via radio occultations.

Opening the doors. Before NASA could authorize the occultation experiment or the science team, it had to establish a procedure for vetting such science proposals. The usual rules did not apply, because the experiment used tracking data rather than an onboard scientific instrument. The selection of instruments, experiments, and scientists normally took place through a process developed shortly after NASA's founding and which Homer Newell, in charge of space science at Headquarters, formalized in April 1960. Any scientist, except those working at NASA Headquarters, could propose an experiment.[38]

Because the occultation did not fit into the types of scientific research addressed by those rules, JPL Director Bill Pickering and Homer Newell discussed a policy, in their words, regarding the "Utilization for Scientific Purposes of Tracking Data from the Deep Space Network," in August 1964. The launch of the first Mars Mariner was only three

---

[35] Asmar and Renzetti, 13; Hamilton, interview, Alonso, 6.
[36] Hamilton, interview, Alonso, 6; Tito, "Trajectory Design," 293.
[37] Nicks, 32; Letter, Homer Newell to Arvydas Kliore, January 21, 1965, F13/B1, JPL421; Letter, Newell to Kliore, June 23, 1965, F13/B1, JPL421; "Preliminary Results," 263-265; *MM64 Operations Report*, 2; "M4 Occultation Experiment," 72; *Report from Mars*, 31; University of Illinois at Urbana-Champaign, "2006 Distinguished Alumni Award Recipient Profiles: Arvydas J. Kliore," http://www.ece.uiuc.edu/alumni/awards/distinguished/kliore.html (accessed July 11, 2008).
[38] John E. Naugle, *First Among Equals: The Selection of NASA Space Science Experiments*, SP-4215 (Washington: NASA, 1991), especially 82-84.

months away. Newell wanted to bring the vetting of such research proposals in line with the procedures already in place for flight experiments as well as, among other issues, to protect the propriety of discovery of the experiment's originator. NASA policy required that proposals be submitted to Headquarters for review by the Space Science Steering Committee, just like other experiments, but it also stipulated input from the DSN by way of the NASA Office of Tracking and Data Acquisition.[39]

Pickering and Newell also discussed the requirement that the distribution of raw tracking data had to have advance approval from NASA Headquarters, the only exception being the flight project itself. Tracking data customarily was not made available to scientists or the public. Mariner 4 became the exception and set a precedent. JPL published the flight tracking data in a form suitable for use by scientists.[40]

The new policy had wide-ranging implications, and not just for implementing future occultations, because it applied to all data acquired by NASA's tracking system. One of the first applications of the new policy was to make the DSN available to radio astronomers. Nick Renzetti played a key role by shepherding the endeavor. Of particular interest to radio astronomers was the exceptionally sensitive 64-meter (210-foot) Mars Station at Goldstone, which was 6-1/2 times more sensitive than the existing 26-meter (64-foot) dishes.[41] The DSN was now an instrument of scientific research.

NASA started approving DSN radio astronomy observations in 1967, a pivotal year in radio astronomy history. Earlier that year, S. Jocelyn Bell, a Cambridge University graduate student working under Sir Antony Hewish, had discovered a scintillating radio source dubbed LGM 1 (for Little Green Men),[42] but known today as a pulsar. The announcement spurred a worldwide hunt for more of these pulsating radio sources using the DSN. Seizing the opportunity, Alan Moffet and Ron Ekers, Caltech astronomers associated with the Owens Valley Radio Observatory, detected pulsars at 2,295 MHz in April and May 1967.[43] In July 1967, Alan Maxwell, Harvard University Radio Astronomy Station in Texas, started measuring the variations in intensity of several galactic radio sources during lunar occultations. Also in 1967, Douglas S. Robertson, Australian Department of Supply's Space Research Group, proposed long baseline interferometry measurements of a number of galactic radio sources using a baseline between the Goldstone and Canberra DSN stations. Alan Moffet joined Robertson as a co-investigator, and their observations began in September 1967.[44]

The volume of radio astronomy requests for DSN antenna time mushroomed. On March 8, 1968, a joint meeting of the NASA Headquarters Office of Space Science

---

[39] Letter, Newell to Pickering, November 8, 1964, and attachment, "Policy on the Utilization for Scientific Purposes of Tracking Data from the Deep Space Network," August 13, 1964, 07 00024 BF, JPLHC.
[40] Appendix A, "Listing of Tracking Data," 57-170 in Mariner IV OD.
[41] DSN: Instrument for Research, 2-1.
[42] Anthony Hewish, S. Jocelyn Bell, John D. Pilkington, Paul F. Scott, and Robin A. Collins, "Observation of a Rapidly Pulsating Radio Source," Nature 217 (February 1968): 709-713; Sir Bernard Lovell, Astronomer by Chance (New York: Basic Books, 1990), 294-295; Benjamin K. Malphrus, The History of Radio Astronomy and the National Radio Astronomy Observatory: Evolution Toward Big Science (Malabar, FL: Krieger, 1996), 137-138.
[43] Moffet and Ekers, "Detection of the Pulsed Radio Source CP 1919 at 13 cm Wavelength," Nature 218 (April 20, 1968): 227-229; DSN: Instrument for Research, 2-1 to 2-2.
[44] DSN: Instrument for Research, 2-2; George D. Nicolson, "The Development of Radio Astronomy at Hartebeesthoek," Astrophysics and Space Science 230 (August 1995): 315-327.

and Applications and the Office of Tracking and Data Acquisition set up a group to establish procedures for scientists to use the DSN. In 1969, the DSN established the JPL Radio Astronomy Experiment Selection Panel consisting of astronomers selected from across the country. Invitations to use the DSN circulated in technical journals in December 1969 throughout Europe, Africa, Asia, Australia, and North and South America.[45]

Occultation at Mars. The Mariner 4 radio occultation experiment was a technical and scientific success. The experiment took place on July 15, 1965, over the dramatically short period of about 100 seconds. When radio contact with the craft ceased, it was just 22,559 km (14,021 statute miles) from the limb (edge) of the planet. The time of the occultation was about 8 minutes later than that predicted by the navigators. Around 53 minutes later the spacecraft emerged from the other side of Mars, and the DSN reacquired the signal.[46]

In order to derive scientific benefit from the experiment, Eshleman and Fjeldbo developed models of the Martian atmosphere under a separate NASA contract. Their models took into account the influence of the Earth's troposphere and ionosphere on the radio waves as well as assumptions about the composition of Mars' atmosphere.[47] From their models and the DSN occultation data, Eshleman and Fjeldbo drew conclusions about the planet's atmosphere, ionosphere, and scale height. The data, for instance, suggested the existence of a second, less-dense ionized layer, yet no indications of an ionosphere were present as the probe left the occultation area. The discovery probably resulted from taking the data when local Martian time was about midnight and atmospheric electron density was at least 20 times lower than daytime levels.[48]

More importantly from NASA's perspective, the data suggested a significantly different range of values for the planet's surface pressure. Previous observations implied that the pressure varied from 20 to 75 millibars, while more recent spectroscopic observations had yielded a range between 10 and 40 millibars, far lower, for example, than the 1,000 millibars more or less at the surface of the Earth. As Donald M. Hunten, Kitt Peak National Observatory, observed: "Until about 2 years ago, the accepted value

---

[45] Thomas V. Lucas, Memorandum for Record, "OSSA/OTDA Meeting on March 8, 1968," March 28, 1968, OTDA Chron Files, NHRC; *DSN: Instrument for Research*, 2-2.

[46] *MM64 Final Report*, 28, 32, 318 & 319; *Mariner Mars 1964 Project Report: Mission and Spacecraft Development*; Vol. I: *From Project Inception through Midcourse Maneuver*, TR No. 32-740 (Pasadena: JPL, March 1, 1965), 572; *Space Flight Operations Plan: Mariner Mars '64*, EPD 122 (Pasadena: JPL, July 15, 1965), Addendum I, VII-6; Null, "Mariner IV Flight Path," 28; Corliss, *History of the DSN*, 110; Mudgway, *Uplink-Downlink*, 39 & 515; *M4 Occultation Instrumentation*, 1-3, 5-6, 8 & 19; Asmar and Renzetti, 14; Nicks, 32-33; "Preliminary Results," 258 & 261.

[47] Stanford Radioscience Laboratory, *The Stanford Study of the Mariner Mars Occultation Experiment, Semi-Annual Report Number 1, for the Period 1 October 1964-31 March 1965* (Stanford: Radioscience Laboratory, June 1965).

[48] *MM64 Final Report*, 319, 321 & 322; Joseph W. Chamberlain and Michael B. McElroy, "Martian Atmosphere: The Mariner Occultation Experiment," *Science* 152 (April 1, 1966): 21; "M4 Occultation Experiment," 72; "Preliminary Results," 262.

of the pressure was higher yet, around 85 mb, and there was a great deal of reluctance about accepting even the 20 or 30 given about that time by the first spectroscopic measurements." The occultation indicated a dramatically lower pressure variation of about 4 to 7 millibars, equivalent to the Earth's atmosphere at an altitude of around 45,720 meters (150,000 feet). Hunten concluded: "The old photometric and polarimetric result of 85 mb is completely discredited."[49] The Mariner 4 occultation had made its mark on science.

The experiment also had an impact on NASA's future Voyager Mars lander. The lower-than-anticipated surface pressure led the agency to believe that designing craft capable of landing on the planet's surface was going to be harder than anticipated. On the other hand, the occultation data suggested that it would be possible to orbit the planet at altitudes lower than those previously planned.[50]

The celestial mechanics experiment. The Mariner 4 mission officially ended on October 1, 1965, when the DSN lost contact with it. The spacecraft was quite popular among JPL employees. One, Jim H. Wilson, even honored it with his "Ode to an Ancient Mariner."[51] The prediction that the probe once again would fly within the range of the DSN stirred the creation of new chapters, Phase II and Phase III, in the mission's history,[52] with further navigation-related activities taking place during Phase II.

Those activities included the acquisition of fresh trajectory data for a number of navigation experiments and for refining certain constants, the chief one being the astronomical unit. George Null was the instigator of these navigational pursuits. Null had been a member of the Mariner 4 Flight Path Analysis and Command Group and had overseen the post-flight analysis. The contribution of that analysis to the advancement of scientific measurement did not go unnoticed by NASA Headquarters. Oran W. Nicks, Director of Lunar and Planetary Programs, declared that, even though the determination of astronomical constants "was not treated as a scientific experiment" on the mission, "George Null and his associates" derived from the tracking data "two classes of results which are of scientific consequence." They were a confirmation of the constants

---

[49] "Mariner to Mars," *Engineering and Science* 29 (October 1965): 13; Gunnar Fjeldbo, Wencke C. Fjeldbo, and Eshleman, "Atmosphere of Mars: Mariner IV Models Compared," *Science* 153 (September 23, 1966): 1518-1523; *MM64 Final Report*, 319 & 321; "Preliminary Results," 257-258; Kliore, Cain, Levy, Eshleman, Fjeldbo, and Drake, "Occultation Experiment: Results of the First Direct Measurement of Mars's Atmosphere and Ionosphere," *Science* 149 (September 10, 1965): 1243-1248; Lewis D. Kaplan, Guido Münch, and Hyron Spinrad, "An Analysis of the Spectrum of Mars," *Astrophysical Journal* 139 (1964): 1-15; Donald M. Hunten, "CO_2 Bands and the Martian Surface Pressure," 240 & 243 in Harrison Brown, Gordon J. Stanley, Duane O. Muhleman, and Guido Münch, ed., *Proceedings of the Caltech-JPL Lunar and Planetary Conference* (Pasadena: Caltech, 1966).
[50] *MM64 Final Report*, 321.
[51] Wilson, "Ode to an Ancient Mariner . . ." December 21, 1967, F47/B3, JPL421.
[52] *MM64 Final Report*, 154 & 212; Corliss, *History of the DSN*, 110; Mudgway, *Uplink-Downlink*, 41; Harold Gordon and John Michel, "Midcourse Guidance for Mariner Mars 1964," AIAA 65-402, AIAA Second Annual Meeting, July 26-29, 1965, San Francisco, CA, 19; *MM64 Operations Report*, 80, 81 & 83.

computed from previous spaceflight data and radar observations and "a dramatic improvement" in the gravitational constant of Mars.[53]

In June 1966, taking advantage of the new policy regarding tracking data, Null submitted 30 copies of his "Proposal for a Determination of the Astronomical Unit and Earth Ephemeris using MIV [Mariner 4] Return Tracking Data."[54] The Celestial Mechanics Experiment, as he termed it, was the first of its kind. He planned to use Mariner 4 Doppler collected between June 1966 and April 1968 to improve "ephemeris information related to Earth-Sun measurements." He argued that the experiment would benefit NASA by "improving future planetary navigational accuracy." It also would have "direct value to astronomers." Null planned to use the third-generation orbit determination program "scheduled for completion in April 1967." His collaborator in the experiment, Ted Moyer, developed the new program's mathematics. When Homer Newell informed him that the agency had selected his Celestial Mechanics Experiment for inclusion on the Mariner 4 mission, Null became the Principal Investigator.[55]

The new policy for the "Utilization for Scientific Purposes of Tracking Data from the Deep Space Network" opened the door for JPL navigators and others to exploit the scientific value of tracking data. Joining a mission science team—or forming one's own science team—became another route for navigators to contribute to science. Of course, there also was the possibility of making a scientific discovery while analyzing Doppler or performing some other navigation-related task.

---

[53] *MM64 TDA Report*, 36 & 39; *MM64 Final Report*, 35, 330, 325 & 328; Null, "Mariner IV Flight Path," 28, 33 & 38; Null, "A Post-Flight Tabulation of the Mariner IV Real-Time Orbit Determination Errors," 24-30 in JPLSPS 37-39:III; Nicks, 34 & 35.
[54] *MM64 Operations Report*, 80; Null, "Proposal."
[55] Null, "Proposal," 4, 13 & 14; Newell to Null, December 27, 1966, F13/B1, JPL421.

Ode to an Ancient Mariner, by Jim H. Wilson, December 21, 1967

ODE TO AN ANCIENT MARINER ....

As it must to all man's institutions, the end came yesterday for the ancient Mariner IV, oldest established permanent floating heliocentric explorer pioneering far out beyond the horizon. It was a quiet peaceful finish for the grand old machine rocketed into the fleecy caribbean sky on Thanksgiving 1964, three scant weeks after its elder sister Mariner III had been hopelessly disabled by a minor technical flaw. After an historic voyage of some six hours unattended in the frosty blackness of deep space, this quarter-ton miracle of modern technology flew by the mysterious planet Mars, transmitting back to Earth the first closeup pictures of another world, in a stunning demonstration of the brilliance of contemporary science and the moxie of the doxy.

Then, after a two-year journey around the sun, America's most-travelled ambassador made another whirlwind tour past the earth in collaboration with the equally dazzling Venus flight of Mariner V.

Finally, years after the end of its usefulness had been predicted by scoffers, the black-and-blue windmill-shaped interplanetary scientific robot finally ran out of gas, lost its orientation, and tumbled bonnet-over-windmill.

Regretfully, NASA spokesmen, project officials, JPL space scientists, spacecraft engineers, and others too numerous to mention, terminated the three-year venture yesterday after vain attempts by the worldwide Deep Space Tracking Network to establish and maintain effective communications with the distant receding automated spaceship.

"We shall not soon forget Mariner IV," said one solemn spokesman.

"I'll miss the old bird," snivelled an engineer.

"Hot damn - home for Christmas!", grinned a representative of the press.

Jim H. Wilson
21 December 1967

Chapter 8

The First Navigators

With the awarding of the Ranger program to JPL, the laboratory underwent a sweeping reorganization that, in the process, created an institutional home for navigation. The organization of the Systems Analysis Section reflected its orientation toward supporting missions and mission design. The navigation area, despite the critical dependence on software development, had no computer programmers. The laboratory organization had placed them all in the Computer Applications and Data Systems Section.

The navigation culture that emerged reflected Caltech's "egalitarian policy" and a "campus" atmosphere meant to cultivate employee competence in preparation for future missions. Participation on flight project teams fostered its own culture at the flight operations center where different navigator groups focused on data reduction, trajectory computation, and midcourse maneuvers in addition to orbit determination. The idiosyncratic culture of the flight center also arose out of the need to provide continual mission coverage and led to such traditions as the Ranger Hotel. The pressure to succeed—or at least not fail—was ever present, but even more so during these earliest space missions. Failure was not an option in the Cold War race into space.

An institutional home. The institutional home of JPL navigation, the Systems Division, came into existence in 1960 in the wake of NASA's assignment of the Ranger program to JPL on December 21, 1959. Previously, the laboratory had just three Departments established in 1956.[1] Technical staff now were spread over seven Divisions. Director Bill Pickering's sweeping 1960 laboratory-wide changes instituted a more formalized structure based on a matrix model. During the two Pioneer flights, the navigation function fell within the Space Sciences Division (Division 32),[2] whose head was Al Hibbs, a JPL employee since 1956. Below him, Manfred Eimer managed the Pioneer orbit-determination operation.[3]

Pickering's matrix scheme soon led to the formation of the Systems Division (Division 31). It had three major flight-related duties, two of which were navigation and mission design.[4] It was at this time that the names of Johnny Gates and Tom Hamilton, long-associated with JPL navigation, came to the Systems Division. Gates became the

---

[1] Hall, *Lunar Impact*, 57; "JPL Research Division Re-Organized," *Lab-Oratory* 5 (July 1956): 3; Koppes, 107; "Robert J. Parks Appointed Planetary Program Director," *Lab-Oratory* 9 (June 1960): 12; "Jet Propulsion Laboratory, California Institute of Technology, Organization Chart," April 1, 1958, F4648 NHRC.
[2] Koppes, 106-107.
[3] "New Role for JPL Indicated," [Pasadena] *Independent Star-News*, September 13, 1959, 1-2; *Tracking Pioneer III and Pioneer IV*, iv; "New Aides Announced at JPL," [Pasadena] *Independent Star-News*, November 13, 1960, 3; "Ten Years Service," *Lab-Oratory* 9 (September 1959), 2.
[4] Hall, *Lunar Impact*, 36.

Systems Division head in 1961 after having been Manager of the Systems Analysis Section since 1957. The Systems Division supervisor was the principal bureaucratic interface between upper management and navigation. Gates remained Systems Division Chief until 1967, when JPL designated him manager of NASA's ill-fated Voyager mission to Mars.[5]

When Gates advanced to Division Chief, Hamilton replaced him as Manager of the Systems Analysis Section (Section 312), the institutional home of navigation. The navigation area had no computer programmers; the laboratory organization had placed them all in the Computer Applications and Data Systems Section (Section 372), whose supervisor initially was Bill Hoover. Navigators also interfaced frequently with the tracking network's Communications Engineering and Operations Section (Section 332).[6]

During Hamilton's long tenure as Section Chief, the size of the Systems Analysis Section mushroomed in step with the escalation of NASA's program of space exploration. The number of Section employees and the creation of new subunits within the Section reflected this growth. Between 1962 and 1965, the number of engineers in the Section doubled from 27 to 54. These numbers only partially reflect population growth as JPL designated employees of the technical divisions as either engineers or technicians depending on rank. From the beginning, the Section included navigators with university-training in astronomy and mathematics. The number of groups constituting the Section evolved over time to fit changing ideas about what constituted navigation.

The core set of navigators made up the Orbit Determination Group. Its head in 1965, Don Trask, joined JPL in 1962. The number of engineers in the Group more than doubled between 1962 and 1965 from 6 to 14. In 1962, the Group included two women, Carolyn R. Mathews and Catherine L. Thomas (later Thornton). The latter had a graduate degree in mathematics from Northwestern University and her first tasks at JPL centered on general error analysis in support of various projects. The others in the Group included Dan Cain and John Anderson.[7]

Another vital navigation unit set up in 1962 was the Systems Analysis Research Group whose Supervisor was Bill Melbourne. Its staff size remained constant between 1962 and 1965, frozen at five engineers and one part-time technician. They included Bill Kizner of B-plane fame, Jack Lorell, Carl Sauer, Jr., Carl Solloway, and Harry Lass. A Caltech graduate, Lass wrote a text on vector and tensor analysis in 1950 and a textbook on mathematics published in 1957.[8]

When he took responsibility for the Systems Analysis Research Group, Melbourne was not only a recent JPL hire, but also chronologically young, being in his late twenties. His degree from Caltech was in astronomy. Melbourne started working at

---

[5] JPL, OrgChart, February 22, 1960, F4648, NHRC; finding aid, JPL235, 1; "Clarence R. Gates" http://crgates.org/Site/Home.html (accessed March 3, 2010).

[6] Hamilton, interview, 1992, 1; "Twenty Years Service," Lab-Oratory, July-August 1972, 23; "Dr. Gates Appointed JPL Division Head," Pasadena Star-News, June 26, 1969, B-1; JPL Telephone Directory, June 1962, viii, NHRC; Ranger 4 OD, 82.

[7] "Ten Years Service," Lab-Oratory, May-June 1972, 21; OrgChart, Systems Analysis Section, February 1, 1962, JPL556; OrgChart, Systems Analysis Section, June 15, 1961, JPL556; Thornton, interview, 1 & 4.

[8] Lass, Vector and Tensor Analysis (New York: McGraw-Hill, 1950); Lass, Elements of Pure and Applied Mathematics (New York: McGraw-Hill, 1957); OrgChart, Systems Analysis Section, June 15, 1965, JPL556.

JPL in 1956 as a summer student and became a full-time employee in 1959. He recalled being selected to head the Systems Analysis Research Group: "I was appointed a group supervisor early on, when I got my PhD. I became head of a research group, of high-powered mathematician types and so on. . . . I was just kind of a novice. I thought they were old geezers . . . It was a lot of fun. I think being the supervisor wasn't all that hard; it seemed like it was a good job from my standpoint."[9]

The organization of the Systems Analysis Section reflected its orientation toward supporting missions and mission design. For example, it included groups for Mission Analysis, Space Guidance Theory, and Trajectories and Performance. Mission design and navigation were in the same Section. In 1965, a new group called simply "Section Staff" reinforced the emphasis on flight projects. Section Staff typically had responsibilities that crossed group boundaries and reported directly to the Section Manager (and Deputy Manager), rather than to a Group Supervisor. The "Section Staff" consisted of "Project Engineers" assigned to specific missions such as Apollo (Dave Curkendall), Ranger 3 (Bill Kirhofer), Mariner (Norm Haynes), and Surveyor (Tom Thornton).[10]

Navigation managers informed their employees that, in accordance with Caltech policy, an "egalitarian policy" would apply to technical staff. Once hired, JPL would judge employees on their work and knowledge, not their academic degree. This policy, George Null explained, was especially important as most of the earliest employees hired had either a bachelor's or a master's degree. A cornerstone of this culture was the "campus" atmosphere. The lab allowed, encouraged, and even funded engineers to use a portion of their own time to work on technical problems of their choosing and to publish their results. The budgetary largesse of the space race enabled the ready availability of funding as well as a degree of relaxed accounting. The basic reason for these practices was to develop employee competence in preparation for future missions. Informal training courses—Tom Hamilton on Ranger mission data weighting, John Anderson on orbit determination theory, and Carl Solloway on orbit determination statistical methods, to name a few—contributed to both this preparation for future missions and the "campus" atmosphere.[11]

Space flight operations. A selected number of navigators participated in flight projects as a team. Don Trask, for instance, led the navigation teams on Mariner 2 and Ranger 6, while Bill Kirhofer was the navigation chief on the next three Rangers.[12] The team's name changed from time to time, from flight to flight. For example, Mariner 2 had the Orbit Determination Group, while the Block III Rangers and Mariner 4 had the Flight Path Analysis and Command Group.[13] The Mariner 4 navigation leader, Norm Haynes,

---

[9] "Twenty Years Service," *Lab-Oratory* 6 (1976): 15; Melbourne, interview, 1.
[10] OrgChart, Systems Analysis Section, February 1, 1962, JPL556; OrgChart, Systems Analysis Section, June 15, 1965, JPL556.
[11] Personal communication, George Null, June 12, 2012.
[12] *Tracking Mariner Venus 1962*, 22, 49, 108, 113 & 119; *Mariner-Venus 1962*, 287.
[13] *Mariner-Venus 1962*, 19, 229, 287, 288 & 289; *Tracking Mariner Venus 1962*, 5, 21, 22, 49, 87, 93, 94, 108, 113 & 119; *Mariner R Project*, 167; Hall, *Lunar Impact*, 89.

enjoyed a long career at JPL that included succeeding Johnny Gates as Manager of the Systems Division in 1980 and as Voyager project manager in 1987.[14]

Serving on a navigation team meant working in a different building reserved for flight operations. The spaceflight operations facility was replete with the computers and other equipment required for analyzing tracking data. There navigators worked alongside other teams such as the Scientific Data Group, the Data Reduction Group, the Trajectory Group, and the Spacecraft Data Analysis Team, which evaluated the tracking data. The Scientific Data Group consisted of the project scientist and certain JPL scientists.

Specialized functions and software came into play during flights. One such specialization was the midcourse maneuver. A dedicated team called the Midcourse Maneuver Commands Group or the Maneuver Analysis Group depended on and intermeshed with the orbit determination crew. The accuracy of all course corrections hinged on the accuracy of the flight path calculations performed in preparation for the maneuver. Of course, maneuver accuracy equally was critical to mission success. If a launch failed to put a probe on the right trajectory, a course change provided by the maneuver team might be able to rectify its flight path and save the mission. Unlike the orbit determination team, the maneuver group also was responsible for generating the commands needed to alter probe trajectories. The maneuver group computed various course correcting options for project managers and generated the requisite commands.[15]

As with Vanguard, visual displays during a flight were standard. The transition to the IBM 7094 mainframe for Mariner 4 navigation gave rise to a software revision whose emphasis was on being compatible with the various display media in the flight operations center.[16] The midcourse maneuver program was no exception. Navigators Hal Gordon and John Michel were in charge of the Mariner 4 maneuver software. Unlike the Rangers and Mariner 2, the Mars Mariners had a modified propulsion system that enabled them to carry out a second midcourse correction, if needed.[17] Future missions would be capable of performing multiple course changes thanks to advances in spacecraft propulsion technology.

Rodney Hamburg, in the Computer Applications Section, wrote the maneuver program so that it would be compatible with the JPL Trajectory Monitor (JPTRAJ). Created in 1963, the Trajectory Monitor was the work of his programmer colleague Skip Newhall, who started at JPL in 1963 after graduating from Caltech.[18] The Trajectory

---

[14] *MM64 Final Report*, 325, 328 & 330; *Tracking Mariner Venus 1962*, 36 & 39; Swift, 181.
[15] *Mariner-Venus 1962*, 19, 81 & 289; *Tracking Mariner Venus 1962*, 49, 93-94, 108, 113 & 119; *Mariner R Project*, 49, 53, 54, 55, 56 & 168.
[16] J. N. James to Donald P. Hearth, October 28, 1966, F50/B3, JPL421; Corliss, *History of the DSN*, 131; Mudgway, *Uplink-Downlink*, 62; Null, "The Mariner IV Flight Path," 23; *MM64 Final Report*, 226.
[17] "Mariner 4 Flight Path," 28; Harold J. Gordon and John R. Michel, "Midcourse Guidance for Mariner Mars 1964," AIAA 65-402, AIAA Second Annual Meeting, July 26-29, 1965, San Francisco, CA, 1 & 2; *Report from Mars*, 24-25; Harold J. Gordon, *Functional Design of the Mariner Midcourse Maneuver Operations Program*, TR 32-1139 (Pasadena: JPL, July 15, 1967).
[18] Newhall, "JPLTRAJ: The New JPL Trajectory Monitor," 481-488 in *American Federation of Information Processing Societies Conference Proceedings* 26 (Baltimore: Spartan Books, 1964); Newhall, "Two New Integral Transforms and their Application," Ph.D. diss., Caltech, Division of Engineering and Applied

Monitor interfaced with the space flight operations facility display system and was a collection of separate links that one could combine with other programs and subroutines in various ways.[19]

In the same way, the latest iteration of the Space Trajectory Program (SPACE) underwent modification to be compatible with the flight display media, but without alterations to its basic structure. This SPACE iteration was the trajectory program for the flights of Pioneer 6, the Lunar Orbiters, and the Surveyors. The adaptation to the flight display media, however, deprived SPACE of some features, so Newhall and others created SFPRO to restore those deficits when operated in combination with SPACE.[20]

An idiosyncratic work culture developed among navigators in the flight operations center that arose out of the requirement to provide continual flight coverage. Ranger flights lasted about 70 hours, but Mariner 2, the first interplanetary mission, lasted for months and had requirements for round-the-clock coverage. The rationale for continuous tracking was to enable JPL to detect spacecraft problems early enough, so that they could take preventive or remedial measures. All in all, the nonstop coverage of Mariner 2 lasted over 109 days between launch and Venus encounter.[21]

The constant tracking meant that navigators had to be on continuous duty. The practices improvised for Ranger became Mariner routines. Bill Sjogren, a member of the Ranger 4 and Ranger 5 navigation teams, remembered staying overnight in the space flight operations center in what his colleagues called the Ranger Hotel. "We'd sleep in the bunk beds. We wouldn't go home for seventy-two hours. The trajectories were something like seventy-two hours long, and we'd just sleep overnight. Sometimes we'd take a break and go down to the Pizza Hut for some pizza, but come right back. We didn't go home for three days. Yeah, those were the good old days."[22]

Snake bitten. Performing a midcourse maneuver was just one of the NASA navigational firsts that occurred on the Ranger and Mariner Venus and Mars missions. The era witnessed many navigation feats performed for the first time for the simple reason that NASA was just starting to explore the solar system. The first "Hohmann transfer orbit," the first trajectory correction maneuver, and the first hard landing on another celestial body were among the milestones that the Rangers attained, while Mariner 2 realized the first planetary flyby. Later voyagers to the Moon—the Surveyors and Lunar Orbiters—added the first soft landing (the Surveyors) and the first orbit around another body (Lunar Orbiter). Mariner 2 visited the farthest celestial body to date, Venus, and required the acquisition of tracking data over unprecedented distances, even greater

---

Science, 1972. He later went by the name X X Newhall, but retained the nickname Skip. "Ten Years Service," *Lab-Oratory*, March-April 1973, 23; biographical information at Internet Movie Database, "Dr. Skip Newhall," http://www.imdb.com/name/nm2663197/bio (accessed December 20, 2009).
[19] *MM64 Final Report*, 226; *MM64 Operations Report*, 578; Gordon and Michel, 3, 6, 8, 11, 12 & 13; "Mariner 4 Flight Path," 28.
[20] *SPACE*, 2; *SFPRO*, 1-4.
[21] *Tracking Mariner Venus 1962*, 6, 18, 23, 76 & 80; *Mariner-Venus 1962*, 262; Corliss, *History of the DSN*, 62, 63 & 70.
[22] Sjogren, interview, 2000, 7.

than those that STL had achieved with Pioneer V. The Venus probe also executed the most distant course correction on September 4, 1962, when its motors fired at a distance of 2,400,000 km (1,492,000 miles) from Earth.[23]

One must not overlook the first parking orbit, performed by the Rangers and several Surveyors, and the first direct ascent, featured in three Surveyor trajectories and all Ranger launches.[24] Four of the seven launches, Surveyor 3 and Surveyors 5 through 7, incorporated a parking orbit, while the first two Surveyors and Surveyor 4 flew directly to the Moon.[25] The parking orbit is a segment of a spacecraft trajectory that begins when the launcher upper stage and its attached probe enter an Earth orbit. It ends usually a few minutes later, when a second engine burst injects the probe into an approximate Hohmann transfer orbit that sends it to the desired target. Direct ascent, as the name implies, entails a direct trajectory from launch pad to space.

An academic exercise. Ranger 3 was the first lunar voyage that JPL navigated. Two Pioneers and the first two Rangers had failed well within Earth's gravity well because of Agena malfunctions. Ranger 3 offered a number of JPL navigational firsts, including the first parking orbit and the first midcourse maneuver. Unfortunately, the probe failed to accomplish its mission,[26] despite navigators' best efforts.

When Ranger 3 left the Cape on January 26, 1962, it appeared to be on track. The only hitch was that the final engine burns had been too long. The craft was traveling too fast, and the Agena was not programmed to compensate. Instead of cruising to the Moon at 39,430 kph (24,500 mph), it was going over 40,235 kph (25,000 mph). The navigation team calculated that Ranger 3 would make the 66-hour trip in just 55 hours and arrive at its destination about 14 hours before the Moon.[27] The situation was tailor-made for a course correction. The original flight plan called for the maneuver to take place 16 hours after launch, but its execution was no longer within the capabilities of the probe. Project management decided to reconfigure the mission from a lunar impact to a flyby. The maneuver team generated the necessary commands, so that Goldstone would be in contact with Ranger as it glided past the Moon. The goal was to have the craft pass the Moon at a distance of about 40,235 km (25,000 miles).[28]

Oddly, after the maneuver commands went to Ranger, Doppler from the Woomera and Johannesburg stations indicated that the spacecraft was moving in a direction that was the mirror image of the one expected. Members of the Spacecraft Data Analysis Team pored over the tracking data in an attempt to determine the cause

---

[23] Wheelock, 65.
[24] Renzetti, *DSN History*, 39.
[25] Theodore F. Gautschi and Victor C. Clarke Jr., *Direct-Ascent vs Parking-Orbit Trajectory for Lunar-Soft-Landing Missions*, TM 33-114 (Pasadena: JPL, December 3, 1962), 1 & 2; Corliss, *History of the DSN*, 159; Byers, 4, 5, 9, 52 & 115.
[26] Corliss, *History of the DSN*, 53.
[27] Corliss, *History of the DSN*, 54 & 55; Hall, *Lunar Impact*, 143; NASA Facts, 8; Scala, 7; Hall, "Project Ranger," 460.
[28] NASA Facts, 5; Hall, *Lunar Impact*, 145; Bill Wilks, "Documentation Error Lost Ranger III Photos," *Missiles and Rockets*, February 19, 1962, 33, F10/B64, Willy Ley; Richard van Osten, "Ranger: What Might Have Been?" *Missiles and Rockets*, February 5, 1962, 12, F10/B64, Willy Ley.

of the unorthodox course change. They analyzed the fault for about thirty hours after the maneuver, but without coming to any conclusion.[29] The navigation team finally realized what had happened.

Tom Hamilton went home at nine o'clock the night of the maneuver—which took place sometime after midnight—because he "didn't want to stay up with the thing. I wanted to be there the next day." He recalled: "A number of our guys stayed up and they looked at the maneuver and were very happy that it was all performed and the engines fired and the bosses changed. Everyone went home with a smile on their face."[30] Actually, a sign had been inverted between the digital maneuver code used in the JPL ground computer and the spacecraft computer. The hardware and software sign conventions were just the opposite of each other. In the words of Jim Burke, the JPL Ranger project director: "The spacecraft spoke a language in which zero was positive and one was negative. The other part of the system, the maneuver computing program, had just the reverse information. In other words the vehicle operated in one system of coordinates and the human operators in another."[31] Preflight tests had checked the magnitude and polarity of the commands, but they had not checked their meaning.[32]

Bill Sjogren recalled his boss's dramatic reaction. Hamilton arrived the morning after the midcourse maneuver and went directly to the Doppler residual printout. After briefly examining them, he exclaimed "Damn, they did it backwards" and ran with the printout to the Project Manager in the control room. Sjogren pointed out that "even though the maneuver was backwards, we would not have been able to encounter the Moon because the push out of Earth orbit was so bad that even if the maneuver had gone perfectly, we still would have missed the Moon by something like 10,000 kilometers. It was just an academic exercise at that point, to make sure we could do a maneuver correctly. Well, it went wrong."[33]

Hamilton's recollection of his reaction was less salty:[34]

> I came in the next morning, seven o'clock in the morning and I said how'd the maneuver go. Somebody said well, over on that machine there is the Doppler data. There's a deck of cards that's got the Doppler data. Okay, how about I print it? So I printed out the Doppler data and I looked at it and I said the maneuver is in the wrong direction. They said surely you are jesting; you do not have the polarities right. I said if there's one thing in this entire world that I know, it's the polarities of the Doppler system; I specified them all. And that maneuver's in the wrong direction. Well, four or five hours later we had assembled enough people to look at it and discovered that yes, what we thought was a plus pitch turn, our spacecraft thought was a minus pitch turn.

---

[29] Hall, "Project Ranger," 460; Scala, 7; Hall, *Lunar Impact*, 146.
[30] Hamilton, interview, 1992, 19.
[31] Quoted in Wilks, 33.
[32] Hall, *Lunar Impact*, 146; Corliss, *History of the DSN*, 54-55.
[33] Sjogren, interview, 2000, 4. His retrospective analysis of the tracking data showed that Ranger 3 would have missed the Moon by some 40,000 kilometers. *Ranger 3 OD*, 2.
[34] Hamilton, interview, 1992, 19.

Following the discovery of the backwards maneuver, a dispute broke out that pitted "unidentified" NASA officials against each other. One group claimed that Ranger 3 might have been able to reach the lunar surface, if only the Johannesburg station had had the ability to send commands. Johannesburg was the first to lock onto the probe after launch and retained contact for the next 11 hours. The transmitter and associated hardware for sending commands already were at the station, but they would not be operational until sometime in April or May, meaning that Johannesburg also might miss the Ranger 4 launch due to occur in late April or early May. Opponents argued that Johannesburg could not have made a useful course change even if its equipment were operational, because the station simply did not have enough time to determine the nature of the problem and to formulate a remedy. Moreover, the spacecraft itself could not have executed such a maneuver, because its high-gain antenna had not yet locked on to Earth, and Ranger 3 had not been designed to make a course change so early in its trajectory.[35]

Ranger flight personnel attempted to rescue the mission in the wake of the backwards midcourse maneuver by means of the terminal maneuver. This maneuver normally occurred just prior to the probe's collision with the Moon. Its purpose was to position the craft to send pictures back to Earth and to receive radar reflections from the Moon. The original Ranger 3 trajectory scheduled the terminal maneuver for about 65 hours after launch or about an hour before impact.[36]

Goldstone sent the terminal maneuver command, and the spacecraft started executing it. However, midway through its prescribed turns, signal strengths at the Echo and Pioneer antennas began to fluctuate. The degree of signal strength deterioration precluded the reception of television pictures. Ranger's high-gain antenna no longer pointed directly toward Earth; the spacecraft was out of control. More commands from the Echo station tried to improve signal reception, but to no avail. Even boosting transmitter power from 200 to 7,000 watts had no apparent effect.[37]

The failure of the midcourse maneuver, Hamilton opined, "didn't harm anything because we had an Agena error that was bigger than we could ever correct on board and what we're just doing was tweaking the flyby conditions for flying by the Moon. We couldn't hit the Moon with the mission."[38] The sign inversion was not the only glitch that had doomed Ranger 3. An electronic malfunction aboard the spacecraft registered at the tracking stations as "an extraneous blip" on a readout. The probe's central computer and sequencer had failed, immobilizing the Earth and Sun sensors that kept the craft in the proper attitude. Its signals to Earth fluctuated as Ranger 3 drifted aimlessly about its axis, even as the television camera began sending back pictures. The video signals were extremely weak, full of noise, and virtually useless to scientists, because the high-gain antenna remained pointed away from Earth. The Moon was nowhere to be seen.[39]

---

[35] Van Osten, 12.
[36] NASA Facts, 6.
[37] "Participation in Ranger III," 51 in JPLSPS 37-14:1; Hall, *Lunar Impact*, 147; Corliss, *History of the DSN*, 55; Wilks, 33; Van Osten, 13.
[38] Hamilton, interview, 1992, 20.
[39] Hall, *Lunar Impact*, 147; Wilks, 33; Van Osten, 13.

The "end of the story," Hamilton reminisced, "should be that we got the pictures of the backside of the Moon. But we didn't. It turns out the spacecraft when we sent it into the turn just for reasons to this day are not understood just kept on spinning. So we lost Ranger at that point. . . . It started to turn and then it never stopped turning. I don't think anybody knows why. So we didn't get the pictures. But that was still a high moment [of my career]."[40]

Puzzle solvers. Both Ranger 4 and Ranger 5, whose mission objectives were essentially the same as that of their predecessor, similarly met with ill fortune. Nonetheless, their flights gave navigators an opportunity to demonstrate the value of their expertise. In the case of Ranger 4, trackers lost its signals just 460 seconds (nearly 8 minutes) after launch, as it was heading toward South Africa. The two South Africa antennas managed to lock onto the transponder signals, but something had gone awry while Ranger was out of touch. One telemetry channel was lost, another was working improperly, and the signal level was fluctuating wildly. Commands sent to switch to the high-gain antenna failed.[41]

The Atlas-Agena had performed nearly perfectly and put Ranger 4 on a lunar collision course without the need for a midcourse correction, but the probe's health continued to worsen. Its ability to perform any science at the Moon was doubtful. It remained mute and deaf to ground commands as the Ranger died bit by bit. Once the solar panels failed to open, battery power ran out 10-1/2 hours after launch. The transponder stopped broadcasting. The only sign of life was the signal sent from the seismometer capsule which had its own battery and antenna.[42]

As Ranger 4 approached the Moon, Goldstone managed to track the capsule signal off and on until the craft flew behind the Moon. Project management concluded that the spacecraft had impacted the far side of the Moon on April 26, 1962. The flight was a technical success—Ranger 4 had impacted the Moon—and a scientific failure because of the lack of television images and experimental data.[43]

The real question was whether Ranger actually had crashed into the Moon. The Soviet Union denied NASA's claim that it had crashed into the Moon. Navigators now stepped in to analyze the one-way Doppler from the capsule's transmitter. They were able to confirm that, indeed, impact had occurred.[44]

By running the one-way Doppler from the 65-hour voyage through their software, navigators were able to show that the probe's elliptical path changed when it entered the lunar gravitational field as it took on a Moon-centered hyperbola orbit. Even though the probe had not been observable from Earth, the data and software indicated

---

[40] Hamilton, interview, 1992, 20.
[41] "Program Objectives," 1 in JPLSPS 37-15:l; "Ranger Project," 3 & 4; *Ranger 4 OD*, 11; Corliss, *History of the DSN*, 56; Hall, *Lunar Impact*, 152.
[42] "Ranger Project," 2; Hall, *Lunar Impact*, 153; Corliss, *History of the DSN*, 56; *Ranger 4 OD*, 2.
[43] "Ranger Project," 3 & 4; Hall, "Project Ranger," 460; Hall, *Lunar Impact*, 153.
[44] Corliss, *History of the DSN*, 56.

that it had struck the lunar surface at a location that was at 12.0 degrees south latitude and 231.4 degrees east longitude.[45]

Ranger 5 similarly posed a puzzle for navigators. The Atlas-Agena again appeared to function correctly, but the trackers lost their lock on the probe. Only half an hour into the launch, the mission began to disappoint. The solar panels had deployed, but were not generating sufficient power. The orbit determination program estimated that the probe would miss the Moon by 631 km. Because of the battery's limited life, the Spacecraft Data Analysis Team recommended trying the midcourse maneuver ahead of schedule.[46]

Johannesburg sent the maneuver commands, but it was too late. The batteries had depleted at 8 hours and 15 minutes into the cruise phase in the middle of the maneuver; the burn would have been completed only six minutes later. Was it possible that the course correction *had* taken place? Telemetry received around 10 hours later showed that the probe had not burned the midcourse motor fuel, yet the supply of attitude control gas was completely gone. Because the battery was dead, trackers had to rely again on the one-way Doppler from the seismometer capsule. On October 21, 1962, Ranger 5 tumbled past the trailing edge of the Moon at a distance (more like an altitude) of 720 kilometers (450 miles).[47]

One of the questions that navigators tackled in their post-flight analysis of Ranger 5 was whether the midcourse maneuver had taken place. During the flight, they assumed that no maneuver had occurred and, based on early transponder data, predicted an occultation time, that is, the moment when Ranger 5 flew behind the Moon. They discovered a discrepancy between the calculated and observed times; the calculated time was 8 minutes earlier. Navigators made a systematic study to decide the cause of the time discrepancy. The completely spent attitude control gas denoted the addition of a known force to Ranger's flight path. Had the probe actually accelerated? Was the orbit determination program wrong, and the probe had not sped up?[48]

No single possibility that the navigators investigated offered conclusive proof that a maneuver had taken place. All the same, they concluded that: "All results seem to indicate a maneuver." They checked their software, including the trajectory program, using different software. They reexamined the various models that underlay the software. They sought consistencies with the results from the previous two flights. Their best estimate of the amount and direction of the maneuver was an increase of 0.69 meters per second in a direction 5 degrees off the probe-Sun line directed away from the Sun. It still was not clear when this small velocity change had occurred. It might have happened hours later, when solar power was available. It might have taken place slowly, as the attitude control gas escaped.[49] Escaping spacecraft gases could act as a

---

[45] *Ranger 4 OD*, 3, 5, 10 & 13; Renzetti, *DSN History*, 39; Hall, *Lunar Impact*, 154.
[46] "Tracking Operations," 1, 3 & 4 in JPLSPS 37-18:III; Corliss, *History of the DSN*, 56 & 57; Hall, *Lunar Impact*, 167, 169 & 170; Hall, "Project Ranger," 460; *Ranger 5 OD*, 3; Raymond L. Heacock, "Ranger: Its Mission and Its Results," *Space Log* (Summer 1965): 5, photocopy, F5370, NHRC.
[47] Hall, "Project Ranger," 460; "Tracking Operations," 4; Corliss, *History of the DSN*, 57; *Ranger 5 OD*, 3 & 31; Hall, *Lunar Impact*, 167, 169 & 170; Heacock, 5.
[48] *Ranger 5 OD*, 31.
[49] Ibid., 31, 32 & 34.

small force diverting the probe from its flight path. Indeed, this would not be the last time that such small forces altered a trajectory and frustrated navigation.

Mariner too. Like the later Block III Rangers, Mariner 2 was a success. Mariner 1's launch lasted only a few minutes before the Range Safety Officer destroyed it.[50] Likewise, its Mars-bound successor, Mariner 4, triumphed after its forerunner failed to jettison the shroud protecting its solar panels.[51] The navigational successes of these first Mariners lay in having an accurate value for the astronomical unit. Mariner 4, in particular, also showed once again the vexing character of unexpected small non-gravitational forces acting on a craft's trajectory.

The primary basis for the successful navigation of Mariner 2 was a new, more accurate value for the astronomical unit determined from radar measurements of Venus. Navigators applied this value, although the IAU did not adopt it until May 1963. Nick Renzetti, a long-time supervisor of JPL's tracking network, pointed out that if JPL had used the value determined by Eugene Rabe, an astronomer at the Cincinnati Observatory, based on telescopic observations of Eros between 1926 and 1945, instead of the radar value, Mariner 2 would have passed Venus without acquiring any useful data (like Ranger 3). William R. Corliss, the historian of NASA tracking systems, echoed Renzetti: "If the old optical value of the A.U. [astronomical unit] had been used in Mariner-2 trajectory computations, its flyby of Venus might have been at a much greater distance with an attendant loss of scientific data."[52]

Many institutions, including Lincoln Laboratory and Jodrell Bank, were calculating values for the astronomical unit based on their own Venus radar observations. The DSN's Eb Rechtin pointed out that JPL had "a particular interest in an accurate determination of the distance to Venus in order that we might guide our space probes to that target." In June 1960, a month before NASA Headquarters approved Mariner Venus, Rechtin proposed the radar experiment to NASA, emphasizing not its scientific value, but the "practical, purely project point of view."[53]

The tremendous utility of more precise radar measurements stirred JPL to conduct more Venus trials in support of Mariner 2 between October and December 1962, as the probe was heading toward Venus. The radar tests sought to verify the radar-based value of the astronomical unit and to study the planet's motions.[54] Like navigation, radar served the needs of both astronomy and spacecraft tracking.

---

[50] *Mariner-Venus 1962*, 13 & 87; Corliss, *History of the DSN*, 67; Wheelock, 43-45; Hall, *Lunar Impact*, 160; *Tracking Mariner Venus 1962*, 9.
[51] The ill-fated Mariner 3 also foiled the first attempt at dual-spacecraft tracking. The idea was to follow the two Mariners simultaneously within the same beam-width of a single antenna. *MM64 Final Report*, v, 2, 6, 22, 131, 192 & 325; Corliss, *History of the DSN*, 96 & 104; *MM64 TDA Report*, 3, 4, 6, 8, 9, 31, 32, 49 & 51.
[52] Butrica, *See the Unseen*, 48; *Tracking Mariner Venus 1962*, 9, 17 & 75-76; Corliss, *History of the DSN*, 29.
[53] Quoted in Butrica, *See the Unseen*, 39.
[54] Wheelock, 107; JPL, Office of Public Education and Information, Press Release, October 30, 1962, F5189, NHRC; NASA, News Release, "Mariner II Deep Space Telecommunications," February 26, 1963, F7/B62, Willy Ley.

This early contribution of radar to astronomy presaged the rise of radar astronomy as an integral part of astronomy and a contributor to space science and astronomy, including celestial mechanics. Radar added a new type of data that astronomers had at their disposal for improving their ephemerides as well as their understanding of the theories of motion undergirding them. JPL navigators soon joined their ranks by refining astronomical constants and ephemerides and devising new ways to study distant planets.

Part Three

The Golden Age Of Exploration

Chapter 9

Navigators in the Golden Age

The decade of the 1970s, in the words of Robert S. Kraemer, NASA's director of planetary programs, was "a golden age of planetary exploration," heralded by the Mariner Mars 1971 (Mariner 9) mission.[1] Between 1971 and 1978, the United States launched 12 probes to explore the Sun and the solar system. The missions included the Mariner 10 trek to Mercury by way of Venus, the visits of Pioneers 10 and 11 to Jupiter and Saturn, the Helios solar mission undertaken jointly with West Germany, the pairs of Viking orbiters and landers, the Pioneer Venus radar imager, and the Voyagers that explored the outer planets from Jupiter to Neptune and beyond.

This so-called golden age began inauspiciously. In October 1967, Congress canceled NASA's Voyager program that would have sent orbiters and rovers to Mars to conduct a gamut of scientific experiments including a search for signs of life. Voyager was elaborate, technologically sophisticated, and very expensive, costing at least $5 billion in 1968 dollars. NASA Administrator James Webb, angry that Congress had rejected Voyager, announced that he would emphasize the loss by not proposing any new planetary projects for at least two years.[2]

This threatened loss of projects moved an anonymous JPL employee to write a "Dirge for the staff advancement of a Project Manager," sung to the tune of "Londonderry Air" ("Danny Boy"). The "Danny" reference was apt. Dan Schneiderman was JPL's Mariner Project manager, and his office would have felt the brunt of Webb's proposal. The Dirge lamented: "The flights, the flights are dwindling, From globe to globe, There sail no spacecraft now." But, it also decried that: "paperwork, Is all that we'll allow," and told an unnamed "Executive" that: "You banish with a memo, sir, Your golden past, Cast into space, alone."[3]

A shrinking NASA budget became a hallmark of the 1970s. Planetary exploration struggled for funding. Projects that were expensive per se, such as Pioneers 10 and 11, Voyager, and Pioneer Venus, had to compete against the costliest way to explore: human spaceflight. As President Richard Nixon and Congress slashed NASA's budget, planetary exploration competed against Skylab, an early space station,[4] and the Apollo-Soyuz program.[5] In addition, as Kraemer notes, Space Shuttle development began

---

[1] Kraemer, 3, and the book's subtitle: *A Golden Age of Planetary Exploration*.
[2] Kraemer, 124; Ezell and Ezell, *On Mars*, 83-118; Curtis Peebles, "The Original Voyager: A Mission not Flown," *Journal of the British Interplanetary Society* 35 (1982): 9-15.
[3] Handwritten verse, "Dirge for the Staff Advancement of a Project Manager," nd [probably 1969], F43/B3, JPL421.
[4] On Skylab: Roland W. Newkirk and Ivan D. Ertel, with Courtney G. Brooks, *Skylab: A Chronology* (Washington: NASA, 1977); David J. Shayler, *Skylab: America's Space Station* (New York: Springer, 2001); W. David Compton and Charles D. Benson, *Living and Working in Space: A History of Skylab* (Washington: NASA, 1983).
[5] On Apollo-Soyuz: Ezell and Ezell, *The Partnership*; Walter Froehlich, *Apollo Soyuz* (Washington: NASA, 1976); Yuri Y. Karash, *The Superpower Odyssey: A Russian Perspective on Space Cooperation* (Reston, VA:

consuming a larger and larger portion of NASA's budget. Bruce Murray, JPL Director, and John Naugle, NASA Associate Administrator for Space Science and Chief Scientist, among others, put the blame entirely on NASA's Space Shuttle: there was no longer room in the NASA budget for both Shuttle development and vigorous exploration of the solar system.[6]

The Cold War had driven the civilian space program during the 1960s until the triumphant landing of U.S. astronauts on the Moon. The Nixon Administration oversaw the closing stages of the war in Vietnam and ushered in a policy of détente toward the Soviet Union. The Cold War no longer was a compelling argument to support space exploration of any kind. A sign of the times was the cooperation with the Soviet Union initiated by NASA Administrator Thomas Paine that culminated in the Apollo-Soyuz joint venture. Paine firmly believed that the basis for human spaceflight had to be a motive other than Cold War rivalry.[7]

Complexity begets complexity. Over the course of the 1970s, the kinds of spacecraft that navigators would help to fly became increasingly more complex. Missions almost always involved two spacecraft: Mariner Mars 1971, the outer-planet Pioneers, Helios, Viking, Voyager, and Pioneer Venus. The one notable exception was Mariner 10, which flew to Venus and Mercury in 1973. These craft were larger and heavier, so NASA had to launch them atop a combination of Titan III and Centaur rockets. Probes had the capability to undertake many more course maneuvers, because they carried added propellant. Missions also entailed a multiplicity of probe varieties. Viking, the heir to the canceled Mars Voyager, sent both orbiters and landers, while Pioneer Venus consisted of an orbiter and a separate bus laden with a large probe (Sounder) and three smaller probes (North, Day, Night).[8] Complex missions called for cutting-edge approaches to navigation.

Navigators benefited from the introduction of new data types. On top of Doppler, they could make range measurements whether to the Moon or the planets. The drive toward greater accuracy would see the decade begin with pioneering efforts to guide spacecraft using their own television cameras intended for scientific exploration and end with the triumph of so-called optical navigation on Voyager. While simultaneously hunting down sources of systematic errors, collecting measurements, and improving their models, some navigators started to experiment with a different approach. Rather than attempting to calibrate data in the usual way, they conceived and tested sophisticated new data types that would cancel out or at least minimize error sources. Their holy grail was self-calibrating data.

During this golden era of solar-system exploration, navigators also benefited from the growing precision of DSN tracking data. The most fundamental source of greater precision was the shift into higher frequency ranges. The DSN upgraded its dish hardware to operate in the S band, and now its antennas functioned in the X band as

---

AIAA, 1999), esp. 88 & 95-97.
[6] Kraemer, 222 & 226-231.
[7] Ezell and Ezell, *The Partnership*, 1.
[8] Wong and Guerrero, 2.

well. The synchronization of DSN station clocks exploited the new X-band capability by sending signals from Goldstone to overseas antennas via the Moon. The stability of time and frequency standards improved significantly, specifically in preparation for Voyager, with the installation of hydrogen masers.[9]

The DSN of the 1970s had an entirely new architecture, the so-called Mark III configuration, that could support the variety and complexity of planetary missions. The philosophy behind the Mark III was flexibility—so that the system hardware could handle multiple missions—and the automation of many functions that previously had been manual or at best semi-automatic. The previous DSN architecture relied heavily on equipment tailor-designed for a single mission to perform specific functions. The idea of the Mark III was to have a "multi-mission" approach in which the DSN provided a generic set of equipment capable of operating over a wide range of parameters.[10]

The Mark III also marked the shift from mainframe computers to minicomputers and distributed computing, the arrangement of multiple autonomous computers that enabled them to communicate with each other through a network and interact to achieve a common goal. By the time Voyager neared Jupiter in March 1979 two strings of dedicated (MODCOMP) minicomputers were monitoring telemetry, tracking, and engineering.[11]

Over the course of the 1970s, the DSN added two antenna subnets. One consisted solely of 64-meter antennas built for the Pioneer and Voyager projects aimed at exploring the outer solar system. NASA consolidated facilities by placing new 64-meter antennas near the old 26-meter dishes. NASA dropped its South Africa facility for obvious political reasons and added a new dish outside Canberra (Australia) in the Tidbinbilla Valley at a place known alternatively as Ballima and Booroomba and another near the Robledo antenna in Spain. A second subnet consisted of 26-meter antennas inherited from Goddard which had built them for the Apollo program near the DSN stations at Goldstone, Canberra, and Madrid. The pairing of DSN and Goddard dishes at each location gave rise to JPL naming them affectionately "Mutual" stations. They provided the main tracking support for Pioneer 10.[12]

We are not alone. JPL was not alone in having navigation expertise. As we saw in Chapter 1, by the end of the 1950s, the number and variety of military and civilian organizations involved in tracking launches and computing flight paths was appreciable. The Naval Observatory, the Smithsonian Astrophysical Observatory, Lincoln Laboratory, and the Applied Research Laboratory stood alongside such companies as Lockheed, Convair, and Ford's Aeronutronic Systems. These institutions and businesses were potential sources of know-how that JPL could exploit in developing its navigation resources. At

---

[9] Mudgway, *Uplink-Downlink*, 46, 47, 126, 150 & 156; Burnell, Phillips, and Zanteson, 152-159.
[10] Corliss, *History of the DSN*, 43, 71 & 185; Edwin C. Gatz, "DSN Telemetry System, 1973-1976," 5 in DSNPR 42-23; Stinnett, 5; Mudgway, *Uplink-Downlink*, 73 & 74.
[11] Tomayko, 265 & 267; Mudgway, *Uplink-Downlink*, 150 & 151.
[12] Corliss, *History of the DSN*, 77, 198, 200 & 201; Mudgway, *Uplink-Downlink*, 56, 58, 75, 77, 119, 120 & 121; Sonny Tsiao, *'Read You Loud and Clear!': The Story of NASA's Spaceflight Tracking and Data Network* (Washington: NASA, 2008), 147-149 & 232-235.

the same time, though, they also were potential competitors in providing NASA with navigational expertise, as the Ranger experience demonstrated.

JPL's lack of hegemony meant that deep-space navigators had to be creative and resolute, if they were going to participate in Ranger. The initial failure to assign responsibility for Ranger navigation to a single organization reflected the unsettled state of project management in 1960.[13] A May 1960 meeting attempted to settle the question of who would compute the Ranger flight paths. Both Lockheed's Missile and Space Division (Sunnyvale, California), the prime contractor for the Agena booster, and NASA's Marshall Space Flight Center vied for the right. Lockheed also wanted extra money to perform the computations. In March, Victor Clarke, Jr., the head of the small JPL trajectory group, had proposed a way of meting out shared responsibility for Ranger navigation. Clyde D. Baker, Marshall's representative, objected and suggested that "the subject be referred to higher authority for resolution."[14]

Clarke's plan had provided a role for Lockheed, but not Marshall. He proposed that Lockheed furnish JPL with trajectories of the powered-flight from launch to its injection into the so-called parking orbit for a range of launch windows specified by JPL. JPL's responsibility would be the flight path from that point, including the last Agena burn, to lunar impact. Clarke made it clear that Lockheed was not to compute any portion of the trajectory past that final burn and that JPL was "the final authority on the standard coasting trajectory."[15]

Underlying Clarke's seemingly arbitrary division of trajectory tasks was his intention to prevent Lockheed from calculating the entire flight course from launch to lunar impact. As he explained to his supervisor, Johnny Gates, "it is vital that JPL compute the parking orbit and last Agena burn, otherwise Lockheed must compute the whole trajectory from launch to impact. This alternative is held to be highly undesirable." Clarke's recommended division of responsibilities also required "the minimum of technical information interchange between JPL and Lockheed, while insuring a definitive set of responsibilities." Clarke also argued against paying Lockheed to compute "the coasting trajectories because it will provide no useful service and be an unnecessary duplication of effort." In fact, "to avoid confusion and conflict," Clarke advised discouraging both Lockheed and Marshall from computing the coasting trajectories, "even on an unofficial basis. Their effort would be much better spent on the pre-injection trajectories."[16]

Accordingly, JPL representatives Clifford Cummings and James Burke were not inclined to hand over this important assignment to Lockheed, and they were doubtful of the company's computations. Marshall continued to push for a piece of the action, and JPL began to lean toward them. Indeed, JPL already had told the head of the Marshall Agena Project Office, Frederick Duerr, that they "would be delighted" to have Marshall perform "any computation work they want to relative to the post-injection

---

[13] Hall, *Lunar Impact*, 64.
[14] Memorandum, Clarke to Clifford Cummings, "Agena Trajectory Computation Management," May 6, 1960, F359/B17, JPL421.
[15] Memorandum, Clarke to Clarence R. Gates, "Recommendations for Agena Trajectory Computation Management," March 11, 1960, F359/B17, JPL421.
[16] Ibid.

trajectories." JPL's main concern remained Lockheed. Gates told Duerr to "keep tabs of what Lockheed would do so Lockheed didn't spend too much money doing unnecessary work."[17]

The search for a solution continued into November 1960. The partition of work now had JPL responsible for the flight after orbital injection; Marshall and its contractors—including Lockheed—had charge of everything from launch to that moment. Clarke railed against dividing responsibilities at orbital injection. "The natural technical division," he explained, was "not at injection, but at parking-orbit injection." If JPL could not calculate and choose the parking-orbit and final-burn portions of the trajectory prior to injection into trans-lunar orbit, then Marshall and its contractors would make those calculations and choices and they, not JPL, "would compute and select the overall trajectory." Clarke characterized this alternative as being "totally untenable," because it "thoroughly violates the administrative interface and would materially weaken JPL's design position and result in unnecessary duplication, confusion, and waste of money." Nonetheless, he *did* admit that giving the parking orbit to JPL violated the administrative assignment of trajectory responsibilities.[18]

The dispute came to a conclusion on December 14, 1960. JPL, the Air Force, Lockheed, and Marshall signed an agreement that called for Lockheed to compute the pre-injection trajectory in accordance with specifications supplied by JPL. JPL would calculate the trajectory from the parking orbit to lunar impact, a victory for Clarke and JPL navigation. STL had a role to play, as well. It would integrate the two trajectories and generate accurate firing tables under a separate contract with the Air Force. Later, STL computed near-Earth trajectories for Mariner 2, though under contract to Lockheed.[19] Companies desirous of acquiring prestigious NASA navigation contracts, such as STL and Lockheed, benefited when the agency assigned projects to centers other than JPL.

Lunar Orbiter. Such was the case when NASA assigned management of the Lunar Orbiter program to the Langley Research Center. The main rationale for the assignment was the feeling that JPL was burdened excessively with the Rangers, Mariners, and Surveyors. The mushrooming number of probes—which now included Mariner 4 (Mars), Mariner 5 (Venus), and Pioneers 6 through 9 (launched into heliocentric orbits)—was weighing down the DSN as well.[20]

---

[17] Hall, *Lunar Impact*, 64; TWX, ABMA Redstone Arsenal to JPL, May 17, 1960, F359/B17, JPL421; Memorandum, Gates to Cummings, "Marshall Space Flight Center and Lockheed," September 12, 1960, F359/B17, JPL421.

[18] Memorandum, Clarke to Gates, "A Method of Specifying the Pre-Injection Lunar Trajectory," November 16, 1960, F99/B9, JPL421.

[19] Hall, *Lunar Impact*, 65 & 67; Jerry G. Reid and William R. Lee, "Launch-to-Mission Completion Targeting for Ranger and Mariner Missions," 545-556 in Victor Szebehely, ed., *Celestial Mechanics and Astrodynamics* (New York: Academic Press, 1964); *Mariner-Venus 1962*, 48-50; F. L. Barnes, Willard Bollman, David W. Curkendall, and Thomas H. Thornton, Jr., "Mariner 2 Flight to Venus," *Astronautics* 8 (December 1962): 68.

[20] Corliss, *History of the DSN*, 150, 152-154, 156, 158 & 172-173.

From the start, it was not clear how—if at all—the Langley Program Office would rely on JPL navigation or the DSN. This was the first sizeable space project managed by Langley, and the center was experiencing some difficulty in defining just how JPL would participate. Langley made the obvious choice to rely on the DSN for tracking data. How they might use JPL's navigation expertise remained undecided. All parties were trying to circumvent the shortage of computer time at JPL caused to some extent by Surveyor. In April 1964, JPL made it clear to Langley that it was not in a position to commit the necessary manpower and computer time to become involved in trajectory design and related analyses.[21]

Langley, though, did request Headquarters approval for JPL to evaluate all computer programs and mathematical formulas involved in flight operations. They wanted JPL to "make a definitive study of Lunar Orbiter tracking data requirements, including the accuracy of real-time trajectory determination, considering tracking sites, data types, sampling rates, data noise biases, site errors, etc." They also wanted JPL to "check the Space Flight Maneuver Specifications Tables; i.e., the guidance philosophy for midcourse, deboost, and retro firing, including numerical firing tables which will be used in DSN operations." At the same time, Boeing would carry out a similar study and would review all JPL work.[22]

Langley Director Floyd Thompson presented this expanded request to Victor Clarke and Marshall Johnson, the DSN Tracking and Data Systems Manager. They approved, but insisted that JPL's navigation and computer sections needed a larger workforce to perform the work. Langley and JPL agreed that orbit determination would be in the hands of the Lunar Orbiter Program Office, specifically under the direction of William Boyer, the Lunar Orbiter Project Office Operations Manager, and John B. Graham, who was responsible for operations integration. The two centers further agreed that JPL would perform orbit calculations for the launch phase and that Boeing navigators would handle course calculations during the trip to and around the Moon. This was the reverse of the Ranger arrangement. Matt Grogan was at the head of the Boeing navigation team, known officially as the Lunar Orbiter Flight Path Analysis and Control team.[23]

Change is afoot. During the 1970s, NASA grew increasingly more bureaucratic and imposed a greater degree of formality, procedures, and management levels on project activities. To outsiders, the most obvious sign of these changes was the agency's decision in 1975 to drop the so-called meatball logo that President Eisenhower had approved in 1959. JPL was not immune to the trend. The institutional milieu and culture of navigation reluctantly abandoned the flexibility of the past for the rigidity of the present. Viking became a sort of continental divide separating the culture of the past from that of the future.

---

[21] Byers, 52, 91-93, 94 & 149; J. R. Hall, ed., *Tracking and Data System Support for Lunar Orbiter*, TM 33-450 (Pasadena: JPL, April 15, 1970), 1-1.
[22] Byers, 150 & 151.
[23] Byers, 151 & 152; Mayo, 395 & 396; "Selenodesy Experiment," 22.

One characteristic of NASA's changing culture was the formal recognition of individual accomplishment as opposed to team achievement. Heretofore, navigation had been an overlooked, if not neglected, activity within the space program, even though the agency's loudly lauded successes depended fundamentally on navigators' skills and expertise. In 1969, NASA Headquarters began to recognize both individuals and groups through an ongoing program of employee awards. The Group Achievement Award maintained the established cultural value of team achievement. The entire Pioneer Mission Navigation Team received one of these, as did the Voyager Navigation Team.[24]

Individual accomplishment equally received attention, mainly for scientific or mission success, with the granting of two medals. The Exceptional Scientific Achievement Medal went "to both Government and non-Government individuals" who had made "an unusually significant scientific contribution toward achieving NASA's mission." The Exceptional Service Medal recognized "significant, sustained performance characterized by unusual initiative or creative ability . . . in engineering, aeronautics, spaceflight, administration, support, or space-related endeavors that contribute to NASA's mission." A number of navigators have received these two medals for service on a particular mission or for participation on a science experiment over the years.[25]

These NASA Headquarters honors provide only a narrow window on the changing culture of navigation during the 1970s. Much of the change reflected the nearly constant state of expansion and reorganization. A major 1969 laboratory reordering that emphasized mission planning formed the Mission Analysis Division (39) from the Systems Division (31), the institutional home of navigators' Systems Analysis Section.[26]

Johnny Gates' Mission Analysis Division comprised three sections: Bill Melbourne's Tracking and Orbit Determination Section (391), Elliot Cutting's Navigation and Mission Design Section (392), and Rody Stephenson's Systems Analysis Section (393). The word "navigation" appears in several places, most conspicuously in Tom Hamilton's JPL Navigation Program. Section 391, Tracking and Orbit Determination, was where one found the navigators organized in groups led by Don Trask, Jay Lieske, John Anderson, Frank Jordan, and Dave Curkendall.[27] This arrangement and numbering system did not last long, swept aside by a tsunami of bureaucratic upheavals following the succession of Bruce Murray as JPL Director in 1976.

Murray's dramatic reorganization had several objectives. They included: "(1) streamlining of the organization to reduce burden costs and provide room for adaptive future grown, (2) consolidation of all technical divisions into a matrix reflecting today's new technical priorities, (3) increased effectiveness in identification and development of new flight projects, and (4) reduction of the number of separate offices reporting to the

---

[24] NASAPeople, "Agency Honor Awards," http://nasapeople.nasa.gov/awards/nasamedals.htm (accessed March 23, 2010); "Voyager Navigation Team," [JPL] *Universe* 9 (March 2, 1979): 1; *Pioneer: First to Jupiter*, 274.
[25] NASAPeople, "Agency Honor Awards," http://nasapeople.nasa.gov/awards/nasamedals.htm (accessed March 23, 2010); lists of awards to NASA employees 1969-1978 in Appendix A of *NASA Historical Data Book*, Vol. IV (accessed March 23, 2010) http://history.nasa.gov/SP-4012/cover.html.
[26] Wood, "Chronology"; Brancheau, "Chronology."
[27] OrgChart, Mission Analysis Division 39, June 30, 1970, JPL556.

Director's Office." The overall objective, though, was to reduce "our burden rate target for FY '77 [Fiscal Year 1977]."[28] Cuts in costs and personnel became commonplace.

Murray considered two alternative structures for the System Division, both of which featured Gates and Hamilton at the helm. Option I had seven Sections: Advanced Projects (311), Mission Design (312), Systems Design and Integration (313), Tracking and Orbit Determination (314), Systems Analysis (315), Tracking Systems and Applications (316), and Science Payload (317). Oddly, Jordan's name as manager of 314 was struck out, and Curkendall was split as Assistant Manager of 314 and 316 (Melbourne). Option II had fewer sections: Systems Analysis (311), Mission Design (312), Systems Design and Integration (313), Tracking and Orbit Determination (314), Tracking Systems and Applications (315), and Science Payload (316). Murray chose the latter arrangement.[29]

The new System Division reflected JPL's goal of reducing costs: it relied on a smaller number of staff with 9 managers instead of 10. Despite the reduced number of managers, the Systems Division remained huge: 344 employees in all. JPL's reversion to the old number system (the Systems Division was again 31) carried a sense of familiarity, but the expansion of the Division from three to six Sections was certainly not familiar. Moreover, from 1976, several navigation managers—namely Frank Jordan, Roger Bourke, and Bill O'Neil—had no Assistant or Deputy Manager, while the other Division Sections all had Assistant Managers.[30]

Frank Jordan, who replaced Bill Melbourne as head of the Tracking and Orbit Determination Section (314), had been in both mission design and navigation and sensed some critical cultural differences arising between the two. Navigation, he explained, was "more mathematically demanding then mission design. You could very easily see that mission design began to be dominated with educational backgrounds up through the master's degree in the main, but Ph.D.'s went to NAV [navigation], because they had to have this extra dimension of the statistical and radiometric knowledge that the mission designers didn't have to have." Jordan himself chose navigation over mission design. "I was going to go where the Ph.D.'s went, you know, so I went into navigation." He "noticed that the two sections . . . had different character. They were different kinds of people. I even noted, and I'm not sure why this was, their political leanings were different in the two groups." "I also noticed that even though I was in NAV [navigation], most of my friends were in the other group."[31]

Furthermore, Jordan reflected, "There was a certain—I wouldn't call it an arrogance—in navigation. But others have commented on this to me. There is a bit of an elitism in navigation, that other engineers at JPL have commented to me about, because it is a very sophisticated mathematical field. There are things navigators know that nobody else knows, and they know it. Whereas mission designers are more the salt of the Earth, navigators are a little bit the academic. I'm exaggerating, but I used to

---

[28] Director's Letter.
[29] Reorganization proposals.
[30] Reorganization proposals; OrgChart, Division 310, January 3, 1977, JPL556; Director's Letter.
[31] Jordan, interview, 15.

notice this as a youngster. I don't know how it occurred that I was in one and had more friends in the other."[32]

Clash of cultures. Murray's laboratory reorganization occurred while NASA was in the midst of the Viking project. From the perspective of JPL navigators, Viking marked a memorable watershed, if for no other reason than its complexity. NASA Headquarters assigned Viking management to Langley, as it had done with the Lunar Orbiter. This time, Langley chose a different firm for navigation expertise, the Martin Marietta Aerospace Company.

The Viking Navigation Working Group undertook initial trajectory and orbit determination work from 1970 until its absorption in 1973 by the Viking Flight Path Analysis Group. JPL's Bill O'Neil and his Deputy Richard Rudd led this multi-agency team composed of members from JPL, Langley, and Martin Marietta as well as General Electric and Analytical Mechanics Associates. They continued the work of designing trajectories and developing navigation strategies, procedures, and operational software.[33]

The project distributed navigation responsibilities among several crews that handled different aspects of the flights of the twin Viking spacecraft from launch to Mars orbital insertion as well as the setting down of the landers on the planet's surface. With two exceptions, the teams were a blend of JPL, Martin Marietta, and Langley specialists. Consequently, JPL navigators played a larger role on Viking than on Lunar Orbiter.

For instance, the Interplanetary Orbit Determination Team heads, Ken Rourke and Carl Christensen, as well as members Chuck Acton, Neil Mottinger, and Bryant Winn, were from JPL. This team did orbit determinations until the engine burn that inserted the spacecraft into a Mars orbit. The Satellite Orbit Determination Team took care of navigation after orbit insertion—when the Vikings were "satellites" of Mars— and determined the landing site of each of the two Landers. The team leaders, Claude Hildebrand and Ed Christensen, and team members Mohan Ananda, Dale Boggs, George Born, and Frank Jordan, were all JPL navigators. The Orbiter Maneuver and Trajectory Team—responsible for developing course corrections and initiating the Lander descents—was the only team made up entirely of JPL navigators. The interface between the Orbiter and the Lander, the so-called separation orbit, was the joint responsibility of the latter all-JPL team and the Lander Flight Path Analysis Team, which had no JPL navigators. Its members worked for General Electric and Analytical Mechanics Associates; the Team Leader was Edward Euler of Martin Marietta Aerospace.[34]

"Somehow," Frank Jordan recalled, "everybody worked together and made it work." The project, nonetheless, "was as much of a political challenge as it was a technical challenge."[35] Managing Viking was a bureaucratic challenge that the Langley program manager, James S. Martin, Jr., met head on. Martin previously had been Lunar Orbiter assistant project manager. Because Viking was the first NASA planetary project in which several centers and contractors participated in the design, development, and

---

[32] Jordan, interview, 15-16.
[33] O'Neil and Rudd, 1-2.
[34] O'Neil and Rudd, 2 & 3.
[35] Jordan, interview, 33.

operation of major spacecraft systems, Martin undertook a number of steps to manage these new complexities.[36]

One step was the creation of the Viking Project Management Council to "facilitate common understanding of the overall project objectives and provide a forum where technical and management problems can be freely discussed." Martin also arranged for a series of project-wide quarterly reviews starting in October 1969. Each systems manager had 90 minutes to summarize progress in his area. He also introduced a system of mass documentation, formal reviews, telecoms, and informal conversations. At the same time, Martin centralized responsibility and authority for Viking, a measure later deemed to be a key factor in the project's success.[37]

Martin soon realized that the Viking and JPL management styles were at odds with each other. JPL engineers functioned within Divisions responsible for specific engineering activities or disciplines, such as the Systems Division. The JPL Viking program office, run by Henry Norris, allocated funds, prepared plans and schedules, assigned tasks, and received progress reposts. However, the Divisions carried out the actual work of designing and developing the spacecraft and experimental hardware, preparing and operating the DSN and the flight operations facility, and planning and performing navigation tasks. Each Division chief and his subordinates supervised their personnel and selected the engineers who represented their Division.[38] This was the JPL matrix at work.

The matrix structure did not mesh well with the Langley approach. Langley personnel worked through a more centralized organization in which everyone was directly responsible to the project director. Langley's project orientation differed significantly from JPL's matrix and line management. The Langley Viking Project Office was uneasy with JPL's way of doing things. A clash of management cultures erupted. Martin went to Pasadena to learn firsthand about how JPL worked. In the end, he was still "not entirely comfortable" with the organization, but at least he had experienced the JPL approach and the people who worked there. Likewise, JPL personnel began to appreciate the sources of Martin's concerns.[39] All the same, Viking was on its way to reshaping JPL navigation culture.

The cultural divide. Some navigators noticed that, over the course of Viking, JPL had changed. The change was strikingly apparent when one contrasted Viking with the way that JPL was managing the concurrent Mariner 9 project. Bill O'Neil, who later managed the Galileo project, was in charge of navigation on both Viking and Mariner 9. Jordan worked for him on both missions and witnessed the differences.

On Mariner 9, Jordan recalled, O'Neil "could run and hide this little crew. We could lock the door, and nobody could bother us. Oh, no, not on Viking. It was a more public affair with all kinds of Headquarters and Martin Marietta people and Langley people all over the place." Viking was driven from NASA Headquarters and far more

---

[36] Ezell and Ezell, *On Mars*, 128.
[37] Ezell and Ezell, *On Mars*, 173, 182 & 193.
[38] Jordan, interview, 34; Ezell and Ezell, *On Mars*, 193 & 194.
[39] Ezell and Ezell, *On Mars*, 194.

bureaucratic with a lot more "management overhead" and many more reviews. In Jordan's words: "Everything had layers of management and review. You just couldn't find more contrast than between the ways those two projects were managed. I never even met the [Viking] project manager. But I sure knew who the document project manager was, and it was a personality of real control there. . . . The number of tests went up. The number of reviews went up. The number of management layers went up."[40]

Mariner 9, though, was not immune entirely from interference by Viking outsiders. Jordan found the "Martin Marietta people" to be particularly bothersome for those on Mariner 9 navigation. The company "had all their independent analysts looking at orbit determination," and they decided that the project would not be able to compute orbits successfully. "Big scare tactic," Jordan recalled. NASA Headquarters Viking management "went marching down to the poor little Mariner 9 management" and told them: "You're going to fail next month." Earl W. Glahn, the NASA Headquarters Mariner 9 manager, was shocked, because "he was trusting JPL." As a result, Jordan recalled being "drug into Viking reviews before I even encountered this thing. I had to put them down and shut them up, and tell them we can do it. And we did it, and we did it without any of their help."[41]

In retrospect, Jordan mused, "Viking worked. I have always believed that this changed the lab forever. It changed the nature of who gets promotions. It was basically a wholesale movement from can-do to can't-fail attitudes. It has changed the lab forever. In a sense, it's when the lab grew up, to have to do things the way that our agency does things. We've never gotten out of it."[42]

"The whole cultural attitude of what is valued here shifted with this mission," Jordan continued. "All of a sudden the guy with the quickest equation derivation was not the top; it was the guy who had the procedures written out." "It was really different. It really cut into the morale of my workforce. I remember that, because people were spending only 70 percent of their time on productive work. The rest of it was just trying to make the boss look good."[43]

On and off again. Perhaps one of the most outward signs of the new navigation culture was the movement of navigators around campus, then off campus, as a result of the frantic pace of construction to keep up with JPL's explosive growth. "You know, we moved around so many times," Bill Sjogren recalled. "First we were on-lab, then we were off lab. Then we came back on lab. Then we moved off." JPL moved navigation out of their building to have more rooms for administrators. "We were up on about the fifth or sixth floor," Sjogren related. "They kicked us out of there. I think that's when they put us off lab actually, when they built building 264, because the administrative area was growing. You know what I mean? They had to have more administrators. Public relations and all these other places had to have more offices."[44]

---

[40] Jordan, interview, 33, 34 & 36.
[41] Jordan, interview, 34-35.
[42] Jordan, interview, 35.
[43] Jordan, interview, 35-36.
[44] Sjogren, interview, 2008, 31.

During the 1970s, navigators initially worked off campus in the Angeles Crest Building located next to the Angeles Crest Highway in La Cañada, a few miles from the JPL campus. Because navigation was so dependent on computers, which remained on campus, working in the Crest Building had its challenges. "There was no operational support done there," Neil Mottinger explained. "I don't know what the logic was for moving us there. Like all good people that resist change, we screamed bloody murder, and we didn't want to go. They installed a card reader and printer there that could connect us to the mainframe, the Univac in those days, so we could submit jobs and that stuff." Putting "some distance between a card reader and a mainframe," he judged, "was quite a trick."[45]

The Angeles Crest Building in La Cañada (Author)

Gaining access to the Crest Building for Mariner Mars 1971 other than during the daytime was problematic. Computer time was cheaper at night, so Mottinger and Bill Zielenbach frequently did computer runs to develop Mariner station locations at that time. But, building policy stipulated that its doors be locked after 6pm and on weekends, that is, when about 20% of all navigational software development work was

---

[45] Sjogren, interview, 2008, 31; Wood, interview, 60; Mottinger, interview, March 10, 2009, 11.

to take place. The lack of access impacted other navigational software, such as the new double-precision trajectory program. The solution was to issue keys to employees.[46]

Still, the Crest Building did offer some advantages as a workplace. "We came to like it very much," Mottinger remembered. "We could park right outside the building, and we just ate in the local restaurants or brought lunches." The Hill Street Café, where retired navigators now enjoy a weekly breakfast get-together, was then known as Lloyd's. Also, there was Berge's, "a little deli sandwich shop. "He and his wife opened up this shop in a little tiny building. It's no longer there." Min's Restaurant, another "tiny restaurant with maybe six booths," featured "home cooking." Today it's a Thai restaurant known as Min's Kitchen.[47]

The next off-campus location was less pleasant, to say the least. Mottinger described the building, located at 133 North Altadena Drive in Pasadena, as "an old Bekins warehouse" that had "no windows, and that's where our section was." "We were there until about 1978. Then we came back on lab." "We didn't want to come back particularly," Mottinger explained, "mainly because of the parking." In 1978, navigation was again on campus, this time in building 264, where it remained into the 1980s.[48]

---

[46] Mottinger, interview, March 10, 2009, 11 & 12; Letter, Schneiderman to W. Hansen, "Evening and Weekend Closure of the Crest Building," July 27, 1970, F40/B3, JPL421.
[47] Mottinger, interview, March 10, 2009, 12.
[48] Wood, interview, 60; Mottinger, interview, March 10, 2009, 13.

NASA Group Achievement Award given to the Navigation Software Development Team in 1978. The names listed are: Daniel Alderson, Peter Breckheimer, John Ekelund, Jordan Ellis, David Hilt, Ahmad Khatib, Victor Legerton, Susan McMahon, Robert Mitchell, Patricia Molko, Theodore Moyer, Nick Panagiotacopulos, Theodore Pavlovitch, Thomas Talbot, and William Zielenbach. (Courtesy of JPL)

Chapter 10

Navigation Becomes Astronomy

The relationship between navigation and astronomy has been multifaceted. Space navigation differed from earlier types of navigation in that navigators' observations aided in the improvement of the tables (data files) that furnished the positions of celestial objects (the ephemerides). The desiderata of navigation and astronomy differed as well, but accuracy and consistency were of equal importance to both. For astronomers, it was imperative that a set of ephemerides be internally consistent to the maximum extent possible. Absolute accuracy was secondary, because the main purpose of the ephemeris was to define a *system*, not to describe the positions of celestial bodies. For navigators, positional accuracy was of the highest importance, though without subordinating internal consistency.[1] Also, both astronomers and navigators sought consistency in data format types, an issue that grew in importance as the types of observations that both communities subjected to computer analysis proliferated.

JPL navigation, initially a consumer (and adapter) of Naval Observatory data and ephemerides, was, by the end of the seventies, the source of constants, models, and ephemerides for the world's almanac offices. The key to this dramatic success was the replication of the celestial mechanics paradigm constructed by Brouwer and others within the walls of JPL. The paradigm, to use the terminology of philosopher of science Thomas Kuhn, consisted of the astronomers' research agenda in dynamical astronomy (a set of problems) plus their common research tools, such as punched-card and computer technologies and numerical integration.

From astrology to celestial mechanics. The first leader of the JPL ephemeris group was Neil L. Block, who had a reputation unknown to JPL as an astrologer working under the pseudonym Gary Duncan. Ken Seidelmann, the former chief of the Naval Observatory's Almanac Office, characterized Block as "a very competent sort of numerical person" who was the "principle person out at JPL initially."[2] One éloge of the astrologer describes him as "a computer scientist and professional astronomer." Another relates: "He lectured in celestial mechanics, numerical analysis and statistics. He claims to be the first astrologer ever to use a computer and to have performed statistical work on these machines."[3] JPL's Neil Block wrote as a colleague to such contemporary celestial

---

[1] Mulholland, "Corrections," 4.
[2] Seidelmann, interview, 20-21.
[3] Michael Erlewine, "Gary Duncan (Neil Llewellyn Bloch) 1931-1988," 2001, http://www.solsticepoint.com/astrologersmemorial/duncan.html (accessed February 8, 2009); Erlewine, "Remembering Gary Duncan (1931-1988)," January 7, 2009, ACT Astrology, http://actastrology.com/viewtopic.php?f=30&t=25 (accessed February 8, 2009).

mechanics experts as Samuel Herrick, Wallace Eckert, Donald Sadler, and George Wilkins (the latter two being at Greenwich Observatory).[4]

Chuck Lawson took over the ephemeris development team from late 1964 until 1967, when Block left JPL to return to school. Lawson was the individual most responsible for transforming the ephemeris effort and saw to the hiring of individuals trained in astronomy. Between 1964 and 1966, a number of Yale students spent their summers with the JPL ephemeris group, including Carol Williams, Jay Lieske, Brian Marsden, and David Dunham. Lieske became a regular staff member after completing his dissertation and took over leadership of the ephemeris group after Lawson. In 1972, Lieske hired fellow Yale astronomy graduate E. Myles Standish, who succeeded Lieske as the group's head. Doug O'Handley, a Yale graduate employed by the Naval Observatory, joined the team at that time as well.[5] The permanent ephemeris workers also included Herget graduates from the University of Cincinnati, such as J. Derral Mulholland, and a good number of Herrick astrodynamics doctorates from UCLA. The first of the Herrick students were John Anderson, Sam Dallas, Ted Moyer, and John Ondrasik.[6]

The tapes. The Development Ephemeris Project was the primary effort of the Celestial Mechanics Group. They were not alone in devising and improving ephemerides to meet the demands of the Space Age. For example, Lincoln Laboratory had its own ephemerides program known as the Planetary Ephemeris Program or PEP. Its significance at least in part was its reliance on a combination of traditional optical observations (Naval Observatory data back to 1850) and radar measurements (ongoing radar observations of Venus, Mars, and Mercury). The radar readings came from the laboratory's own facilities at Millstone and Haystack in Massachusetts and Cornell University's gigantic dish outside Arecibo, Puerto Rico, which also relied on PEP for pointing their radars. PEP eventually analyzed lunar laser range and VLBI measurements as well. Irwin Shapiro, who ran the project, exploited Lincoln Laboratory's computing power and designed the program so that it numerically integrated the equations of motions of the planets. While Arecibo planetary radar astronomers continued to use PEP, pulsar observers and other radio astronomers shifted to the JPL ephemeris. In the end, a lack of funding left PEP just able to keep up with the Arecibo ephemeris work.[7] In contrast, the JPL ephemeris program enjoyed a more abundant source of manpower and funding, because it supported NASA missions at a time when the space agency's budget was ever on the increase.

---

[4] Letter, Block to Eckert, November 23, 1960; Letter, Block to Sadler, November 30, 1961; Letter, Wilkins to Block, December 11, 1961, all in F2-2/B2, Eckert Papers, copies provided courtesy of Allan Olley.
[5] Ekelund, "History of ODP at JPL," 4; O'Handley, Standish, and Fliegel, interview, np; "JPL Ephemeris Development," 3; Standish, interview, 3-4; *Astronomy at Yale*, 199.
[6] Dallas, interview, 1; "Ten Years Service," *Lab-Oratory*, September-October 1973, 21; "JPL Ephemeris Development," 2 ; Herrick students' dissertations, B4 & B5, Herrick Papers; OrgChart, Systems Analysis Section, November 1, 1964, JPL556; OrgChart, Systems Analysis Section, June 15, 1965, JPL556.
[7] Butrica, *See the Unseen*, 124-125; Shapiro, interview, September 30, 1993, 18-22.

Goddard and the Naval Surface Weapons Center also had ephemeris projects that addressed a niche need peculiar to their mission. The Naval Surface Weapons Center served the missile-related needs of the Navy, while Goddard concentrated on Earth-orbiting satellites, ceding deep space to JPL, a division that reflected the different types of missions supported by the two NASA centers. The JPL ephemeris program differed from these in its commitment to providing ephemeris tapes to industry and government. The distribution of its ephemeris tapes, though, put JPL in direct competition with the Almanac Office.

The chief difference between JPL and the Almanac Office was that the latter offered products intended for general scientific use, while the former addressed the more limited needs of aerospace users. The basis of the Almanac Office products was a fixed, self-consistent set of astronomical constants that rarely underwent revision, while the JPL ephemeris tapes were rooted in the latest and most precise measurements. The astronomer's need for *consistent* values was set against the navigator's craving for the newest and most *precise* values. In a sense, one had a dual system of constants: (1) the astronomer's set of self-consistent constants not subject to speedy change and available for general scientific use, and (2) the navigator's set of continually-updated constants prepared ad hoc for a specific mission.

In 1967, JPL standardized how it designated its ephemerides. The original tapes, discussed in Chapter 5, simply went by the names E9510, E9511, and E9512, where "E" stood for ephemeris. Ever since 1967, regardless of delivery format, JPL has used the Development Ephemeris nomenclature in which the letters "DE" followed by a number indicate a particular iteration. The first in the "DE" series was the DE19 version; the "E" tapes became DE3 at the same time. JPL also issued a specialized lunar ephemeris, labeled "LE" plus a numeral, to function in tandem with the Development Ephemeris. JPL began distributing its ephemeris tapes outside NASA in 1964. A not insignificant number of copies went out to governmental and industrial clients. A December 1964 Interoffice Memorandum listed 18 companies using the JPL tapes. Later versions were available from the NASA Computer Software Management and Information Center at the University of Georgia.[8]

Although the JPL ephemeris effort did not suffer the funding issues faced by Lincoln Laboratory, financial backing for the Development Ephemeris was rather piecemeal until the arrival at JPL of Irwin Shapiro's student, Robert Preston. Preston enlisted the help of navigator Dave Curkendall and approached the JPL office that supported flights (the Flight Project Support Office). Myles Standish recalled Preston telling them: "You guys are going to need an ephemeris now and on into the future." They had two options. "You either have to support it here or get it somewhere else. You can't get it somewhere else." Standish remembered a viewgraph that said: "MIT won't, and USNO [U. S. Naval Observatory] can't." It meant that "MIT didn't want anything to do with supporting navigation; they just didn't want to get into that game. The Naval Observatory did not have the expertise technology. There was no one else."[9]

---

[8] *LE4*; *DE19*; "JPL Ephemeris Development," 2; *JPL Ephemeris Tapes*, 1-2; *DE69*, 1; *DE96*, 7.
[9] Standish, interview, 25.

During the 1970s, Skip Newhall made the Development Ephemeris available to astronomers all over the world by focusing on data formats. "There are a lot of different types of computers, and everybody had his own format and stuff," explained Standish. To overcome the multiplicity of formats, Newhall converted the files into different basic formats including binary. "On the Univac," Standish recalled, "he actually would make a tape in the format that an IBM machine would have written. He actually pushed the bits around. I mean, it was pretty clever. We would send an IBM guy an IBM format tape, and a VAX guy would get a VAX tape, and so forth. Skip Newhall did that."[10] Today, astronomers can acquire a copy of the latest JPL Development Ephemeris on a laser-readable disk or download it in a variety of formats including Java, ASCII, and even FORTRAN.[11]

The paradigm. The Celestial Mechanics Group was a mirror image of the celestial mechanics paradigm—to use Thomas Kuhn's expression—represented by the collaboration of Clemence, Brouwer, and Herget discussed in Chapter 2. They focused on problems of dynamical astronomy, the technology of computers, and the technique of numerical integration. However, although navigators shared the same problems, they approached them from a different, idiosyncratic perspective.

According to Kuhn, before evolving into normal science, scientific activity passes through a developmental phase in which the problem-solving consensus that characterizes normal science does not exist yet. In this "preconsensus" or "pre-paradigm" phase, and immediately before a phase of normal science, groups of investigators addressing roughly the same problems but from different, mutually incompatible standpoints compete with each other. As a consensus emerges, members of the competing schools join the group whose achievements are better as measured by scientific values.

In this case, the establishment of the JPL celestial mechanics unit threw the navigators and dynamical astronomers into competition. The greatest difference between the two groups during this "preconsensus" or "pre-paradigm" phase—what made them mutually incompatible—was the type of data each used. JPL navigators used both tracking (Doppler and range) and radar data, while celestial mechanics, especially those responsible for the world's almanacs and ephemerides, used telescopic observations mainly.

A similar divergence existed in planetary radar astronomy, but the divergence was one of complementary, not competing, groups. The so-called bistatic radar approach of Von Eshleman at Stanford University complemented the research of ground-based planetary radar astronomers at Arecibo and Haystack. Eshleman had intended that complementarity; ground-based planetary radar astronomers distinguished themselves from the Stanford approach. The establishment of turf lines is clear from a 1973 article written by Tor Hagfors and Donald B. Campbell of Cornell that reviewed

---

[10] Standish, interview, 23.
[11] "JPL Planetary and Lunar Ephemerides: Export Information," November 29, 2011, ftp://ssd.jpl.nasa.gov/pub/eph/planets/README.txt (accessed March 16, 2010).

the field. They wrote: "We have, however, chosen to omit this work [space-based radar] here since it is our opinion that it properly belongs to the realm of space exploration rather than to astronomy."[12] The dividing line thus was one that separated space-based research from ground-based astronomy.

One could argue similarly that navigation belonged to "the realm of space exploration rather than to astronomy" based at least partially on the different data collecting methods of almanac astronomers and JPL navigators (optical measurements versus Doppler and range). Navigators also reaped the benefits of radar observations of the Moon and planets as well as the highly precise observations made by the Naval Observatory's Six-Inch and Nine-Inch Transit Circles, while the International Astronomical Union was struggling with the controversial question of adopting a radar-derived value for the astronomical unit. The differences between the two scientific communities of navigators and astronomers also extended to how they went about confronting the most fundamental—and most challenging—problem in celestial mechanics: the motions of the Moon.

Corrections to corrections. Astronomers had been working the problem for centuries, but despite significant progress, an acceptable solution remained elusive for a number of reasons, the chief of which was the complexity of the Moon's movements. The IAU had assigned responsibility for wrestling with this eely problem to the Greenwich Royal Observatory.[13] Wallace Eckert led a parallel effort at Columbia University. Their common goal was to correct the theory of the Moon worked out earlier by Yale Professor Ernest W. Brown whose *Tables of the Motion of the Moon*, published in 1919, represented literally the work of a lifetime.[14]

Astronomers typically solved the problem of lunar motion in two steps. The first step, which they called the "main problem," treated the Sun, Earth, and Moon as point masses and assumed that the mutual center of gravity of the Earth and Moon (the barycenter) moved in a fixed ellipse around the Sun. Once a solution for the "main problem" was in hand, astronomers introduced, as small variations to this solution, various perturbing forces such as the effects of variations in the Sun's elliptical orbit caused by the Moon and planets, the direct influences of the planets on lunar motion, the elliptical shapes of the Earth and Moon, and the effects of relativity.[15]

Brown had followed the methods introduced by George W. Hill (1838-1914) of the Naval Observatory. The expressions for the lunar coordinates generally in use after 1923 were those of Brown.[16] The so-called Hill-Brown method used rectangular coordinates oriented in a specific fashion that Eckert and others used to compute their lunar ephemeris,[17] making the coordinate system part and parcel of the ephemeris. For

---

[12] Hagfors and Campbell, "Mapping of Planetary Surfaces by Radar," *Proceedings of the IEEE* 61 (September 1973): 1219-1225, quote 1224.
[13] Letter, Block to Sadler, November 30, 1961, F2-2/B2, Eckert Papers.
[14] Brown, *Tables of the Motion of the Moon* (New Haven: Yale University Press, 1919).
[15] Eckert, "Improvement," 415.
[16] Dick, *Sky and Ocean*, 285; Eckert, "Improvement," 415; Chapront-Touzé, 53.
[17] Eckert, "Improvement," 416.

the 1952-1972 ephemeris, Eckert replaced Brown's tables with new coordinates that were integral to what Eckert termed the Improved Lunar Ephemeris.[18] He computed these coordinates without reference to the tables built from Brown's theory and indicated a means for correlating the two coordinate systems. Errors arose, nonetheless, when sufficient accuracy was not maintained in transforming from the original Brown coordinates to Eckert's.[19]

Despite eliminating the inaccuracies of the old tables, the Improved Lunar Ephemeris retained some of the loss of precision that Brown had introduced in his transformation from rectangular to polar coordinates and elsewhere in his lunar model.[20] The space age demanded greater exactitude. In November 1958, Eckert admitted that the expressions for the coordinates were "not sufficiently accurate, however, for the new methods of observation coming into use."[21] The space era's reliance on highly precise radio, radar, and eventually laser measurements demanded more accuracy.

From 1957, Eckert and his astronomer colleague Miss Rebecca Jones began applying numerical methods, rather than algebraic expansions, in the hopes of developing a more accurate theory of the Moon. They tested their computations by running them in parallel on an IBM 650 mainframe, but adapting them to the computer was neither an easy nor a straightforward task. The process required what Eckert characterized as "shoe-horning" to cut down computer running time and to avoid an excessive amount of card handling. In the end, the entire calculation, he reported, was "done with a very moderate number of hours of machine time and comparatively few cards."[22]

Eckert returned to the Sisyphean labor of improving the accuracy of the lunar ephemeris in 1966. His second-generation corrections improved the precision of the solar part of the ephemeris. The so-called Improved Brown Lunar Theory with 1966 corrections now represented the main improvements in the lunar ephemeris available to astronomers and navigators.[23]

Numerical integration triumphant. By then, JPL navigators were busy updating their lunar ephemeris. Its accuracy was essential not just for flying to the Moon, but also for synchronizing DSN clocks via the Moon, and that synchronization in turn impacted navigational accuracy. Lunar Ephemeris Number 4 (LE4), prepared for the Surveyor and Lunar Orbiter, was the work of Neil Block and Derral Mulholland. Available in 1967, LE4 differed from previous iterations by virtue of several corrections to Eckert's lunar theory. Block and Mulholland intended LE4 to be the most accurate theoretical lunar ephemeris available at the time of its release. They also wanted it to be perfectly

---

[18] Eckert, Jones, and Clark.
[19] Chapront-Touzé, 53.
[20] Eckert, Walker, and D. Eckert, 314.
[21] Eckert, "Improvement," 415.
[22] Eckert, "Improvement," 416.
[23] Eckert, Walker, and D. Eckert, 315; Chapront-Touzé, 53.

consistent with the IAU System of Astronomical Constants.[24] Unfortunately, only after completing LE4 did they discover that "an archaic value" that accounted for the flattening of the Earth was "buried deeply in the structure" of the Improved Brown Lunar Theory. Consequently, perturbations in lunar motion caused by the shape of the Earth were wrong in both the Improved Brown Lunar Theory and the LE4.[25] How was one to achieve the necessary level of precision?

Mulholland and the JPL lunar ephemeris group—Charles Devine, Doug Holdridge, and Bryant Winn—along with Raynor Duncombe and Thomas Van Flandern at the Naval Observatory collaborated in addressing problems in the lunar ephemeris. Mulholland believed that the ultimate solution was to apply numerical integration. Electronic computers provided both the means to treat lunar motion using numerical integration as well as the need to apply numerical integration. "The emergence of solar system exploration has posed demands for accuracy in the solar system ephemerides that cannot be satisfied with the present analytic theories," he concluded.[26]

The JPL navigators achieved a major breakthrough by applying numerical integration to the Moon. The issue initially was to determine a "meaningful time span" over which to extend their numerical integrations. The utility of numerical integration increases with the length of the time span of data treated and, consequently, with the sheer amount of data integrated. To treat the long-term characteristics of the motion properly, Mulholland and Devine regarded the nodal period (18.6 years) as a practical minimum interval. The interval they actually adopted was slightly more than 20 years and began at the initial time of the JPL ephemeris tapes.[27]

Mulholland and Devine were not really satisfied with the results. They felt that the underlying model was in need of revision. So, they switched to a model that allowed empirical (observation-based) adjustments to the theory. This was the same approach, for example, that navigators used to model and calibrate for media effects. The need for these adjustments to the theory arose from such known factors as the tides and "perhaps some yet unknown cause." Mulholland also performed numerical integrations with the planetary data and lunar theory contained in Development Ephemeris 19 (DE19).[28]

When Mulholland and Sjogren numerically integrated Lunar Orbiter range data, the results gave residuals that often far exceeded those expected from known effects. When they incorporated Eckert's *original* Improved Brown Lunar Theory plus several modifications devised by Mulholland's team, they realized an order-of-magnitude reduction in the size of the range residuals. They compared their revised ephemeris with the old one by measuring the root mean square (RMS) of the residuals, a statistical measure of the magnitude of a varying quantity. With the old ephemeris, the RMS residual was about 800 meters with a maximum near 1,700 meters. The corrections

---

[24] *LE4*; *LE6*, 1.
[25] *LE6*, 1.
[26] Mulholland, "Numerical Studies," 249; Mulholland and Sjogren, 74; Garthwaite, Holdridge, and Mulholland, 1133 (quote).
[27] Mulholland, "Numerical Studies," 247.
[28] Mulholland, "Numerical Studies," 247; *DE19*.

reduced the RMS residual to about 110 meters.[29] JPL now had a new and more accurate ephemeris for the Moon. This news boded well for the planetary ephemerides to which it was coupled.

Longer and longer integrations. Consistency had become the quest of the JPL ephemeris group: consistency between the lunar and planetary ephemerides as well as within the theoretical models that went into the ephemerides. Attention to the correctness of those theories had come to the forefront. "An inconsistent model is an incorrect model," Mulholland wrote, "just as an inaccurate model is an incorrect model." Navigators needed better astronomical constants derived from tracking and radar data, but it also was "absolutely necessary to use as consistent and as accurate an ephemeris as can be found. Every improvement in either of these properties is an improvement in the degree of confidence with which we can improve our basic data."[30]

If the ephemeris of a single body lacked consistency with the ephemerides of other bodies, it still could be perfectly adequate for computing flight paths that lay entirely within the vicinity of that body provided that all observations were made from the Earth or the probe. On the other hand, if a spacecraft trajectory passed near two or more bodies whose ephemerides contained inconsistencies, then navigators would have to make a significant amount of empirical (observation-based) adjustments in order to offset the inconsistencies in positional information. In other words, they would have to rely on creating an ephemeris, as it were, on the fly. But, if they eliminated the internal consistencies in all of the ephemerides, navigators could rely less on this empirical approach.[31]

The quest for consistency began in 1964, when Brian Marsden laid out a series of steps for navigators to take in that direction. He focused on the ephemeris for Mars and found inconsistencies in the various theories for that planet's motion (and their associated corrections). These inconsistencies became especially evident when he compared the ephemeris to the results of a numerical integration of the planet's orbit. He proposed conducting a consistent numerical integration of the nine principal planets' motions over the period 1700 to 2000 incorporating both optical and radar observations. Marsden believed that only after performing these computations could one "think seriously about changing the constants of integration and planetary masses and produce a new set of more accurate ephemerides."[32]

The achievement of a simultaneous integration of the planets' motions had to wait until 1967 and the completion of the Solar System Data Processing System (SSDPS) under the direction of Chuck Lawson. The SSDPS was actually a series of programs that determined the motions of the planets through simultaneous numerical integration. The first such simultaneous integration, carried out in the spring of 1968, covered the period from 1967 back to 1950. By 1971, they were integrating from 1969 back to 1910. The integrations processed a massive amount of observational data, over 37,600

---

[29] Mulholland and Sjogren, 74 & 76; Eckert, Jones, and Clark.
[30] Mulholland, "Corrections," 5.
[31] Mulholland, "Corrections," 4.
[32] Brian G. Marsden, "On a Consistent Ephemeris," 61-65 in JPLSPS 37-29:IV.

observations of the planets. They included photographic transit-circle measurements of all planets except Pluto and radar bounces from Mercury, Venus, and Mars collected between 1964 and 1969. Tracking data, of course, was in the mix, including range and Doppler acquired during the encounter segment of the Mariner 5 flyby of Venus in October 1967.[33]

A pivotal factor in the move toward simultaneous integrations of increasingly larger data sets was the hiring of Jay Lieske, a Yale graduate student. One of his more far-reaching contributions came out of his doctoral dissertation. Dubbed DE28, Lieske's ephemeris ran from 1800 to 2000, a longer time span than any previous JPL ephemeris. The incorporation of the treasure trove of past astronomical data into the JPL ephemeris database marked a major milestone in the program's evolution.[34] Lieske argued that large-scale, high-speed digital computers made it relatively easy to generate more precise ephemerides than had been possible just a few decades earlier. In fact, one could perform numerical integrations of planetary orbits over a time span that was longer than the period of time for which actual observations were readily available. The result of this impasse between the capabilities of digital computers and the dearth of usable data made it both necessary and conventional to fit the numerical integrations to the classical theories rather than to actual observations.[35]

DE3 and DE19 already contained numerical integrations, but they were solutions for a single body. Lieske wrote his own software to adapt an existing N-body (multiple-body) computer program developed for asteroids and comets by Joachim Schubart and Peter Stumpff of the Heidelberg Astronomisches Rechen-Institut.[36] Here again cometary astronomy had an impact on navigation history. For Lieske, the next logical step toward a multiple-body solution was to perform numerical integrations simultaneously on two and more bodies using JPL computers and software.

Fortunately, the Solar System Data Processing System (SSDPS) was now available. It generated DE40, the "first dynamically consistent planetary ephemeris produced at the laboratory." Created for Mariner Mars 1969, DE40 incorporated the recently adopted IAU masses, LE4, Venus radar data, and meridian circle observations made from 1950 to 1967. It featured better representations of the outer planets' motions, but navigators felt a need to obtain more definitive ephemerides by fitting a much longer arc of data.[37]

The result was DE69 issued in February 1969 again for Mariner Mars 1969. It benefited from both the integration of large quantities of data and the SSDPS. DE69 was a "special purpose" ephemeris not intended to replace the existing DE19. It covered a short time span (October 28, 1961, to January 23, 1976) and integrated 60 years of observational measurements: over 34,000 optical observations of the planets (except

---

[33] "JPL Ephemeris Development," 3; Standish, interview, 6 & 8; *DE69*, iv & 3-5; "Simultaneous Solution," 233-234 & 236-241.
[34] Lieske, 1; "JPL Ephemeris Development," 3; Standish, interview, 8; *DE69*, iv & 3-5.
[35] Lieske, 1.
[36] Lieske, 1-2; Joachim Schubart and Peter Stumpff, *On an N-Body Program of High Accuracy for the Computation of Ephemerides of Minor Planets and Comets* (Karlsruhe: Verlag G. Braun, 1966).
[37] "JPL Ephemeris Development," 3; Standish, interview, 8; *DE69*, iv & 3-5; O'Handley, "Ephemeris Development," 2-2.

Pluto) and the Sun obtained from the 150-mm (six-inch) and 230-mm (nine-inch) transit circles of the Naval Observatory between 1910 and 1968; nearly 800 radar observations of Mercury, Venus, and Mars made from 1964 to 1968 by JPL, Lincoln Laboratory, and the Arecibo Observatory; and Mariner 5 range measurements made around Venus encounter (June 21 to November 12, 1967). DE69 also was the first to incorporate relativistic coordinates and the first to include planetary range data (collected in 1967 from the Mariner 5 mission to Venus).[38]

Each successive Development Ephemeris integrated a larger and larger mass of optical, radar, and tracking data. DE96 was the first to integrate observations of the asteroids Vesta and Pallas and the dwarf planet Ceres taken over the entire twentieth century. The number of parameters handled by the SSDPS also increased. Creating DE71, for instance, entailed finding a simultaneous solution for 63 parameters including such factors as the motions of the eight planets (excluding Pluto), the trajectory of Mariner 5, the mass of Venus, the astronomical unit, and the radii of Mercury, Venus, and Mars.[39] The multiple-planet, simultaneous integration of long-term observational data furnished a far more accurate representation of the positions of the planets than those given by the planetary and lunar theories.[40]

Coming together. The differences between navigators and astronomers narrowed over the course of the 1960s. Astronomers were beginning to appreciate the advantages of radar astronomy for refining and defining the astronomical unit. Both communities wanted to benefit from radar and telescopic observations. Increasingly, too, ephemerides users included radio astronomers. Everyone—radio and radar astronomers, optical and almanac astronomers, and navigators—reduced data on large-scale digital computers. For a number of years, astronomers had been distributing ephemerides and star catalog information in a form readable by computers. Her Majesty's Nautical Almanac Office at Greenwich, the *Astronomisches Rechen-Institut* in Heidelberg, and the Naval Observatory in Washington were three institutions supplying such information in a computer-usable format.[41] If one could express multiple data types in a single, machine-readable form, navigators, almanac astronomers, radio astronomers, and radar astronomers could profit from access to the same observations. The establishment of a common format was an important step in bridging the gap between astronomers and navigators.

The institution of this comprehensive data format was the work of the Ephemeris Working Group. Bill Brunk, the Chief of the NASA Headquarters Planetary Science Branch, initiated the effort following informal discussions among individuals at NASA, the Naval Observatory, and elsewhere involved in making and using optical and radar observations of the Moon and planets. The 24 members of the Ephemeris

---

[38] Moyer, *Formulation of DPODP*, 22; "Interplanetary OD," 22; O'Handley, "Ephemeris Development," 2-3; DE69, iv, 4 & 5; "JPL Ephemeris Development," 2; DE96, 2 & 5.
[39] DE69, iv, 4 & 5; "JPL Ephemeris Development," 2; DE96, 2 & 5 ; O'Handley, "Ephemeris Development," 2-6 & 2-11.
[40] Mulholland, "Numerical Studies," 248.
[41] "Dynamical Astronomy," 149; O'Handley, *Card Format*, 4.

Working Group represented NASA—Headquarters, Goddard, JPL—the Naval Observatory, and the Naval Weapons Laboratory (Dahlgren, VA). The Naval Observatory and JPL contributed 13 members. JPL representatives included Bill Melbourne, Chuck Lawson, Roger Broucke, Derral Mulholland, and Doug O'Handley. Other notable participants were Paul Herget and, from Lincoln Laboratory, Irwin Shapiro, Michael Ash, and Bill Smith.[42]

The idea was to coordinate the exchange and formatting of both optical and radar data types and to make the optical data format uniform from 1750 to the present. The Working Group agreed on a single card format of 80 columns (the standard for IBM punched cards) for both data types. Herget was responsible for standardizing the code numbers that designated the various observatories to allow the interchange of data with other IAU organizations.[43]

The Ephemeris Working Group enabled radar, radio, and optical astronomers to exchange observations and to integrate them into their computations. Optical astronomers gained the added precision of radar measurements, while radar astronomers now had access to the treasure trove of optical observations made over the centuries (or at least since 1750). Navigators benefited from access to both radar and optical data. In addition, the common format enabled a closer collaboration between navigators and astronomers. A paradigm shift was taking place. Astronomers and navigators would operate within a single paradigm, one in which the JPL Celestial Mechanics Group was in charge of the world's almanacs and ephemerides.

The IAU's decision to adopt the JPL ephemerides as *their* ephemerides came after a long search by astronomers for a better set of physical constants. The IAU was a forum for astronomers to agree on a consistent set of constants and other issues related to the formulation of ephemerides. It counted among its members the astronomers who created ephemerides and almanacs for the United States, Great Britain, France, and other countries. The IAU had been searching for better constants since the end of the previous century. A 1950 IAU meeting convened to discuss the adoption of new constants postponed the question.[44]

The IAU met again in Paris in May 1963 to discuss, among other issues, an appropriate value for the astronomical unit. According to Gerald Clemence, the recent radar-determined value of the astronomical unit and the precise value of the Sun-Venus mass ratio computed from Mariner 2 tracking data "provided the immediate incentive" for holding the meeting. Had it not been for the radar value, Mariner 2 would have passed Venus without acquiring any useful data.[45] The IAU adopted a radar-derived value expressed in kilometers for the astronomical unit. The movement of knowledge

---

[42] Notes, telephone conversation, Seidelmann, February 16, 2009, F326, DSN Files; Dick, *Sky and Ocean*, 532; "Dynamical Astronomy," 149; O'Handley, *Card Format*, iii.

[43] O'Handley, *Card Format*, iii, vii, 1 & 4; Dick, *Sky and Ocean*, 532; "Dynamical Astronomy," 149; Alan D. Fiala, "Astronomical Data Files at the U. S. Naval Observatory: Ephemerides, Star Catalogues, and Observations," *Bulletin of the American Astronomical Society* 6 (1974): 217.

[44] Clemence, "System of Constants," 96-97; Clemence, "On the System, 169-179.

[45] Clemence, "System of Constants," 97-98; *Tracking Mariner Venus 1962*, 9, 17 & 75-76; Corliss, *History of the DSN*, 29.

between celestial mechanics (theory) and navigation (practice) started to flow from practice to theory for the first, but not for the last, time.

In 1964, the IAU adopted the so-called 1968 IAU System of Constants—the values to be incorporated in ephemerides effective 1968—and the Catalog of Fundamental Stars FK4 as the fundamental reference system. Many in attendance believed that some work remained unfinished, the planetary masses for example. Astronomers thought that fresh information about planetary masses would become available in the coming years mainly, but not totally, from tracking spacecraft. The next meeting to address planetary masses and other questions was the Colloquium on the IAU System of Astronomical Constants held in Heidelberg in August 1970. Both Ken Seidelmann and Ray Duncombe from the Naval Observatory contributed to the discussion on planetary masses, as did JPL's Bill Melbourne.[46]

Melbourne told the gathering that JPL already had deemed the 1968 IAU System of Astronomical Constants to be "out of date" and "of insufficient accuracy for astrodynamic purposes." Of all the constants, he stressed, the planetary masses were "the most seriously out of date." He highlighted the "dramatic improvement in astronomical constants" realized by JPL navigators in refining the accuracies of the mass values of Venus, Mars, the Earth, and the Moon by two or three orders of magnitude thanks to direct measurements of their gravitational fields by spacecraft.[47]

Clemence argued for deferring the question until at least 1973, when better values would be available. IAU members voted their agreement. They also approved a motion to update planetary masses at a time closely linked to the production of the next fundamental star catalog (FK5) due in 1978 thereby making several basic changes at the same time. To prepare for this radical revision, the IAU recommended setting up working groups, including one for the purpose of specifying "in time for consideration in 1973, the basis for the planetary ephemerides to be published in almanacs for 1980 onwards." The Working Meeting on Constants and Ephemerides took place in October 1974 at the Naval Observatory to draft the working group's report. The chairmen of all working groups met again in September 1975 and June 1976 in Herstmonceux, England, and Washington, DC, respectively. Finally, in August 1976, the IAU Sixteenth General Assembly adopted their report and recommendations.[48]

The cumulative result of these calculated and incremental deliberations and resolutions was a major fundamental readjustment of the quantitative and theoretical basis for the ephemerides. Astronomers endorsed the IAU (1976) System of Astronomical Constants along with the FK5 star catalog. National almanacs would incorporate both simultaneously beginning with ephemerides published in 1984. The IAU also adopted the International Systems of Units—the meter, kilogram, and second—for all astronomical measurements. This was the same unit system used in

---

[46] Jean Kovalevsky, "Introductory Remarks," *Highlights of Astronomy* 3 (1974): 209; Emerson and Wilkins, 128 & 136.

[47] *Constants and Related Information*, 33 & 35.

[48] Emerson and Wilkins, 128-149, 138-139, 144 & 147-148; Seidelmann, 99 & 101; Dick, *Sky and Ocean*, 430 & 538.

navigation. Additionally, the 1984 almanacs of the world would feature new theories replacing current ones that were from three to eight decades old.[49]

In order to base the almanac ephemerides on the latest and most accurate constants and the best theories, the IAU chose to make JPL Development Ephemeris 200 (DE200) the basis for the constants, theories, and tables printed in the Naval Observatory's *American Ephemeris and Nautical Almanac*, renamed *The Astronomical Almanac* with the 1981 edition. DE200 was a high-precision ephemeris created by numerical integration of multiple data types spanning more than 44 centuries, from 1410 BC to 3002 AD. Seidelmann had recommended adopting DE200 to his working group, because he concluded that it was the most accurate available, and nobody else was going to have that level of accuracy.[50] Navigators and astronomers now had a common paradigm. They shared a research agenda in dynamical astronomy; research tools, such as digital computing technologies and numerical integration; and multiple data types expressed in a single, machine-readable form. JPL navigation had become astronomy.

---

[49] Seidelmann, 99, 106, 108 & 111; Dick, *Sky and Ocean*, 539.
[50] Standish, "JPL Planetary Ephemerides," 181-186; "DE102," 150-167; Standish, "Observational Basis for DE200," 252-271; Seidelmann, interview, 2009.

Chapter 11

Scientists and Explorers

Celestial mechanics was only one field in which navigators were scientists. The so-called golden age of planetary exploration opened up new opportunities to be project scientists and to expand the kinds of experiments performed. Navigators continued their celestial mechanics observations and studies of gravitational fields. The results enriched the JPL ephemerides and improved gravity models, thereby raising navigational precision on future flights. Navigators continued to collaborate with Stanford in the field of radio occultations. A significant departure from the past was a venture into theoretical physics and General Relativity.

The many flights of the 1970s also afforded fresh challenges from a number of directions. Sometimes challenges arose from the design of the spacecraft and its scientific objectives, as in the case of Pioneer Venus. In other instances, the navigation team dealt with various factors that degraded Doppler quality, while at other times they faced some unusually problematic effects of solar pressure and attitude control leaks. Small, unmodeled forces continued to be the navigators' bane. Mars probes over time have suffered a far higher failure rate than those aimed at any other solar-system target. Each unsuccessful mission has had its idiosyncratic technical reasons for going awry, but the Mariner 7 anomaly gave rise to a space legend, the Great Galactic Ghoul, that lives on in space lore.

Putting Einstein to the test. Navigators continued as project scientists into the 1970s conducting celestial mechanics and radio occultation experiments. John Anderson and Arv Kliore were Principal Investigators on two of the fourteen outer-planet Pioneer science teams. The S-band Occultation Experiment, led by Kliore, also involved his JPL colleagues Gunnar Fjeldbo (formerly at Stanford), Dan Cain, and Boris L. Seidel plus S. Ichtiaque Rasool from NASA Headquarters. The radio occultations mainly examined the atmosphere and ionosphere of Jupiter, as Pioneer 10 swung behind the planet for about one hour. When it flew behind Io, additional occultations attempted to discover whether that satellite had an atmosphere.[1]

Anderson led the Celestial Mechanics Experiment team which consisted of himself and George Null. Consistent with past practice, the experiment used the spacecraft as an instrument for studying the gravitational fields of Jupiter and its large moons. The gravitational force of those bodies was in direct proportion to their masses. By teasing gravitational values out of the two-way Doppler, Anderson and Null achieved a five-fold improvement in accuracy over Earth observations in determining the mass of

---

[1] Kraemer, 66; *Pioneer Odyssey*, 58, 59 & 202; Alfred J. Siegmeth, "Pioneer Mission Support," 6 in DSNPR:VII.

Jupiter. They also calculated more exactly the masses of Jupiter's four large moons, Io, Europa, Ganymede, and Callisto.[2]

Mariner Mars 1971 marked a milestone of sorts in navigators' conduct of science. First, because the probes orbited Mars, navigators could study and map the planet's gravitational field at close range rather than from Earth, as they had done with the Lunar Orbiters. Accordingly, the celestial mechanics experiment incorporated gravity researchers. A second experiment represented a departure from previous scientific inquiry into the field of theoretical physics, namely, relativity.

The Mariner Mars researchers formed two separate groups: one for gravity, one for relativity. The investigators came from JPL and MIT. JPL's Jack Lorell shared Principal Investigator status with MIT's Irwin Shapiro. Lorell's gravity group consisted of John Anderson and Frank Jordan from JPL and Robert Reasenberg and Shapiro from MIT. Reasenberg and Shapiro shared Co-Investigator status on the relativity team. The remainder of the latter team included Anderson, George Born, Pat Esposito, Jordan, Philip Laing, Warren Martin, and Bill Sjogren all from JPL.[3]

The Mariner Mars 1971 relativity test was not the first of its kind. It was the Fourth Test. Three others had preceded it in attempting to ascertain the validity of the theory of General Relativity. After announcing his Special Theory of Relativity in 1905, Albert Einstein spent another ten years developing the theory of General Relativity.[4] That theory accounted for the precession of Mercury's perihelion, the point at which Mercury is closest to the Sun. It predicted that a gravitational field would bend or deflect the path of light rays and that it would cause the speed of a light wave to slow ("red shift"). Support for the theory historically came from experiments conducted over several decades that confirmed the precession of Mercury's perihelion, the deflection of light rays in a strong gravitational field, and the red shift.

Irwin Shapiro conceived a new test that entailed measuring the impact of the Sun's gravitational field on radar signals sent from Earth to either Mercury or Venus. He performed his so-called Fourth Test of General Relativity with Lincoln Laboratory's Haystack facility and Cornell's Arecibo Observatory with some success. However, he realized that an experiment carried out with a spacecraft would increase the accuracy of these measurements significantly.[5]

The first such relativity test took place during the extended portion of Mariner Mars 1969.[6] Its goal was to measure the retardation of tracking signals caused by the Sun's gravitational influence during superior conjunction, as Mariner 6 and Mariner 7

---

[2] Kraemer, 66; *Pioneer Odyssey*, 58 & 201; George W. Null, "Gravity Field of Jupiter and its Satellites from Pioneer 10 and Pioneer 11 Tracking Data," *The Astronomical Journal* 81 (December 1976): 1153-1161.

[3] "Celestial Mechanics Experiment," 13.

[4] The following discussion of Einstein's theory of General Relatively draws from Banesh Hoffmann, *Relativity and its Roots* (New York: W. H. Freeman and Company, 1983); Peter G. Bergmann, *The Riddle of Gravitation*, revised and updated (New York: Charles Scribner's Sons, 1987); Mendel Sachs, *Relativity in Our Time: From Physics to Human Relations* (Bristol, PA: Taylor & Francis, 1993).

[5] Irwin I. Shapiro, Michael E. Ash, R. P. Ingalls, William B. Smith, Donald B. Campbell, Rolf B. Dyce, Raymond T. Jurgens, and Gordon H. Pettengill, "Fourth Test of General Relativity: New Radar Results," *Physical Review Letters* 26 (1971): 1132-1135; Butrica, *See the Unseen*, 126-129; Shapiro, "Fourth Test of General Relativity," *Physical Review Letters* 13 (1964): 789-791.

[6] MM69 Final Report, Vol. III, 17; Butrica, *See the Unseen*, 128-129.

swung behind the Sun. The Sun's gravitational pull would increase the probe's speed, but also decrease the time that radio signals took to travel because of the distortion or stretching of the fabric of space by the intense gravitational field. General Relativity predicted a delay of a few thousandths of a second (about 200 microseconds) in the two-way signal return time of up to about 45 minutes.[7]

The experiment obviously required an extremely precise and sensitive ranging system. Warren Martin therefore made a number of changes to the tracking system to accommodate the experiment. As a result, the relativity test verified the approximate 200 microsecond time delay with an uncertainty of no more than 3 percent. Non-gravitational forces acting on the spacecraft largely caused the failure to gain greater precision. The impact of free electrons from the Sun's corona on the tracking data also complicated measurement of the delay though to a far lesser extent.[8] Although Mariner Mars 1971 was the next flight opportunity to run the Fourth Test, Anderson and Shapiro took advantage of the delay to gather additional radar data, so that they could make more precise measurements during those next flights.[9]

In addition, another opportunity arose to conduct Shapiro's experiment, but using a radio interferometer and substituting quasar emissions for spacecraft radio signals. The quasar pair selected, 3C279 and 3C273, were relatively close in the sky (about 8° apart). Every October the Sun passes in front of 3C279 and creates an ideal opportunity to measure the bending of the quasar's radio signals by comparing the apparent angular separation of the two sources as the one signal grazed the Sun. An added advantage of using quasars instead of spacecraft was their high frequency (about 23 GHz), which dramatically decreased the effects of the solar corona.[10]

In October 1969, two groups at two different locations used this pair of quasars to perform Shapiro's relativity test with a short baseline interferometer. Dewey Muhleman led one group at Goldstone, while radio astronomer George Seielstad oversaw a second at his home institution, the Owens Valley Radio Observatory. The two groups' results were consistent with Einstein's predictions.[11] Subsequently, in

---

[7] *MM69 Final Report*, Vol. II, 252; *MM69 Final Report*, Vol. III, 17; Wilson, 38-39.
[8] Warren L. Martin, "Radio Tracking With The Deep Space Network," 158-165 in Davies; Anderson, Esposito, Martin, Catherine L. Thornton, and Muhleman, "Experimental Test of General Relativity using Time-Delay Data from Mariner 6 and Mariner 7," *Astrophysical Journal* 200 (August 15, 1975): 221-233; John D. Anderson, Pasquale B. Esposito, Martin, and Duane O. Muhleman, "Measurement of General Relativistic Time Delay with Mariner 6 and 7," 1623-1630 in Sidney A. Bowhill, L. D. Jaffe, and Michael J. Rycroft, ed., *COSPAR Space Research XII*, vol. 2 (Berlin: Akademie-Verlag, 1971); Anderson, Esposito, Martin, and Muhleman, "A Measurement of the General Relativistic Time Delay With Data From Mariners 6 and 7," 111-135 in Davies; David W. Curkendall, Susan G. Finley, Melba W. Nead, V. John Ondrasik, and Thornton, "The Effects of Random Accelerations on Estimation Accuracy With Applications to the Mariner 1969 Relativity Experiment," 148-157 in Davies; Wilson, 38 & 39.
[9] Anderson, "General Relativity Data Analysis," 56 in JPL Science Office, *Annual Progress Review, July 1, 1972-June 30, 1973* (Pasadena: JPL, August 30, 1973).
[10] Durham and Purrington, 188; Shapiro, "Testing General Relativity: Progress, Problems, and Prospects," 143 in Davies.
[11] Muhleman, Ronald D. Ekers, and Edward B. Fomalont, "Radio Interferometric Test of the General Relativistic Light Bending Near the Sun," *Physical Review of Letters* 24 (June 15, 1970): 1377-1380; George A. Seielstad, Richard A. Sramek, and Kurt W. Weiler, "Measurement of the Deflection of 9.602-GHz Radiation from 3C279 in the Solar Gravitational Field," *Physical Review Letters* 24 (June 15, 1970): 1373-

October 1970, researchers from JPL, MIT, and Goddard carried out another experiment with the same quasars, but using a very long baseline interferometer that stretched from California (the DSN Mars dish) to Massachusetts (Lincoln Laboratory's Haystack antenna), both of which now had state-of-the-art hydrogen masers to ensure greater accuracy. Data analysis took place at Goddard on an IBM 360 computer. The results again indicated success.[12]

Einstein more precisely. The Mariner Mars 1971 Fourth Test, the next flight opportunity, saw Shapiro, Lorell, Anderson, and Jordan obtain results that were more precise than those realized during the previous Mars mission. Mariner Mars 1969 time delays of approximately 200 microseconds had confirmed General Relativity to within 3 percent. The Mariner 9 test realized time delays of just 180 microseconds and corroborated Einstein with a precision of about 2 percent.[13]

Navigators subsequently proposed methods for improving the accuracy of future tests. Rather than focus on ways to account for error sources more accurately, they considered approaches that obviated the effects of certain error sources (the "direct" method; see Chapter 13). For example, Dan Cain pointed out that one could alleviate the effects of non-gravitational forces arising from solar radiation pressure and spacecraft attitude control systems by "anchoring" the spacecraft. A lander on a distant moon or planet would not experience such forces. Perhaps with Viking in mind, Cain specifically suggested a Mars lander. He also argued that existing uncertainties in the planet's ephemeris would not matter, because the orbiter would be undergoing the same motions.[14]

A different proposal attacked the influence of interplanetary free electrons on radio signals. The relativity test required that navigators distinguish between the relativistic delay and the signal delay caused by free electrons. These plasma effects varied with the inverse square of the signal's frequency, while the relativistic delay was independent of frequency. In principle, therefore, one could minimize the plasma effects by ranging at two different frequencies.[15]

The introduction of dual-frequency transponders on probes starting with Mariner Venus-Mercury 1973 empowered such a relativity test. Viking was an excellent candidate for the next relativity test. The Viking spacecraft featured a dual-frequency radio system, and, in line with Cain's proposal, the mission would set landers down on the planet's surface. The Viking relativity test entailed analyzing the round-trip travel

---

1376; Marshall H. Cohen, "A History of OVRO: Part II," *Engineering & Science* 70 (2007): 35-36, 37, 39-40 & 42.

[12] Takeshi Sato, Lyle Skjerve, and Donovan Spitzmesser, "Radio Science Support," 151-153 in JPLSPS 37-66:II; Corliss, *History of the DSN*, 61, 90, 126; Renzetti, *DSN History*, 65; A. R. Whitney, Shapiro, Alan E. E. Rogers, D. S. Robertson, C. A. Knight, Thomas A. Clark, G. E. Marandino, N. R. Vandenberg, and Richard M. Goldstein, "High-Accuracy Determination of 3C273-3C279 Position Difference from Long-Baseline Interferometer Fringe Phase Measurements," *Bulletin of the American Astronomical Society* 3, no. 4, pt. 1 (1971): 465.

[13] Will, *Einstein*, 127 & 157.

[14] Cain, "Anchoring Spacecraft to Planets," 245-248 in Davies.

[15] Anderson and Lau, 431 & 432.

time of radio signals transmitted from Earth to the two Viking spacecraft. Shapiro's MIT researchers and JPL navigators, including Cain, looked for a time delay of up to about 250 microseconds. A preliminary analysis of the Viking data acquired near the 1976 superior conjunction of Mars indicated agreement with Einstein's predictions to within the estimated uncertainty of 0.5 percent.[16]

Subsequent analysis of 14 months of Viking Mars Orbiter and Lander data acquired in 1976 and 1977 verified General Relativity to an estimated accuracy of 0.1 percent. The data accuracy benefited from a combination of approaches. Improved calibration of tracking data and precise knowledge of the position of the landers combined with the self-calibrating dual-frequency downlinks. As a result, this was the most accurate test of the theory of General Relativity to that point.[17]

Later repetitions of Shapiro's Fourth Test took place with various spacecraft traveling to Mars and beyond. For example, Anderson and others used the Voyager 1 craft during its encounter with Saturn in 1980 and Voyager 2 during solar conjunction in December 1985.[18] More relativity tests occurred during the Helios mission as part of its celestial mechanics experiment.[19] Anderson and Eunice Lau, along with Giacomo Giampieri (Imperial College London) and their Italian colleagues Luciano Iess (University of Rome) and Bruno Bertotti (University of Pavia), planned two relativity experiments to be conducted during the solar conjunction periods of the Cassini mission to Saturn in 2002 and 2003. The increased accuracy in the relativistic measurements this time resulted from the use of the higher-frequency Ka band (26.5-40 GHz), which, with a wavelength of less than one centimeter, was affected less by interplanetary media than the S-band or X-band signals used previously.[20]

Still, the best measurements of the relativistic effect remained those realized during the Viking mission and those obtained with lunar laser ranging equipment. Between 1969 and 1972, Apollo astronauts placed four laser retroreflector arrays on the Moon (the Lunar Laser Ranging Experiment or LURE). Later, the Soviet Luna 17 and Luna 21 missions added two French-built reflector arrays. In 1969, various observatories began ranging to these reflectors. The precision of range readings between 1970 and

---

[16] Anderson and Lau, 432; Shapiro, Reasenberg, P. E. MacNeil, Robert B. Goldstein, Joseph P. Brenkle, Cain, Thomas A. Komarek, Arthur I. Zygielbaum, Edward Cuddihy, and William H. Michael, Jr., "The Viking Relativity Experiment," *Journal of Geophysical Research* 82 (September 30, 1977): 4329-4334.

[17] Reasenberg, Shapiro, MacNeil, Goldstein, Breidenthal, Brenkle, Cain, Kaufman, Komarek, and Zygielbaum, "Viking Relativity Experiment: Verification of Signal Retardation by Solar Gravity," *Astrophysical Journal, Part 2: Letters to the Editor* 234 (December 15, 1979): L219-L221; Mudgway, *Uplink-Downlink*, 517; Will, "Confrontation," 40.

[18] Timothy P. Krisher, Anderson, and James K. Campbell, "Test of the Gravitational Redshift Effect at Saturn," *Physical Review Letters* 64 (March 19, 1990): 1322-1325; Mudgway, *Uplink-Downlink*, 517-518.

[19] Anderson, William G. Melbourne, Cain, Lau, Wong, and Wolfgang Kundt, "Relativity Experiment on Helios: A Status Report," AAS 75-102, AAS/AIAA Astrodynamics Specialist Conference, July 28-30, 1975, Nassau, Bahamas; *Helios Solar Probes Science Summaries*, TM-82005 (Greenbelt: GSFC, August 1980), 83-84.

[20] Bertotti, Iess, and Tortora, "A Test of General Relativity using Radio Links with the Cassini Spacecraft," *Nature* 425 (September 25, 2003): 374-376; Iess, Giampieri, Anderson, and Bertotti, "Doppler Measurement of the Solar Gravitational Deflection," *Classical and Quantum Gravity* 16 (1999): 1487-1502.

1975 hovered around 25 cm. Later, the McDonald Laser Ranging System drove the precision toward 3 cm and even as low as 2 cm.[21]

Among other projects, the 24 years of lunar laser ranging data accumulated by this self-described continuing legacy of Apollo aided in the refinement of the JPL ephemerides in the hands of Skip Newhall, Myles Standish, and others. Navigators also used the lunar laser ranging to conduct relativity tests. Instead of the Shapiro Test, they relied on a new relativity test tailored to lunar ranging data that Kenneth Nordtvedt, Jr., a professor in the Physics Department, Montana State University, had developed in 1968.[22]

Not so pioneering. In addition to being scientists, navigators also were explorers on the many missions that flew during the 1970s. New problems and puzzles came from a number of directions. Pioneer Venus, for example, was complicated from a mission and spacecraft design perspective. It carried four probes designed to impact the planet at various times and locations, because each carried a variety of instruments to measure the characteristics of the atmosphere from high altitudes down to the surface of Venus. The bus, designed not to survive entry, held independent scientific instruments to measure the atmosphere above 130 km. One of the other factors complicating Pioneer Venus navigation was the motion of the spacecraft itself. For instance, the polarization of the antenna combined with the craft's rotation degraded the quality of the Doppler. Because of the rotation, the antenna center point was not on the probe's spin axis, and its spin axis did not coincide with the Earth line of sight. All of these factors introduced peculiar signatures into the Doppler and lessened its quality.[23]

Guiding the Pioneer Venus orbiter and probes to their targets required eight different trajectory changes some of which dealt specifically with the release of the probes. The first two maneuvers occurred 7 and 20 days after the May 20, 1978, launch; a third targeted the large probe (called the Sounder) 30 days before encounter. Then, 24 days before encounter, a fourth maneuver separated the Sounder from the spacecraft and sent it into the planet's atmosphere. At 23 days before encounter, the bus executed a trajectory correction to target the small probes. The sixth maneuver, carried out 20 days before encounter, separated the small probes from the bus and sent them downward toward the planet. The bus released the clamps that held the probes in place, so that mechanical springs and the centrifugal force of the spacecraft ensemble

---

[21] Will, "Confrontation," 47; Standish, interview, 18-20; "Lunar Laser Ranging," 482-490; Thomas W. Murphy, Jr., Jana D. Strasburg, Christopher W. Stubbs, Eric G. Adelberger, Jesse Angle, Kenneth L. Nordtvedt, James G. Williams, Jean O. Dickey, and Bruce Gillespie, "The Apache Point Observatory Lunar Laser-Ranging Operation (APOLLO)," http://www.physics.ucsd.edu/~tmurphy/apollo/doc/matera.pdf (accessed June 12, 2009).

[22] "Lunar Laser Ranging," 485; Bender, Robert H. Dicke, David T. Wilkinson, Carroll O. Alley, Douglas G. Currie, James E. Faller, J. Derral Mulholland, E. C. Silverberg, Harry E. Plotkin, William M. Kaula, and Gordon J. F. MacDonald, "The Lunar Laser Ranging Experiment," 178-181 in Davies; Williams, Newhall, and Dickey, "Relativity Parameters Determined from Lunar Laser Ranging," *Physical Review D* 53 (1996): 6730-6739; Nordtvedt, "Testing Relativity with Laser Ranging to the Moon," *Physical Review* 170 (1968): 1186-1187.

[23] Wong and Guerrero, 2 & 8-10.

could send the probes on their way. The bus performed a trajectory correction maneuver at 18 days before encounter to target the bus for a specific planetary impact. The final maneuver trimmed the flight path of the bus 9 hours before its entry into the Venusian atmosphere.[24]

Pioneer Venus navigators Sun Kuen Wong and Helen Guerrero computed the spacecraft's trajectory almost exclusively from two-way Doppler. Their navigation strategy entailed addressing the usual major errors that affected the accuracy of orbit solutions. However, modeling the effects of solar pressure and attitude control leaks was more problematical than usual because of the craft's motion. Three-axis stabilized craft, such as those in the Mariner series, presented the same surface toward the Sun, but spin-stabilized Pioneers continually changed the surface exposed to the effects of solar forces. The changing orientation relative to the Sun required altering the solar-pressure model. To address this issue, Wong and Guerrero made use of five different models, each representing the solar-pressure effect over a maximum span of only 18 days. They also used four different filter models to deal with solar pressure radiation and attitude control leaks.[25]

Two-way Doppler was not available for navigating the probes' descents toward the planet. So, Wong and Guerrero relied on a combination of onboard hardware and one-way Doppler. Each probe had a coast timer and an onboard sequence programmer that together controlled the entry sequence. The probes remained mute until 22 minutes prior to entering the atmosphere, where entry was defined as 200 km above the planet's surface. At that point, they started to transmit radio signals. The only tracking data available for both the large and small probes therefore was one-way Doppler.[26]

This was the first time that navigators had depended on one-way Doppler so completely in determining a flight path. The main drawback to one-way Doppler is the drift caused by variations in the oscillator frequency. The Pioneer Venus navigators treated this problem by modeling the frequency change in the Orbit Determination Program. The time during which the DSN tracked the probes was very short. Wong and Guerrero, for example, computed the Sounder orbit with data collected over a span of less than 12 minutes.[27]

Meet the Ghoul. The greatest navigation challenges probably are those that arise from the action of small non-gravitational forces. Models constructed *a priori* as mathematical representations of solar pressure and attitude control effects often can fail to cope with the unexpected. These unmodeled forces are the bane of navigation. In the case of Mariner Mars 1969, they gave rise to a space legend.

Navigators knew that Mariner 6's infrared spectrometer was going to influence its flight path, but they were unprepared for what actually happened. The instrument operated in a cryogenic environment created by releasing extremely cold hydrogen and

---

[24] Wong and Guerrero, 3, 5 & 6.
[25] Wong and Guerrero, 9, 18 & 21.
[26] Wong and Guerrero, 5 & 6.
[27] Wong and Guerrero, 8, 9, 34, 35 & 37.

nitrogen gases under pressure. The navigation team knew in advance that the release of those gases imparted a known quantity of force to the craft and changed its trajectory velocity between 0.1 to 0.2 meters per second. The mission design called for the gas to start discharging about 35 minutes before Mars encounter and to continue through encounter for about 80 minutes. The unused gas then escaped into space over a period of around 5 hours. Unfortunately, the spectrometer system failed to operate as designed. The resulting thrust took place over a period of not minutes but for about six days before encounter, a critical navigational period. The malfunction also frustrated John Anderson's attempts to compute a value for the mass of Mars, because it effectively destroyed the Doppler data taken at that time.[28]

A far more troubling malfunction occurred on Mariner 7, the other Mariner Mars 1969 craft. It experienced a number of major anomalies that collectively became known formally as the Mariner 7 pre-encounter anomaly. The malfunctions, some said, were the work of the Great Galactic Ghoul, a mythical creature spun into existence by the mysterious nature of the anomalies themselves and now a part of space lore. A water color of the Ghoul painted by contractor artist G. W. Burton portrayed it as a hideous spotted creature floating around Mars with one hand snatching the Mariner 7 probe.[29]

The anomalies occurred on July 30 and 31, 1969, 5 days before the Mariner 7 encounter and just 7 hours before the Mariner 6 encounter. The sequence of events began when Johannesburg lost the Mariner 7 signal. The Woomera antenna recovered the signal some 13.5 hours later. What had happened? A failure review board convened two weeks after the occurrence investigated the matter.[30]

The immediate concern during the event was whether the spacecraft was locked onto the Sun or was operating solely, or at most partially, on battery power. That stroke of bad luck had not happened. Once Woomera reacquired the Mariner 7 signal, tracking personnel realized that its star tracker had acquired not Canopus but probably Jupiter. This time the infrared spectrometer system was not acting up.[31] What, then, had happened?

A subsequent investigation determined that there was no Great Galactic Ghoul, just a failed spacecraft battery. In reality, a series of malfunctions had taken place. First, four days before the anomaly had become noticeable, the battery began to fail while being recharged. Next, a telemetry loss occurred on three separate occasions over a period of about 13 hours. That anomaly disturbed the craft's course by approximately 130 km. Then, 20 of the probe's 97 engineering telemetry channels were put out of action permanently. Those channels related the position and attitude of the spacecraft. Once the battery failed, Mariner temporarily lost its roll reference. Finally, of the 19

---

[28] John Anderson, "Mariner Mars '69 Celestial Mechanics Experiment," 257 in Carl Sagan, Tobias C. Owen, and Harlan J. Smith, ed., *Planetary Atmospheres* (Dordrecht, Holland: Reidel, 1971); Anderson, "Mariner Mars 1969 Celestial Mechanics Experiment," 5 & 6 in JPLSPS 37-44:IV; "Celestial Mechanics," 130.
[29] Invention of the name Great Galactic Ghoul is attributed to *Time* reporter Donald Neff. The watercolor (JPL129) measures 21" by 31" and has been reproduced in Westwick, 259; Burrows, 730; and elsewhere.
[30] *MM69 Final Report*, Vol. II, 110 & 111.
[31] *MM69 Final Report*, Vol. II, 111 & 112.

subsystems aboard the spacecraft, nine experienced various effects of what appeared to be major electrical transients on the spacecraft. All of these breakdowns probably took place because of a structural failure of the spacecraft battery, with one of the 18 cells in the battery failing by short-circuiting its positive and negative plates. Somehow, the failed cell spread failure to other battery cells. The trajectory perturbation most likely resulted from the release of battery gas inside the spacecraft bus through openings in the thermal blanket.[32] These calamities certainly seemed ghoulish, and many missions to Mars have failed in what seemed to be mysterious ways, but there is no Galactic Ghoul.

A surprise discovery. The following mission to the red planet, Mariner Mars 1971, was, in the words of one DSN engineer, "the most complex unmanned space mission ever undertaken." It was NASA's first mission to orbit, not fly past, another planet. The need to approach the planet close enough to be pulled into orbit, but not crash, made it the agency's least simple flight to date. Mariner 9, its mate lost in the Atlantic Ocean near Puerto Rico, succeeded in orbiting the red planet and narrowly beat the Soviet Mars 2 and Mars 3 probes both of which arrived about a month later.[33] Once again navigators felt the pressure of Cold War rivalry, as Mariner raced to beat the Soviet probes in imaging the planet's surface. What awaited navigators were both frustration and a surprise discovery.

Mariner 9 continued the imaging work of Mariner Mars 1969. It mapped over 70% of the Martian surface from the lowest altitude (1,500 kilometers) at the highest resolutions (1 kilometer per pixel to 100 meters per pixel) of any previous Mars mission. To realize that success, project managers had to wait out months of dust storms. Their patience was rewarded. The Soviet flyby probes, on the other hand, failed to send back pictures because of the large-scale dust storms. This unforeseen occurrence made a strong case for conducting planetary studies from orbiters rather than flyby probes. The storm forced major changes in the television experiment. To aid in obtaining better camera pictures after the dust cleared, the navigation team used the craft's limited maneuver capability to raise, rather than lower, its orbit.[34]

The storm, however, was less surprising to navigators than the roughness of the planet's gravitational field. Navigation team member Frank Jordan remembered it well. After the successful insertion of Mariner 9 in orbit around Mars, he recalled that: "Everybody went home. Or they went to the bars. I don't know. They went partying. All the scientists went partying. [Laughs] And there were like two people in that building left. I was one of them because I was doing the orbit." Jordan began by fitting two hours of data, then four hours of data. At the time, the probe was in a 12-hour orbit. Then he fit six hours of data. "And everything was just such a boring night. Everything was so

---

[32] *MM69 Final Report*, Vol. II, 113 & 114; Harold J. Gordon, Sun Kuen Wong, and V. John Ondrasik, "Analysis of Mariner VII Pre-Encounter Anomaly," *Journal of Spacecraft and Rockets*, 8 (1971): 931-937.
[33] Hofmann, 136; *MM71 Final Report*, Vol. I, 56 & 70; Ezell and Ezell, *On Mars*, 288 & 291; Hartmann and Raper, 25 & 30 & 31.
[34] NASA, "The Mariner Mars Missions," http://nssdc.gsfc.nasa.gov/planetary/mars/mariner.html (accessed June 5, 2009); Ezell and Ezell, *On Mars*, 289-292; Hartmann and Raper, 38 & 40-42; "Celestial Mechanics Experiment," 15.

easy compared to the simulations that we had been going through for six months," but not for long.[35]

Mariner 9 Mars Orbit Insertion, November 13, 1971 (Courtesy of JPL)

When Mariner 9 approached its periapsis (the point of closest approach), something went wrong. "And I'm fitting data. And all of a sudden I get close. And all hell broke loose; it wouldn't fit." "What happened," Jordan explained, "was that Mars was four times rougher for the equivalent size than the Moon." The residuals "were just all over the map." The topography was dramatically different from that of the Moon. In planning the celestial mechanics experiment, they had thought that Mars would be similar to the Moon, but "We missed it by a factor of four."[36]

The rest of the story, Jordan added, was "very humorous." "Immediately I panicked, wondering whether something had happened on the spacecraft. Of course, now we're talking about three a.m. in the morning. Nobody there. Now it's probably more like five or six in the morning, and people are starting to come back in. I'm sitting there knowing something nobody else knows in the whole universe. Something strange! I got my gravity guys in, and I said: 'We have to see what's going on.' Immediately we thought: 'Well, that gravity's just very harsh.' By the time we got two orbits, we had a gravity map. And guess what? We looked at Mars, and there was this great big bulge that

---

[35] Jordan, interview, 24.
[36] Jordan, interview, 24.

later became known as the Tharsis Bulge. It dominated, for a number of years, the way people described Mars in the literature."[37]

For Jordan, the funny thing was that he and the gravity team were "the only people who saw it." "There was a dust storm. You couldn't see what was down there. Two months later, you know what we saw. We saw these enormous caldera volcanoes, including Olympus Mons. We saw that there had been so much activity on that end of the planet, that it had come out and flowed over and created a gravity anomaly four times greater than the Moon and many times greater than that of the Earth. It was so enormous! You see these pictures of Olympus Mons, and it covers California! But we saw it that night. It was the most amazing thing to go through, sitting there, knowing you'd seen something nobody else had seen before. If you just figured it out, you were going to make a dent. And we did. Within a day we had a gravity map that was pretty good. Within a week we knew exactly what that planet looked like, but nobody had seen it except us through gravity through the Doppler."[38]

---

[37] Jordan, interview, 24-25.
[38] Jordan, interview, 25.

Chapter 12

To Be More Precise

The quest for greater navigational precision has taken several routes. One route was via technology: the computers and software that played a major role in shaping the evolution of JPL navigation. During the 1960s, limited computing power constrained the parameters, models, and accuracy of navigation. The 1970s, the last decade of mainframe computing, witnessed a quantum leap in computational power that enabled a major, long-awaited software advancement. Linking two (and later multiple) computers to each other enabled new capabilities. One mainframe arrangement, for example, allowed the DSN to track two spacecraft at the same time, a necessity on such dual-probe missions as the 1969 and 1971 Mars Mariners.

Fundamental change—a revolution in navigational capabilities—came with the acquisition of the Univac 1108. That machine for the first time enabled navigators to make computations in double precision. The term "double precision" did not refer to a doubling of precision, that is, more digits to the right of the decimal point. The "double" came from the fact that a double-precision number used twice as many digital computer bits as a regular ("single") floating-point number. For example, if a single-precision number required 32 bits, its double-precision counterpart would be 64 bits long. Because computers essentially used only integers (so-called whole numbers), they needed complex codes to represent numbers with decimal points. The term floating point came from the fact that there was no fixed number of digits before and after the decimal point. Rather, the decimal point floated.

By allowing navigators to generate double-precision trajectories, the Univac 1108 enabled a software revolution: the Double Precision Orbit Determination Program (DPODP). This new generation program replaced all previous iterations of the single-precision ODP that JPL had been using since the days of the lunar Pioneers. Its development was a painstaking and slow process that took five years to complete and more than another year to become operational. Nonetheless, the DPODP became the basic JPL navigational software and remained so for decades. As of 1979, the program had some 300,000 lines of code written in FORTRAN V.[1] The Univac 1108 and the DPODP together signaled a new era in navigation and spurred fresh efforts to achieve greater navigational accuracy.

Computers for navigation. Before the arrival of the Univac 1108, navigators and the DSN experimented with a variety of new computing configurations. Incoming DSN data went to an IBM 7040 computer for processing. The data went directly from that machine to an IBM 1301 54-megabyte hard disk for storage. The IBM 1301 drive also served the IBM 7094 computer, so that either machine could access data at identical locations (that

---

[1] Wood, "Nav Evolution, 1962-1989," 290; Ekelund, "JPL OD Software," 79-88.

is, at identical addresses) on the disk. The concept was not unlike a network file server. Such configurations were coming into use by airline reservation database systems at about the same time.[2] Because of this arrangement's novelty, especially for navigation, Dan Schneiderman, writing to JPL Director Bill Pickering, related that one of the "Key Lessons Learned" during Mariner 5 was "Developing experience with huge computer complexes, i. e., computers talking to computers."[3]

The interlinked computers and IBM 1301 disk unit were able to handle two spacecraft data streams simultaneously, even though the data came from different DSN stations. This capability was a necessity for both the 1969 and 1971 Mars Mariners.[4] A test of the configuration took place using the Mariner 4 Mars probe, which was still orbiting the Sun, and Mariner 5. Their trajectories were such that the two probes would be in view of the same antenna (the Goldstone Mars Station) at the same time. The networked 7094-7040 mainframes successfully processed data concurrently from the two spacecraft as well as simultaneously from Mariner 5 and one of the Surveyors, though not without some difficulty. Consequently, the Surveyor project added a DEC (Digital Equipment Corporation) PDP-7 as a front-end computer placing it between the tracking stations and the IBM 7040 and other equipment.[5] Navigation on the Mariner Mars 1971 mission, however, benefited from an entirely different computing arrangement with the arrival of the Univac 1108. Introduced in 1964, the Univac 1108 featured integrated circuits (as opposed to vacuum tubes), a faster processing speed, and double-precision calculations. This last feature meant that navigators finally could achieve the long-sought goal of double-precision navigation software.[6]

The Univac 1108. Before reaping the benefits of the Univac 1108, navigators saw the replacement of the IBM 7094-7040 computer strings with model 75 IBM 360 mainframes (the so-called IBM 360/75). They were not new, but rather hand-me-downs from NASA's Manned Space Center near Houston.[7] Although they were a step up, the IBM 360/75s were a source of headaches just hours before the launch of Mariner 9. The trouble already had become apparent on the eve of the launch of its ill-fated twin.

On May 6, 1971, just two days before the launch of Mariner 8, navigators in the Tracking Analysis Group learned that the IBM 360/75 was generating erroneous predicts. Their accuracy was crucial, because they told the DSN antennas where to point. The error source was an incorrect value for a constant known as DUT that

---

[2] Tomayko, 263; MV67: Final Report, 156; Renzetti, DSN History, 34.
[3] Memorandum, Schneiderman to Pickering, Alvin R. Luedecke, Robert J. Parks, and Jack N. James, "General Observations Concerning the Mariner IV, Mariner V," July 12, 1967, F32/B2, JPL421.
[4] Renzetti, DSN History, 34; MM69 Final Report, Vol. I, 589 & 596; Mudgway, Uplink-Downlink, 42; MV67: Final Report, 159 & 160; Tomayko, 262 & 263.
[5] Letter, Schneiderman to Glenn A. Reiff, April 14, 1967, F16/B2, JPL421; Memorandum, Schneiderman to MV67 Project Staff, "Activation of Phase III of the Mariner IV Project," July 19, 1966, F51/B3, JPL 421; MV67: Final Report, 87, 166, 182 & 204-205; Renzetti, DSN History, 46; Corliss, History of the DSN, 166; JPL, Surveyor Final Project Report, Vol. I: Project Description and Performance, TR 32-1265 (Pasadena: JPL, July 1, 1969), 142.
[6] Paul Ceruzzi, A History of Modern Computing, 2d ed. (Cambridge: The MIT Press, 2003), 165.
[7] Corliss, History of the DSN, 194.

related Ephemeris Time to Universal Time. The DUT value changed slowly over time. The IBM 7094 predict system had updated values once a month, but the 360/75 had no method for updating the DUT.[8]

The next day—the day before launch—navigators and DSN engineers tried to determine the nature of the problem. They ran the computers as late as three hours prior to launch, but to no avail. The loss of Mariner 8 shortly after takeoff rendered the exercise academic. The launch of Mariner 9 saw the same problem return. Tests pointed to software problems among other concerns. As a backup measure, navigators used an IBM 7094 to generate tracking data tapes for their use. The backup worked smoothly during launch; the IBM 360 also performed correctly. So, about eight hours after takeoff, the project decided to terminate use of the backup computer and to rely entirely on the IBM 360.[9]

Meanwhile, navigators kept the Univac 1108 busy. By July 1969, key programs had been rewritten for the 1108 including the DPODP; the Satellite ODP, a version of the DPODP tailored to handle the orbital insertion maneuver; and the Double Precision Mission Matrix Program, which interfaced with the DPODP and served quality control and other purposes. The rewriting of software had been going on for six months and remained yet incomplete. JPL managers proposed that navigators instead use the 360/75 computers which still had not arrived at JPL. That decision meant even more software revisions, and those would take even longer to finish, potentially too late for the mission. In order to save time—as well as to have a backup computer for flight navigation—JPL needed to buy a second Univac 1108.[10]

From January 1971, navigators spent thousands of hours of Univac 1108 computer time practicing and training during weekdays, weeknights, and weekends for the Mariner Mars mission. Following the loss of Mariner 8, the Univac's time transferred to the Mariner 9 extended mission. In May 1972, the JPL Mariner program came under fire, especially from Gerald M. Truszynski, the Associate Administrator for the Office of Tracking and Data Acquisition (OTDA), because of the costs incurred by running the 1108 for Mariner 9 navigation.[11]

Truszynski's cost concerns reflected an administrative change that had taken place years earlier. Prior to December 1, 1964, the JPL Space Flight Operations Facility operated under the aegis of the NASA Headquarters Office of Space Science and Applications (OSSA), which was Homer Newell's bailiwick. Because of the intimate relationship between mission control and the DSN, Bob Seamans, NASA's Associate Administrator, asked Edmond Buckley, then head of OTDA, to assume responsibility for

---

[8] *TDA for MM71*, Vol. I, 219.
[9] *TDA for MM71*, Vol. I, 219, 220 & 248.
[10] Memorandum, Donald P. Hearth to Deputy Associate Administrator for Space Sciences and Applications, "Use of Univac 1108 for Mariner Mars '71 (MM '71) Flight Path Analysis," July 1, 1969, F20/B2, JPL421.
[11] Charles J. Vegos and William J. O'Neil to W. E. York, "1108 Compute [sic] Usage for MM '71 Navigation Team Training and Operation," July 30, 1971, F27/B2, JPL421; R. T. Hayes to W. P. Spaulding, "Estimate of 1108 Utilization Requirements for MM '71 Navigation," May 1, 1972, F27/B2, JPL421; Schneiderman to Pickering, "Documentation on 1108 Problem, as requested," May 3, 1972, F27/B2, JPL421.

the flight operations center. As a result, between 1964 and 1972, that facility was part of the DSN.[12]

The 1972 transfer back to the Office of Space Science reflected the dramatic rise in the processing of data, especially scientific data, at the space operations facility. The DSN was responsible for providing computer time, while the flight projects that demanded computer access had no responsibility for funding the computers. Therefore, in October 1971, Truszynski and John Naugle (Office of Space Science) decided to transfer the flight operations center and its computers back to the Office of Space Science effective July 1, 1972. Henceforth, flight projects would bear the responsibility and costs of acquiring computers and processing data.[13]

The next generation ODP. The acquisition of the Univac 1108 revolutionized JPL navigation. It enabled navigators to realize capabilities only imagined previously with their single-precision programs. The new navigation software had been years in preparation, but certification for regular operational use had to wait until Mariner Mars 1971. This "next generation" (as navigators originally called it) program evolved into the DPODP. By Ranger 9, the term Double Precision Orbit Determination Program had come into use and stuck.[14]

The need for better navigation software became clear as early as the flight of Ranger 6. Navigators hoped that the new software would overcome the limitations of the current, though upgraded, software in navigating planetary flights, and that it would extend navigators' ability to cull additional scientific knowledge from tracking data during post-flight analysis. One of the basic needs was simply to change the way in which the navigation computer performed calculations.

The building up of computing errors from rounding off and other mathematical operations plagued the navigation software and continued to thwart the realization of what navigators termed "the full potential" of the tracking data. One of the greatest barriers was the continued computation of trajectories and orbits in single precision. The "next generation" navigational software would address this limitation by performing all calculations in double precision.[15]

Navigators foresaw the "next generation" software as being able to handle solutions with more than 20 parameters, the limit of the existing program. The new one would be able to handle up to 50 parameters, depending on the nature of the run. The increased capability would allow navigation to take into account additional parameters in reaching solutions. These additional parameters included the effects of the ionosphere, the troposphere, and the motion of the Earth's polar axis. The single-precision program essentially ignored ionospheric effects and utilized a preset, unvarying empirically-derived standardized correction for tropospheric effects, the same correction for all DSN antennas. The "next generation" software would have to include a separate, more

---

[12] Corliss, *History of the DSN*, 92.
[13] Corliss, *History of the DSN*, 194; J. R. Hall, "Network Control System," 5 in DSNPR:XI.
[14] "Interplanetary OD," 23; *Ranger 6 OD*, 59; *Ranger 9 OD*, 51.
[15] *Ranger 6 OD*, 59 & 60; *Ranger 7 OD*, 48; *Ranger 9 OD*, 52.

sophisticated tropospheric model for each station that would take into account local conditions as well as ionospheric corrections.[16]

The new software additionally would contain a model for the wandering of the polar axis; the current program had no such provision.[17] It also would take into account relativistic effects. One of the earliest impacts on spacecraft travel that JPL navigators tackled were those predicted by Albert Einstein's General and Special Theory of Relativity, namely the problem of a spacecraft moving in four-dimensional space-time and the so-called clock paradox. Jack Lorell and John Anderson broke down the mathematics of relativity into Newtonian equations of motion plus additional terms that described the perturbing forces on the Newtonian motion.[18] It was the subject of Ted Moyer's master's thesis at UCLA as well.[19] Their combined efforts allowed navigators to incorporate the perturbing acceleration from this aspect of General Relativity into the more classical mechanics of space navigation.

Harry Lass addressed Einstein's so-called twin or clock paradox from the Special Relativity Theory. One of two identical clocks travels in space, returns to Earth, and finds that it has aged less than its identical twin that remained on Earth. From the perspective of the navigator, the paradox required Lass to create a coordinate system that spanned both the inertial frame of reference of the Earth and the non-inertial relativistic accelerating system of the spacecraft.[20] In addition, with NASA funding, other experimenters studied the relativistic effects on spacecraft clocks.[21]

Ready or not. The completion of the DPODP not only took years to achieve, but its validation and adoption for operational navigation required even more time. The initial analysis that formed the basis for the software began in 1964. Writing code for the IBM 7094 computer started the following year. Programmers had to rewrite the code for the Univac 1108 computer, when it became the main navigation computer. That effort ended early in 1970.[22] Thus, although started in 1964, the software was not ready in 1967 for Mariner 5. Three years later, Mariner Mars 1969 included it, but just on an experimental basis and then only after a great deal of effort on the part of navigators.[23]

A November 1966 memorandum established the Mariner Mars 1969 project's policy regarding use of the double-precision program. The factor that rationalized using

---

[16] *Ranger 6 OD*, 59 & 60; *Ranger 9 OD*, 52; Corliss, *History of the DSN*, 126.

[17] *Ranger 9 OD*, 40, 52, 53 & 55.

[18] John D. Anderson and Jack Lorell, "Orbital Motion in the Theory of General Relativity," *AIAA Journal* 1 (1963): 1372-1374; Anderson, "Twins, Clocks and Geometry: General Covariance Principle for Operation and Measurement Requirements of Space-Time Models," *Transactions of the New York Academy of Sciences*, Series 2, 26 (June 1964): 934-946.

[19] Moyer, "Relativistic Equations of Motion," UCLA Masters Thesis, 1965, F385/Box 30, JPL127 (done under Samuel Herrick).

[20] Lass, "Accelerating Frames of Reference and the Clock Paradox," *American Journal of Physics* 31 (April 1963): 274-276.

[21] See, for example, Robert F. C. Vessot and Martin W. Levine, "Measurement of the Gravitational Redshift Using a Clock in an Orbiting Satellite," 54-64 in Davies.

[22] Moyer, *Formulation of DPODP*, 1.

[23] Curkendall, "Precision Navigation Project: Introduction," 2-2.

it was the highly accurate pointing accuracy required by the spacecraft science instruments, especially the ultraviolet spectrometer experiment. In a presentation to project management on October 27, 1966, Tom Hamilton argued that the only way to meet the pointing requirements for the science experiments was to use either the DPODP or an onboard sensor with a closed-loop control system. Project managers decided against the onboard equipment and in favor of the DPODP, but only tentatively. Before firmly committing to the project, Bud Schurmeier, the JPL Mariner program manager, wanted "a more complete picture" of what was involved in implementing this decision. In particular, he wanted to know how the program would be used during the flight. He put Philip Eckman, from the Mission Analysis Division, in charge of the overall effort, so that he could make a final decision "about the end of February 1967."[24]

In his DPDOP presentation, Hamilton suggested using it only on an experimental basis. The software, he pointed out, had "No operational experience" and had been designed with the limitations in mind of the single-precision program. During the flight of Mariner Mars 1969, the software would run in "research mode." Its more sophisticated models, Hamilton pointed out, included compensations for the wandering of the Earth's pole and the effects of charged particles in the ionosphere. It also would benefit from the improved lunar and planetary ephemerides currently in development. The new DSN lunar-bounce clock synchronization scheme, Hamilton believed, would enable the determination of station location accuracies in both latitude and longitude "good to 1 meter during post flight analysis." Using the double-precision program during the mission's Mars encounters, he concluded, would be "extremely advantageous" to navigation.[25]

Development of the DPODP now focused on serving the Mars mission. In November 1967, navigator Sun Kuen Wong outlined how Mariner Mars 1969 might use it. He assumed that the program would be available, that is, certified, by March 1, 1969. The navigation (Flight Path Analysis and Command) team in the flight operations center would not have access to the software. Instead, that navigation team would prepare data tapes and hand them over to their colleagues in the Precision Navigation Project, who, with the help of the Computer Support Group, would handle DPODP operations.[26]

Wong proposed reserving blocks of computer time for DPODP runs during both cruise and encounter phases. During encounter, he wanted an 8-hour block on each of the five days leading up to encounter, and on the day of encounter he wanted to use two computers. Project management rejected this request. In addition, one day before any maneuvers occurred, Wong wanted a special 4-hour block of computer time. The purpose of the cruise computer runs was to develop confidence in the operation of

---

[24] Memorandum, Schurmeier to Hamilton, "Use of Double Precision Orbit Determination Program for MM '69," November 14, 1966, F38/Box 2, JPLDSN.

[25] Memorandum, Hamilton, David C. Grimes, and Donald W. Trask to Harris M. Schurmeier, Philip K. Eckman, and Marshall S. Johnson, "Mariner'69 Encounter Phase Orbit Determination Accuracies," January 12, 1967, F37/Box 2, JPLDSN.

[26] *DSN Operations Plan for the Mariner 1969 Project, Volume III: DSN Systems Descriptions and Special Procedures*, EPD 605-45 (Pasadena: JPL, October 10, 1967), 14.

the DPODP and to detect any serious trajectory discrepancies computed by the single-precision software.[27]

By May 1968, Dave Curkendall, the navigator responsible for getting the DPODP operational in time, was overseeing engineers and technicians from two separate sections of the Systems Division (Bill Melbourne's Systems Analysis Research Section and Elliott Cutting's Systems Analysis Section) and the DSN (the Flight Operations and DSN Programming Section) as well as contract programmers. The amount of computer and personnel time devoted to the project changed from month to month. Among the navigators and programmers involved were Clifford Cary, Theodore Elconin, Ted Moyer, Richard Schumann, Michael Sykes, and Bill Zielenbach.[28]

By November 1968, six months later, navigators were a full year into development of the "third-generation" navigation program for Mariner Mars 1969. The original schedule had called for completion of a working version on July 1, 1968, and the flight version on March 1, 1969. Development was taking much longer than expected. By the July 30, 1968, review, the due date for the working version had become November 15, 1968, with the flight version ready by March 19, 1969. In the end, programmers delivered the certified flight version on May 1, 1969, thus culminating a development period of over five years.[29]

One of the requirements imposed on the double-precision software was that it had to be compatible with the linked IBM 7040 and 7094 computers, neither one of which was designed to handle the DPODP. This requirement meant that, at least in part, program users would have no access to a backup computer.[30]

As a result of these and other limitations, not to mention the untested nature of the DPODP, navigators relied on the single-precision software during the flights of Mariner 6 and Mariner 7 with the uncertified "third-generation" software operating experimentally. Navigators ran the DPODP immediately before the trajectory correction maneuver and during encounter, but, again, they could not rely on its results to resolve operational issues because of its lack of certification. The new software performed satisfactorily within the limits imposed by the IBM mainframes. One such limitation was that the double-precision program was unable to use range information, because the program that converted the single-precision data tape to a double-precision tape handled the range measurements incorrectly.[31]

Inside the DPODP. Ted Moyer began laying out the theoretical and mathematical basis for the "third-generation" orbit determination software in a series of reports starting in early 1966.[32] Its mathematical models were significantly more accurate, and the program

---

[27] Memorandum, Clifford N. Cary to Sun Kuen Wong, "Use of DPODP in MM'69 Operation," November 30, 1967, F36/Box 2, JPLDSN.
[28] *MM69 QR8*, 30, 33 & 34; "Mariner Mars 1969 Project Functional Organization," January 15, 1969, F49/B2, JPLDSN.
[29] *MM69 QR9*, 19; Curkendall, "Precision Navigation Project: Introduction," 2-2.
[30] *MM69 Final Report*, Vol. I, 589 & 596.
[31] *MM69 Final Report*, Vol. II, 19.
[32] The first report was: Moyer, "Theoretical Basis for the Double-Precision Orbit Determination Program

was far more flexible than its predecessor. Those benefits were a direct result of the fact that the new program (and the mainframe computer that ran it) could handle many more parameters. Some of those parameters were physical constants, such as the astronomical unit and the speed of light. The DPODP also treated non-gravitational forces, such as solar radiation pressure and spacecraft motor burns, that affected the motion of a spacecraft. Yet other parameters addressed timing issues, including transformations from Universal Time to Ephemeris Time, the degree of departure of Atomic Time at each tracking station from the broadcast Universal Time (Greenwich Mean Time as broadcast by the National Bureau of Standards), and conversions among different time systems.[33]

The double-precision program also took into account tracking station information (latitude, longitude, distance from the Earth's spin axis, height above the equator). It continued to use the mean 1950.0 equatorial Cartesian coordinate system in all computations. But, because it processed data in other coordinate systems, the program could convert among an extensive set of coordinate systems. It also adjusted for relativity factors. In fact, the General Theory of Relativity heavily influenced the program's formulation.[34]

Many of these advances were possible because of the DPODP's ability to handle as many as 70 parameters, a dramatic advance over the 20 parameters of the older program. The DPODP, like its predecessors, was written in Fortran IV and consisted of a chain of smaller programs commanded by the control program. One of the most important program links in the navigation software set was called Regres for regression analysis. It was one of the major advances in the transition to double-precision software. Regression analysis is a powerful tool for modeling and analyzing several variables, when one wants to know the relationship between a dependent variable and one or more independent variables. It has wide applications in predicting and forecasting, in this case, in estimating the future flight path of a spacecraft. Regres computed the residuals, that is, the differences between what the tracking stations observed as the spacecraft's position and velocity and their predicted values. Each subsequent subroutine program or link operated in response to the control program. Its job was to execute the links in a sequence that was orderly and meaningful, so that the entirety of the links could determine a probe's flight path.[35]

The DPODP, when run on the double-precision Univac 1108, allowed navigators to construct and apply models that more truly—more realistically—represented the four-dimensional relativistic universe through which a given spacecraft was flying. Equally important was the fact that, to a certain degree, the more powerful software and its computer could bring these models to bear during actual flights rather than only through post-flight analysis. Still, in order to navigate spacecraft successfully and with exactitude, the mathematical models written into the software had to copy nature as closely as possible. That was the very nature of the "fitter's universe."

(DPODP)," 24-26 in JPLSPS 37-38:III.
[33] *DPODP*, 1-2, 1-3, 1-4, 2-14 & 2-15; Moyer, *Formulation of DPODP*, 1-18.
[34] Moyer, *Formulation of DPODP*, 1-18; *DPODP*, 1-2, 1-3, 1-4, 2-1, 2-2 & 3-1.
[35] *DPODP*, 1-1, 1-2, 1-4, 1-5, 3-1, 3-2 & 3-3.

Chapter 13

The Quest

Navigators' basic approach to determining a spacecraft's position began to change during the 1970s. The customary way was to fit the probe's flight path to the mathematical models that navigators constructed. Essentially, navigators made an "indirect" measure of the craft's position by relying on measurements to make corrections to mathematical models. Navigators adapted and created a number of tools to help them to extract spacecraft position information from Doppler. This was the fitter's universe. A new approach that was gaining ground was to design one's observation of the spacecraft in such a way as to cancel out or at least minimize error sources, such as the effects of electrically charged particles in the ionosphere and the interplanetary plasma, rather than attempt to calibrate the models for those effects. Navigators tested several of these so-called self-calibrating techniques with some success, but ultimately the quest would lead to the adoption of radio astronomy techniques.

Devils and angels. Navigators live in the "fitters' universe."[1] They construct mathematical models of the real universe expressed in computer code that attempts to encompass all the influences, gravitational and non-gravitational, acting on spacecraft trajectories and on tracking data. The challenge, from one flight to the next, is to eliminate, or at least compensate for, sources of systematic errors. Achieving an increasingly tighter fit between the mathematical representation of reality in the computer and the reality experienced by the spacecraft required that navigators analyze the tracking data in order to winnow the wheat, the desired information, from the chaff. Tom Hamilton, who led the JPL navigation enterprise during its formative years, understood the fitter's universe. He illustrated his talks with a black-and-white print created by the Dutch graphic artist M. C. Escher titled "Angels and Devils."[2]

> If you looked at it, it turned out that the white figures were angels and the black figures were devils. I said this represents the radio tracking data that we use for orbit determination for our missions. The angels are the information that we're getting that enables us to know where we're going. But the devils are the fact there are things about the rotation of the Earth, where our stations are and what the ephemeris of the target is, and the Sun, that make it so that we're limited in how well we can do. Those from the point of view of a navigator are the devils.

---

[1] *Constants from Tracking Lunar Probes*, 1.
[2] Hamilton, interview, 1992, 5. The original name of the 1960 Escher print was "Circle Limit IV." The University of Texas provides an image here: http://www.ma.utexas.edu/users/radin/reviews/escher.html.

In this fitter's universe that Hamilton described, navigators had several techniques at their disposal. One of the most common methods deployed to deal with systematic errors that crept into computations was the filter. Filters are a class of electronic and mathematical techniques for handling data that vary predictably from an established norm. When navigators are uncertain about the magnitude of a certain force, such as the solar pressure exerted on a spacecraft, they might loosen the models that characterize its flight path. Small non-gravitational forces—such as leakages from attitude control jets—are predictable but can be irregular and cause small, but fatal errors in orbit determinations. The size of those errors quickly swells as navigators process larger and larger arcs of data. In these and other cases, filters come to the rescue.

Catherine Thornton, a long-time JPL navigator, explained how a filter works: "It's analogous to saying that the spacecraft is wandering down the road, and you're trying to say it's on a straight line and determine what that straight line is. You can get way off." As a result, she added, "we had to account for the potential noisiness of the dynamics, the uncertainty of the dynamics. You know, take more frequent measurements, kind of put an uncertainty factor in your equations for the forces that are acting on the spacecraft, so that it doesn't lock into the wrong answer real quickly."[3]

The filter best known outside deep-space navigation is probably the so-called Kalman filter developed by Rudolf Kalman with Air Force funding. The Kalman-Schmidt filter resulted from collaboration between Kalman and Stanley Schmidt, an employee of NASA's Ames Research Center. Its first application was the Apollo project.[4] Batch filters are an important class of filter. These workhorse filters have undergone improvement and adaptation over the decades to meet mission requirements. Navigators choose which filter they feel is most appropriate to the mission. The experience of Al Cangahuala, a navigator on Ulysses, illustrates this point. He started with a Kalman filter but found that it was not necessary, because he could wait a few minutes between estimates. Therefore, he just applied "a straight batch filter" of the type he had learned in school. Other filters "always need care and feeding, and then they need tuning," Cangahuala explained. "With the batch filter, if you can afford the time to just run a batch filter, those real-world things go away. I just don't have to worry about tuning it so much." He added: "I was getting the same performance out of the batch, and it was easier to implement and validate. Let's just do the batch."[5]

The Hamilton-Melbourne equations were yet another fundamental tool in the fitter's toolkit. Tom Hamilton and Bill Melbourne derived these formulas and published them in a 1966 internal periodical. "It was everybody's little bible when I started,"

---

[3] Thornton, interview, 7.
[4] Kalman, "A New Approach to Linear Filtering and Prediction Problems," *Journal of Basic Engineering* 82 (1960): 35-45; Stanley Schmidt, "The Kalman Filter: Its Recognition and Development for Aerospace Applications," *Journal of Guidance and Control* 4 (January-February 1981): 4-5; Gerald L. Smith, Schmidt, and Leonard A. McGee, *Application of Statistical Filter Theory to the Optimal Estimation of Position and Velocity Onboard a Circumlunar Mission*, TR R-135 (Washington: NASA, 1962); John D. McLean, Schmidt, and McGee, *Optimal Filtering and Linear Prediction Applied to a Midcourse Navigation System for the Circumlunar Mission*, TN D-1208 (Washington: NASA, March 1962).
[5] Cangahuala, interview, 18.

recalled Frank Jordan. The paper "was an attempt to look at the basics of how Doppler navigation was working." Jordan would tell every new employee: "Read that paper. Learn it cover to cover." He added: "You have to know it, or you're not going to get very far in planetary navigation. I learned it real, real thoroughly. It was part of growing up."[6]

The paper, to use its own language, dealt with the "information content" contained in "a single pass" of Doppler data, that is, the Doppler collected during a single passage of a spacecraft over a tracking antenna. "Information content" was what Hamilton meant by the angels in his Escher-based analogy. The paper resulted from "many conversations with members of the Systems Analysis Section," but especially Jay Light, who worked in Hamilton's Section and had the key idea of the "velocity parallax." In September 1966, just months after the appearance of the Hamilton-Melbourne paper, Light left JPL for the Harvard Business School, convinced that he needed more management skills. He subsequently became a Harvard Business School professor and in 2006 that school's Dean.[7]

Light concluded that the rotation of the observing stations around the Earth's center—caused by the Earth's spin—contributed the greatest to changes in the Doppler.[8] Put another way, the Earth's 24-hour spin dominated Doppler signatures. Hamilton was thinking about Light's discovery and how to extract navigational information from Doppler, when Melbourne walked into his office. Melbourne characterized Hamilton as "one of these guys who will just talk your ear off. He goes and goes and goes. And in those days he was much less focused than he is now; he's more to the point now. We used to call it the Hour of Mystery when you'd go in there. You couldn't figure out what the hell this guy was talking about half the time. And he's so smart; he's a very brilliant guy."[9]

Melbourne suggested that the 24-hour signature could indicate something about the declination of the spacecraft. Declination is a measure of the distance the spacecraft is above the celestial equator analogous to latitude in space. Hamilton had figured out the right ascension, the analog of longitude. "And he said, no, no, no; he kind of waved me off. I didn't argue with him there," Melbourne recollected. "But I went home that night and started fooling around on a piece of paper and convinced myself that I had something here."[10]

The next day Melbourne showed his work to Hamilton, and "he just went crazy." "This is it!" Hamilton exclaimed. "Between the two of us we put this puzzle together. I never would have thought of it if he hadn't been looking around in his usual muddled way of approaching problems which was half-assed backwards and me coming

---

[6] Hamilton, interview, 1992, 1; Melbourne, interview, 1; Jordan, interview, 42.
[7] Hamilton and Melbourne, 23; Roger Thompson [Harvard Business School], "New at the Helm: A Talk with HBS Dean Light," October 18, 2006, http://hbswk.hbs.edu/item/5546.html (accessed April 2, 2009); Harvard Business School, "Jay O. Light: Biography," http://drfd.hbs.edu/fit/public/facultyInfo.do?facInfo=bio&facEmId=jlight (accessed April 2, 2009).
[8] Jay O. Light, "An Investigation of the Orbit Determination Process Following the First Midcourse Maneuver," 8-17, esp. 12-14 in JPLSPS 37-33:IV; Hamilton and Melbourne, 18.
[9] Melbourne, interview, 16.
[10] Ibid., 17.

in and just seeing something he hadn't seen. That's all. When you put the two pieces together you get something that's greater than the whole, greater than the sum of the parts."[11]

The Hamilton-Melbourne equations showed a clearer way to interpret Doppler using three major parameters: the probe's mean geocentric radial velocity, its right ascension, and the cosine of its declination. For many situations, the right ascension and declination were extremely important for achieving navigational accuracy. Hamilton and Melbourne also concluded that the determination of right ascension and declination angles was not straightforward and that the precision of those determinations rested on the accuracy of station locations.[12] Their formula subsequently came to be a fundamental navigational tool, and one that highlighted the critical importance of obtaining the most precise measures possible of station locations.

Inherent accuracy. Navigators' efforts to deal with, in Hamilton's words, the things "that make it so that we're limited in how well we can do," became formalized following the DSN upgrade from the L band to the S band. The S-band frequencies were higher and meant an increase in tracking data precision. Shifting from one band up to the next was like being able to measure Doppler in millimeters instead of centimeters. This increased precision, however, unmasked just as many devils. The DSN had been making piecemeal improvements in navigation quality. As a March 1965 report stated: "It is the policy of the DSN to continuously conduct research and development of new components and systems and to engineer them into the DSN to maintain a state-of-the-art capability."[13] The advent of S-band capability stimulated the establishment of a systematic approach to improving navigation.

At its July 1965 meeting, the DSN Executive Committee, under the direction of Eb Rechtin, formally established the Inherent Accuracy Project. Its goals were to assess the intrinsic accuracy of the DSN for tracking lunar and planetary missions and to formulate means for making it as accurate as practicable. The project brought together experts from both the DSN (the Telecommunications Division) and navigation (the Systems Analysis Section) who met monthly to coordinate and approve relevant studies and experiments.[14]

Hamilton recalled how the Inherent Accuracy Project came about:[15]

> Along about this time—I can't pin down the exact date—Eb Rechtin asked me to form an inherent accuracy project for the DSN since I was the navigation customer of the DSN and some of my people also worked on improvements and analysis that supported the DSN. I put together a group, which was co-chaired with [Richard] Dick Goldstein from the radio systems division [Communications Systems Research, Section 331].

---

[11] Ibid., 18.
[12] Hamilton and Melbourne, 18 & 22-23.
[13] "Introduction," 1 in JPLSPS 37-32:III.
[14] Corliss, *History of the DSN*, 124; Hamilton, "Introduction," 8 in JPLSPS:III.
[15] Hamilton, interview, Alonso, 3.

We held meetings and reviewed all of the limiting error sources, charged particles, atmosphere, water vapor, the whole works. What are the limiting accuracies in the system? Which ones can we do something about, which ones will we just have to live with, and where is the point of diminishing return in improving the performance of these systems? These turned out to be very productive meetings between basically the designers of the hardware to do the job and the users of that hardware.

Some of the earliest such studies looked at accuracy as a function of Doppler sampling rates and Doppler counting methods. One considered the stability of station transmitter frequencies as an error source, while others focused specifically on the quality of Mariner 4 Doppler and the contribution of the navigation software itself to errors. Future studies promised to examine the influences of the troposphere, the ionosphere, and interplanetary plasma on radio waves, tracking station locations, the synchronization of time among antennas, the ephemerides, the impact of non-gravitational forces ("miscellaneous forces") on spacecraft trajectories, and issues dealing with time scales (Universal Time and Ephemeris Time). Specifically, investigations would consider "other model errors in the 'fitters universe' assumed by the Orbit Determination Program."[16]

In June 1966, a year later, Hamilton's Systems Analysis Section and the Systems Analysis Research Section formally set up the Navigation Technology Project. This was a navigation-centered effort intended to act as a companion to both the Inherent Accuracy and the Ephemeris Development Projects. The new initiative, under the leadership of Dave Curkendall, received funding from flight programs, the DSN, and the JPL Supporting Research and Advanced Development Office.[17]

The general purpose of the Navigation Technology Project was to organize and coordinate the development of "technologies" needed for accurately guiding spacecraft. These "technologies" included, for example, building statistical models to describe the non-gravitational forces affecting spacecraft trajectories and to elaborate navigational approaches for estimating flight paths influenced by such forces. Curkendall's group also looked into methods for computing navigational accuracies in advance of a flight, facilitating the selection of orbits, processing long spans of tracking data, navigating pairs of orbiting and landed space vehicles (Viking), and planning advanced projects such as the anticipated Mars Voyager mission. Yet other proposed studies considered how to navigate probes better as they approached target bodies, including the possibility of combining Earth-based radiometric data and optical observations acquired by the spacecraft, that is, what came to be known as optical navigation, as we shall see in Chapter 15.[18]

---

[16] Curkendall, "Orbit Accuracy as a Function of Doppler Sample Rate for Several Data Taking and Processing Modes," 20-24 in JPLSPS 37-38:III; Trask and Hamilton, "Tracking Data Inherent Accuracy Analysis: DSIF Two-Way Doppler Inherent Accuracy Limitations," 8-13 in JPLSPS 37-38:III; Trask, "Tracking Data Inherent Accuracy Analysis: Quality of Two-Way Doppler Tracking Data Obtained During the Mariner IV Mission," 13-20 in JPLSPS 37-38:III.
[17] Curkendall, "Navigation Technology Project," 15 in JPLSPS:II.
[18] Ibid.

In December 1968, the Inherent Accuracy Project grew into a much broader endeavor known as the JPL Navigation Program. It continued to investigate the same issues including the measurement of such non-gravitational forces as solar radiation pressure. In his history of the DSN, William Corliss wrote of the Inherent Accuracy Project: "not only were the practical results impressive, but the scientific value of the effort was unexpectedly high."[19]

The Inherent Accuracy Project and its successors exemplified the fitter's approach. Fitting a spacecraft's flight path to sophisticated mathematical models required gathering large amounts of measurements tailored to local conditions in order to adjust those models for errors arising from media effects and tracking station locations. A different approach was to make spacecraft observations that canceled or abated error effects.

During the 1970s, navigators went on a quest to find such self-calibrating techniques. Two viable techniques became available to obviate the effects of electrically charged particles in the ionosphere and interplanetary plasma. One required differencing range and Doppler, while the other entailed receiving signals transmitted at two different frequencies. Navigators studied and compared the two to determine which method was superior to the long-standing methods based on ionosonde readings and ground-based instruments. Navigators also pursued techniques that averted, or at least lessened significantly, errors arising from station locations. The most promising scheme was to adapt a radio astronomy practice known as Very Long Baseline Interferometry (VLBI). However, using VLBI for navigation was far from simple, as we shall see in the next chapter. In the meantime, navigators focused on mimicking VLBI by differencing Doppler signals acquired simultaneously by two dishes.

Taking a difference. According to DSN historian William Corliss, the most important of the self-calibrating techniques devised between 1968 and 1975 was Differenced Range Versus Integrated Doppler (DRVID). Nonetheless, writing in 1976, Corliss added that DRVID was "not considered the best approach to charged particle correction."[20] The competing method, dual-frequency calibration, vied with DRVID as both a complementary and opposing technique.

Navigators used DRVID on both the 1969 and 1971 Mars missions, but its development started much earlier. In 1966, Tony Liu undertook the first trials of what eventually came to be known as DRVID. Range had become available with the advent of the Lunar Orbiters, and Liu studied the influence of the ionosphere on range readings for the Inherent Accuracy Project. He was interested in the prospect of sending probes to Venus. Ionospheric effects were more severe for such missions, because they always involved taking observations more nearly in the direction of the Sun. The DSN consequently followed spacecraft through a much thicker ionosphere than, say, on Mars

---

[19] Corliss, *History of the DSN*, 124-127.
[20] Corliss, *History of the DSN*, 185.

flights, which entailed acquiring signals in a direction away from the Sun for about half the flight's duration.[21]

Liu decided to compare range and Doppler to achieve a variety of ends but not, at least initially, to create a self-calibrating technique. He compared two ways of determining the change in distance between Earth (a DSN antenna) and a craft over time. One measure came indirectly from continuously counted Doppler integrated over a given time interval. The other was a direct measure realized by taking the difference between range measurements made at the start and end of the same time interval. The desired datum was the difference between the Doppler and the (differenced) range measurements. More importantly, because the changes to the Doppler and range caused by charged particles along the signal paths were equal and opposite to each other, they canceled out each other.[22]

In April 1967, Liu tested his theory with range and Doppler collected at the Goldstone Echo station from Lunar Orbiters 2 and 3. He expressed the range in terms of round-trip time called range units. Each range unit corresponded to *exactly* 16 Doppler counts and to *approximately* 1.06 meters because of the antenna's S-band operating frequency. After converting the Doppler into range units, Liu took the range-Doppler difference (the residual) at each measurement time. In theory, the residual value should have been zero. In reality, because the Doppler and range passed through different hardware both on the ground and aboard the spacecraft, the signals experienced different time lags. Still, as Liu had predicted, the ionospheric effects on the range and Doppler were nearly equal in magnitude, but opposite in sign.[23]

Liu's study entered a second round with more Doppler and range taken at the Echo station from Lunar Orbiters 2 and 4. The results were even more disappointing because of an accidental battery discharge aboard Lunar Orbiter 4, and other error sources clearly existed. Liu concluded that, until navigators understood these error sources, the effects of the ionosphere would remain cloaked.[24]

In 1968, as Tony Liu became involved in other projects, Bryant Winn took over. Winn collected planetary range and Doppler during the flight of Mariner 5 to Venus. In his own words, Winn hoped to take advantage of the "self-contained calibration inherent to the data pair, ranging and Doppler," but which so far was "unusable." His chief advantage over Liu was better hardware, namely the far more precise Mark II planetary ranging system. That advantage, though, turned out to be not so advantageous. All efforts to correlate the range-Doppler differences with ionospheric phenomena failed. A hardware investigation revealed that, in fact, some of the system's modules were subject to phase drift.[25] DRVID's success rested on the quality of the hardware used to collect data.

---

[21] Anthony S. Liu, "Range & Angle Corrections Due to the Ionosphere," 38 in JPLSPS 37-41:III.
[22] Liu and Robert L. Motsch, "Doppler-Ranging Calibration Experiment Proposed for Lunar Orbiter," 28-33 in JPLSPS 37-44:III; Liu, "Results, Phase I," 23-28.
[23] Liu, "Results, Phase I," 27-28.
[24] Liu, "Results, Phase I," 28; Mudgway, *Uplink-Downlink*, 71; Renzetti, *DSN History*, 67; Liu, "Results, Phase II," 30 & 34.
[25] Winn, "Mariner V Differenced Range-Integrated Range Rate Experiment," 25 & 34 in JPLSPS 37-52:II.

The next trials took place on Mariner Mars 1969. The technique was the same; the experimenters were new: Peter MacDoran and Ray Wimberly. The designation Differenced Range Versus Integrated Doppler (DRVID) came into use for the first time. The causes of previous failures were not unknown. Rather than troublesome, MacDoran and Wimberly viewed this history as being "encouraging."[26] They compared their results against the analysis carried out by Brendan Mulhall using data from the Environmental Science Service Administration and discovered that their measure of ionospheric effects was considerably lower than the anticipated amount. What made that lower figure all the more puzzling was the fact that 1969 was a year of high ionospheric concentrations as a result of solar flare activity. However, the period during which they made their measurements, July 1 to August 5, 1969, was one of relative inactivity. MacDoran and Wimberly realized that, once again, the ranging hardware appeared to be at fault.[27]

Rather than give up, MacDoran continued to work on DRVID's mathematical foundations and practical applications. In early October 1969, he began collaborating with Warren Martin. Martin and radar astronomer Dick Goldstein had developed a new variety of planetary ranging called Mu. Its first use was in 1970 for the General Relativity experiments carried out with Mariners 6 and 7. Mu differed from the earlier Tau planetary ranging system devised in 1967 by Bob Tausworthe, a DSN software specialist.[28] When MacDoran and Martin tested DRVID during the Mariner Mars 1969 Extended Mission, they used a new Mu system that generated DRVID information automatically. On November 24 and December 11, 1969, MacDoran acquired Doppler and Mu range from Mariner 6. DRVID finally worked.[29]

Success led to inclusion of MacDoran's DRVID study on the Mariner Mars 1971 celestial mechanics experiment. He compared the advantages and disadvantages of both Mu and Tau range, and his comparisons of DRVID with Faraday-rotation analysis were very good. DRVID provided an accurate measure of the rate of change of electrons in the signal ray path through both the ionosphere and the solar plasma.[30]

DRVID now became a navigation aid on later flights. In 1972, the extended Mariner Mars 1971 mission used Super Mu ranging. Viking relied operationally on DRVID to calibrate ionospheric effects as well, but the days of DRVID were numbered. Already, it was clear that the technique was incompatible with VLBI, the holy grail of self-calibrating data.[31] Moreover, another calibration method was waiting in the wings.

---

[26] MacDoran and Wimberly, 73-74; Hamilton and Trask, "Introduction," SPS 37-58, 66.
[27] MacDoran and Wimberly, 74 & 77; Mulhall, Ondrasik, and Mottinger, "Results of Mariner Mars 1969 In-Flight Ionospheric Calibration of Radio Tracking Data," 76-77 in JPLSPS 37-60:II; MacDoran and Martin, 34.
[28] Corliss, History of the DSN, 133 & 183; Renzetti, DSN History, 67; Mudgway, Uplink-Downlink, 42, 43, 70 & 71.
[29] MacDoran, "A First-Principles Derivation of the Differenced Range Versus Integrated Doppler (DRVID) Charged-Particle Calibration Method," 28-34 in JPLSPS 37-62:II; MacDoran and Martin, 34.
[30] "Celestial Mechanics Experiment," 14; George A. Madrid, "The Tracking System Analytic Calibration Activity for Mariner Mars 1971: Its Function and Scope," 9 & 11 in TSAC for MM71; Madrid, "Charged Particles," 43, 47, 48, 50, 53 & 58 in TSAC for MM71.
[31] TDA for MM71, Vol. IV, 1, 7 & 22; Mudgway, "Viking Mission Support," 40 in DSNPR:III; Mudgway, "Viking Mission Support," 42 in DSNPR:IV; Von Roos, "Analysis of Dual-Frequency Calibration for Spacecraft VLBI," 52 in DSNPR:VI.

Dual-frequency calibration. That was dual-frequency calibration. It required that the DSN transmit signals to craft at two different frequencies one of which was substantially different from the other. The dish receiving the returned transponder signals analyzed the differences between the received and transmitted signals. That information, in turn, allowed one to estimate the number of charged particles in the signal path.[32]

As with radio occultations, the Stanford Center for Radar Astronomy played an important role in initiating dual-frequency studies of the ionosphere. Its director, Von Eshleman, proposed adding a dual-frequency experiment to Mariner 5. It almost did not take place because of cuts in the project's science budget. The Dual-Frequency Radio Propagation Experiment *was* approved and became the "First use of dual-frequency radio propagation technique for quantitative measurement of the ionosphere of another planet."[33]

The experiment involved transmitting from the Stanford Center for Radar Astronomy's 150-foot-diameter antenna in Palo Alto, California. The experiment required Mariner 5 had to have a dual-frequency receiver. The Stanford researchers hoped to use the information to learn about the Venusian ionosphere and charged particles in interplanetary space. Their goal was not to study the terrestrial ionosphere, which was the purpose of dual-frequency calibration. The Mariner 5 experiment was a repeat of one performed with the Pioneer interplanetary probes. Four Pioneers, numbered 6 through 9, launched between December 1965 and November 1968 and embarked on long-duration flights designed to measure interplanetary fields and particles along their trajectories. Writers sometimes portrayed them as interplanetary weather stations.[34]

Brendan Mulhall and Ray Wimberly realized that the Stanford dual-frequency studies directly measured the charged particles (both ionospheric and interplanetary) between the transmitter and the probe. With very little adjustment to DSN hardware, they believed that they could apply the Stanford technique to calibrate Goldstone tracking data because of the relative geographic proximity of the Goldstone and Stanford antennas.[35] They explicitly foresaw dual-frequencies as a new tool for

---

[32] Mulhall, "Analysis," 13-15.
[33] Letter, Dan Schneiderman to Glenn A. Reiff, February 2, 1966, F13/B1, JPL 421; telex, Homer W. Newell to William H. Pickering, February 7, 1966, F13/B1, JPL421; memorandum, Schneiderman to Pickering, Alvin R. Luedecke, Robert J. Parks, and Jack N. James, "General Observations Concerning the Mariner IV, Mariner V," July 12, 1967, F32/B2, JPL421.
[34] MV67: Final Report, 3, 12 & 13; OrgChart, Mariner 1967 Project, March 31, 1967, F56/B4, JPL421; Nicholas A. Renzetti, *Tracking and Data System Support for the Pioneer Project, Vol. V: Pioneer VI Extended Mission, July 31, 1966-July 1, 1969*, TM 33-426 (Pasadena: JPL, February 1, 1971) 11; Renzetti, *Tracking and Data System Support for the Pioneer Project, Vol. IX: Pioneers VI-IX Extended Missions, July 1, 1969-July 1, 1970*, TM 33-426 (Pasadena: JPL, August 15, 1971), 2 & 6; "Mariner Venus 67 Project," 2 in JPLSPS 37-38:II; Horen Chang, "Analysis of Dual-frequency Observations of Interplanetary Scintillations Taken by the Pioneer 9 Spacecraft," Ph.D. diss., Stanford, Department of Electrical Engineering, 1976; Richard L. Koehler, *Interplanetary Electron Content Measured Between Earth and the Pioneer VI and VII Spacecraft Using Radio Propagation Effects*, SU-SEL-67-051 (Stanford: Stanford Electronics Laboratories, May 1967).
[35] Mulhall and Wimberly, 58.

measuring and correcting for the effects of charged particles, but its deployment required that both the DSN and the spacecraft possess dual-frequency transmitters and receivers.

The prospect of such a dual-frequency calibration technique motivated Tom Hamilton and Don Trask to propose it as early as 1969 for the Pioneers flying to Jupiter. They pointed out that the chief determinant to navigational accuracy over such great distances would be the effects of charged particles. They proposed a dual-frequency (S-band/X-band) calibration method to overcome this problem. John Anderson argued that putting an X-band receiver on the two craft would benefit both navigational accuracy and his own celestial mechanics experiment.[36]

Although the Jupiter Pioneers were promising candidates for the dual-frequency approach, other probes carrying both S-band and X-band equipment furnished earlier opportunities for tests, such as Viking and Helios. Chronologically, though, the next launch was the 1973 Mariner flight to Mercury (Mariner 10). Consequently, Mariner Mars 1969 and 1971 were the last missions to use just DRVID for ionospheric calibrations (ground-based methods aside), and Mariner 10 was the first flight to have dual-frequency equipment both on the spacecraft and on the ground thereby allowing the DSN to calibrate for ionospheric effects using both DRVID and dual-frequency methods. The two complemented one another for the time being. In 1972, for example, Pioneer Venus was considering both calibration techniques,[37] but dual-frequency was gaining ground among navigators.

Which method was the best? Brendan Mulhall argued in favor of S/X calibration over all other approaches. Goldstone Faraday rotation measurements, he pointed out, came from a polarimeter that was "only an experimental unit of low reliability." Another drawback was obtaining data from an outside agency, namely the University of New England in Australia.[38] A 1972 study by Oldwig Von Roos and Mulhall evaluated all three charged-particle calibration techniques: dual-frequency, DRVID, and Faraday rotation. They concluded that dual-frequency was by far the most promising scheme for calibrating both range and Doppler for charged particle effects whether in the ionosphere or interplanetary space. Faraday rotation helped to remove only the ionospheric impacts on range and Doppler. The main drawback to DRVID was the need for good signal-to-noise ratio. And, despite its utility, DRVID was, by nature, not applicable to tracking with two antennas,[39] which was the future of navigation.

VLBI-ish. Data acquired from two antennas promised to provide not just information about the impacts of ionospheric and interplanetary charged particles on signals, but also to reduce significantly errors arising from polar motion, inaccurate station locations, imprecise timing, and other station-related sources. Two-antenna solutions potentially

---

[36] Hamilton and Trask, "Introduction," *SPS 37-58*, 66; John D. Anderson, "Implications of Adding X-Band Tracking to a Pioneer F/G Spacecraft," 77 & 78 in JPLSPS 37-58:II.
[37] Corliss, *History of the DSN*, 186; Siegmeth, 49; Mudgway, *Uplink-Downlink*, 47 & 48; Paul S. Goodwin, "Helios Mission Support," 18-21 in DSNPR:II; Mulhall, "Analysis," 15-16 & 18.
[38] Mulhall, "Demonstrations," 40.
[39] Von Roos and Mulhall, 43, 44 & 45.

could deliver the holy grail of self-calibrating tracking data. Navigators put forward several such schemes, but few succeeded and even then only after prolonged effort. One example was three-way range, the ranging equivalent of three-way Doppler. Its demonstration was far in the future, in 1988, when, in preparation for the Voyager 2 Neptune encounter in August 1989, navigators and DSN engineers prepared to gather three-way range.[40]

During the 1970s, navigators also considered so-called alternate ranging. Two tracking stations made range measurements of a spacecraft alternately, sending different range codes. Ranging could not take place simultaneously, because the craft's transponder could not respond to two signals received at the same time. The first station ranged for about an hour, the second station ranged for another hour, the first station ranged again for an hour and so on until the probe was no longer in view. The collective term for these approaches was simultaneous ranging.[41]

A more practicable substitute for real VLBI was a technique that gathered Doppler at two facilities at the same time to create a new self-calibrating data type. Navigators believed that the approach mimicked, but was not quite the same thing as, VLBI. All the same, even the original name of this technique, Quasi Very Long Baseline Interferometry (QVLBI), evoked the goal of adapting VLBI to navigation.

QVLBI involved two dishes and both two-way and three-way Doppler. One antenna obtained two-way Doppler in the usual manner, while a second received the transponder return signal at the same time as the other station. Navigators then took the difference between the two-way and three-way Doppler. An early name for the method was two-way minus three-way Doppler which later shortened to just differenced Doppler.[42]

Navigators tried to give it a better name. "We used to call it pseudo, and I guess Bob Tausworthe complained that pseudo means fake," explained Tom Hamilton. "And so we decided to use Quasi VLBI." The reason it became "Quasi" VLBI, according to Ondrasik and Rourke, was that two-way minus three-way Doppler was "analogous to the VLBI data types, fringe rate, and time delay, respectively," making it "sort of" (quasi) VLBI. Another navigator, Jim Williams, saw the technique as aiding navigators eventually in handling VLBI data. "The similarity of the fringe rate equation to the conventional Doppler equation means that much of our intuition built on experience with Doppler tracking can be applied to VLBI fringe rate data,"[43] he wrote. In any case, QVLBI made sense as a stopgap measure until one could adapt VLBI to space navigation, but it would live on as a viable alternative to VLBI.

---

[40] Ondrasik and Rourke, "Advantages," 61; Duane C. Roth, Anthony H. Taylor, Joseph A. Wackley, and Ralph I. Roth, "Development of Three-Way Ranging for the Voyager Neptune Encounter," AIAA 88-4265, AIAA/AAS Astrodynamics Conference, August 15-17, 1988, Minneapolis, MN, 372-378; Duane C. Roth, Anthony H. Taylor, Robert A. Jacobson, and George D. Lewis, "Performance of Three-Way Data Types during Voyager's Encounter with Neptune," AIAA 90-2878, AIAA/AAS Astrodynamics Conference, August 20-22, 1990, Portland, OR, 129-134.
[41] Von Roos and Mulhall, 43.
[42] Von Roos and Mulhall, 43.
[43] Hamilton, interview, Waff, 24-25; "Interplanetary OD," 51; Rourke and Ondrasik, 50; Ondrasik and Rourke, "Advantages," 62; James G. Williams, "Very Long Baseline Interferometry and its Sensitivity to Geophysical and Astronomical Effects," 51 in JPLSPS 37-62:II.

The expectation was that differenced Doppler would provide its own calibration information, including such persistent error sources as uncertainties in the effects of solar pressure on spacecraft or small accelerations arising from gas leaks in the attitude control systems or any other random small non-gravitational force. Based on a 1971 study by John Ondrasik and Ken Rourke, Brendan Mulhall saw the utility of QVLBI when spacecraft passed through zero declination, when two-way Doppler was far less reliable.[44]

To make differenced Doppler practical, navigators would have to overcome once again a hardware issue, namely oscillator frequency instability, the main error source in three-way Doppler. Replacing the DSN's rubidium oscillators with hydrogen masers appeared to be the solution or at least a good first step. DSN radio astronomy was driving the adoption of the maser. Tests undertaken between October and December 1969 at the Harvard College Observatory radio telescope already had proven the advantages of hydrogen masers for VLBI.[45]

The first QVLBI test took place in October and November 1971. The DSN not having upgraded yet to hydrogen masers, Goddard supplied them to the Goldstone Echo and Woomera facilities. Mulhall, along with Chia-Chun Chao, Donald Johnson, and Bill Zielenbach, undertook these initial trials on Mariner 9. The probe was an excellent candidate for QVLBI, because a spacecraft gas leak was altering its flight path in an unexpected way.[46]

Because QVLBI was a new data type, collecting the data and feeding it into the orbit determination software required some innovation, including creating tailor-made data compression methods and specialized programs. The novelty of QVLBI also complicated post-flight analysis. In the end, the limited amount of data collected and inadequate knowledge of the behavior of the maser-driven frequency system made their results inconclusive, but "promising."[47]

Pioneer 10 presented three separate opportunities to try out QVLBI on a mission to the outer planets. Chao, Sun Kuen Wong, and Anton Lubeley carried out the first trial in 1973, using a short baseline interferometer consisting of the Goldstone Pioneer, Echo, and Mars antennas. The system lacked long-term stability. A second test in 1973 by Chao with Robert Preston and Harold Nance focused on the stability of the rubidium frequency standard and found that its frequency drift rate was inadequate for QVLBI.[48] In 1974, Chao carried out the third Pioneer 10 test with Bruce O'Reilly. This time the goal was to show the superiority of QVLBI over two-way Doppler for providing more accurate orbit computations in the presence of unmodeled forces and for reducing the effects of interplanetary plasma. In this instance, the forces were the gravitational attractions of Jupiter's four largest moons (Io, Callisto, Ganymede, and

---

[44] Mulhall, "Demonstrations," 39; Rourke and Ondrasik, 49-60, esp 54; Ondrasik and Rourke, "Application," 116-132; Ondrasik and Rourke, "Advantages," 61 & 66.
[45] "Interplanetary OD," 50; Ondrasik and Rourke, "Application," 116-132; Martin W. Levine and Robert F. C. Vessot, "Hydrogen-Maser Time and Frequency Standard at Agassiz Observatory," *Radio Science* 5 (October 1970): 1287-1292.
[46] "Two-Station Demo," 27 & 29; "Interplanetary OD," 51.
[47] "Interplanetary OD," 51-52; "Two-Station Demo," 27-40; Chao and Ondrasik, 52.
[48] "Short Baseline QVLBI, Part I," 47-56; "Short Baseline QVLBI, Part II," 20-26; Chao and Ondrasik, 53.

Europa). Chao and O'Reilly concluded that QVLBI had handled the unmodeled forces better than the two-way Doppler. They attributed the success in part to QVLBI's ability to compensate for plasma effects.[49]

The Differenced Doppler Demonstration. By the time Chao undertook this last Pioneer 10 test, QVLBI already had undergone a rigorous demonstration on Mariner Venus-Mercury 1973 as one of the official navigation experiments. In the Differenced Doppler Demonstration, Chao and Ondrasik collected two-way and three-way Doppler over the short baseline between the Goldstone Echo and Mars stations, both of which still had rubidium frequency standards. They had hoped to collect data over a full nine months both before and after encounter. The Venus and Mercury encounter data would not be useful in showing off QVLBI, however, because the primary force acting on the probe, the planets' gravitational fields, would mask any effects from unmodeled forces.[50]

Collecting the differenced Doppler data proved to be tricky for a number of reasons. The level of precision required for navigators to compare QVLBI and customary trajectory estimates was available only during the two weeks prior to encounter, thereby restricting the trial to the two weeks before encounter. Chao and Ondrasik needed to analyze long data arcs that suffered more from error sources.[51]

All things considered, Chao and Ondrasik deemed the demonstration to be a success. They concluded that QVLBI had decreased the effects of unmodeled forces on trajectory computations "by an order of magnitude." The differenced Doppler had reduced the impact of the solar corona on orbit determinations by a factor of five. This was a significant reduction because of Mariner 10's close proximity to the Sun and its ion-filled corona. At one point, the corona effects became so great that dual-frequency calibration was unable to overcome them, so flight navigators relied heavily on orbital estimates based on QVLBI data. The QVLBI solutions proved to be "very useful in making real-time navigation decisions for use by the Mariner 10 project." Chao continued to conduct more QVLBI experiments. Each time, he varied the conditions, using Pioneer 10 and Pioneer 11 as targets and different short baselines—such as one between the Tidbinbilla and Canberra dishes—and a long baseline stretching between the Echo and Cebreros antennas.[52] QVLBI would live on as differenced Doppler, but by the end of the 1970s, the future seemed to belong to VLBI navigation, especially for guiding spacecraft to the outer planets.

---

[49] Bruce D. O'Reilly and Chia-Chun Chao, "An Evaluation of QVLBI OD Analysis of Pioneer 10 Encounter Data in the Presence of Unmodeled Satellite Accelerations," 66-69 in DSNPR 42-22.
[50] Chao and Ondrasik, 53-56.
[51] Chao and Ondrasik, 52 & 53.
[52] Chao and Ondrasik, 53, 55 & 57; Chia-Chun Chao, "QVLBI Doppler Demonstrations Conducted during Pioneer 11 Encounter and Solar Conjunctions," 55 in DSNPR 42-31; Jay H. Lieske, "Improved Ephemerides of the Galilean Satellites," *Astronomy and Astrophysics* 82 (February 1980): 340-348.

Chapter 14

A Page from Radio Astronomy

Very Long Baseline Interferometry (VLBI) promised to provide navigation data that were far more precise than standard Doppler or range. The technique had the potential to become *the* self-calibrating data type. VLBI gave far more accurate information about antenna locations, polar motion, Earth precession and nutation constants, Universal Time, and station timing errors. At the same time, it tackled errors arising from media effects. At first, navigators embraced so-called differential VLBI (also expressed as ΔVLBI or delta VLBI), already in use by the DSN. The name derived from the fact that it entailed two antennas receiving simultaneously from a spacecraft, then turning a few degrees to detect signals from a nearby extragalactic radio source (usually a quasar), and taking the difference between the differenced observations of the probe and the quasar. This reliance on quasars and other extragalactic radio sources added yet another link between navigation and astronomy, not to mention the fact that ΔVLBI itself was a radio astronomy technique. Its use also required that navigators establish links between the established star-based reference frame of the JPL ephemerides and this new radio-based reference frame.

Navigators' chief interest in ΔVLBI was for missions to the outer planets. If Doppler and range remained the only kinds of observations available for those flights, either those projects would be infeasible or their scientific utility would be restricted.[1] The effects of error sources increased as the distance between Earth antennas and spacecraft increased. Two-antenna solutions had the potential to override the impacts of those error sources, and ΔVLBI held great promise. Delta VLBI navigation was a long time in development throughout the 1970s. Its first test on the Voyager outer-planet mission demonstrated the failure of the DSN to keep pace with the needs of ΔVLBI navigation. Not all the hurdles were technical. Navigators had to convince NASA Headquarters of its necessity, and the focus of the DSN's VLBI system was on radio astronomy and geodesy, not navigation.

Interferometry for geodesy. Early radio interferometers were essentially analogs of the optical interferometer pioneered by Albert Michelson in 1890.[2] Two radio telescopes separated by some distance (known as the baseline) received signals from the same source. Combining and comparing the signals from the two telescopes took place

---

[1] V. John Ondrasik, Chia-Chun Chao, F. Bryant Winn, KaBing Yip, Charles H. Acton, and Stephen J. Reinbold, "Demonstration of New Data Types for use in Interplanetary Navigation," AIAA 74-831, AIAA Mechanics and Control of Flight Conference, August 5-9, 1974, Anaheim, CA.

[2] Albert A. Michelson, "On the Applications of Interference Methods to Astronomical Measurements," *Philosophical Magazine* 30 (1890): 1-21; Michelson, "Some Recent Applications of Interference Methods," *Proceedings of the Physical Society of London* 33 (December 1920): 275-285; Michelson, "On the Application of Interference Methods to Astronomical Measurements," *The Astrophysical Journal* 51 (June 1920): 257-262.

electronically. The factor that initially determined the distance between telescopes was the length of cable connecting the two, which was about 1 km. Later, microwave relay links allowed separations up to 100km, the limiting factor being the line-of-sight distance.[3]

Lengthening the baseline distance between the two radio telescopes increased the interferometer's resolution, that is, its ability to see smaller and smaller objects. A major breakthrough was the successful operation of a radio interferometer in Canada over continental distances—between British Columbia and Ontario—in 1967 achieved by the National Research Council, Queen's University, the University of Toronto, and the Dominion Radio Astrophysical Observatory.[4] At the same time, the NRAO (Green Bank, WV), in collaboration with Cornell University, was developing its own VLBI system.[5]

In each case, two radio telescopes simultaneously and separately recorded signals on magnetic tape along with time codes. The electronic cross-correlation of the two sets of recorded signals took place subsequently. The Canadian system was analog, while the United States went digital. The key to eliminating the direct electrical connection between stations (whether a cable or a microwave link) and to obtaining the necessary synchronization between radio telescopes was the frequency stabilization provided by rubidium atomic clocks.[6]

It soon became evident that radio interferometers offered a new and important window on the Earth as well. Signals arrived at the dishes at slightly different times. Signal cross-correlation yielded a measure of the amount of time delay. The time delay arose entirely from the Earth's motion, but related equally to the source location and the baseline between the two antennas. As a result, mathematical analysis of the time delay could furnish information on source locations, the baseline, polar and other Earth motions, and Universal Time (UT1).[7]

At MIT, Irwin Shapiro, Chuck Counselman, and others undertook an improvement in VLBI methods to obtain more accurate Earth and time measurements using a technique developed by MIT radio astronomer Alan Rogers. Specifically, they were interested in determinations of Universal Time, polar motion, Earth tides, and continental drift, which geophysicists were beginning to understand in terms of plate tectonics. They collaborated with Arthur E. Niell at JPL and Thomas Clark at Goddard, where VLBI geodetic work started in 1972 with funding initially from ARPA.[8]

---

[3] MacDoran, "VLBI," 62-63; Benjauthrit, 147-148.
[4] Norman W. Broten, Thomas H. Legg, Jack L. Locke, C. William McLeish, Roger S. Richards, Robert M. Chisholm, Herbert P. Gush, J. L. Yen, and John A. Galt, "Long Baseline Interferometry: A New Technique," *Science* 156 (June 23, 1967): 1592-1593; Galt, "Beginnings of Long-Baseline Interferometry in Canada: A Perspective from Penticton," *Journal of the Royal Astronomical Society of Canada* 82 (October 1988): 242-247.
[5] Kenneth I. Kellermann and Michael H. Cohen, "The Origin and Evolution of the NRAO-Cornell VLBI System," *Journal of the Royal Astronomical Society of Canada* 82 (October 1988): 248-265.
[6] MacDoran, "VLBI," 62-63; Benjauthrit, 148-149.
[7] MacDoran, "VLBI," 65-68.
[8] Hans F. Hinteregger, Irwin I. Shapiro, Douglas S. Robertson, Curtis A. Knight, R. A. Ergas, Alan R. Whitney, Alan E. E. Rogers, James M. Moran, Thomas A. Clark, and Bernard F. Burke, "Precision Geodesy via Radio Interferometry," *Science* 178 (October 27, 1972): 396-398; Clark, "Very Long Baseline Interferometry," 238-247 in GSFC, *Significant Accomplishments*; Irving M. Salzberg, "Very Long Baseline

At JPL, the main thrust of VLBI research was geophysics with an emphasis, not surprisingly, on the study of faults and earthquakes. The idea behind the Astronomical Radio Interferometric Earth Survey (ARIES) Project, carried out in collaboration with the U. S. Geological Survey's National Center for Earthquake Research and the Los Angeles County Surveyor's Office, was to use VLBI to help to predict earthquakes. Two antennas, one mobile and the other stationary, located many miles apart on opposite sides of the San Andreas Fault, received signals from the same quasar, but at slightly different times. The time difference, measured with the help of atomic clocks, allowed one to calculate the distance between the two antennas as well as their elevation to within a few centimeters (an inch) or less. Subsequent measurements made months or even years later would reveal changes that in turn would indicate movement in the Earth's crust.[9] Quasars made excellent reference targets, because, unlike stars and planets, they did not appear to move against the sky. Using a variation of Alan Rogers' VLBI technique, the ARIES project succeeded in measuring variations in the Earth's rotational rate (UT1) with what JPL called a "state-of-the-art precision of 2 milliseconds." By 1973, a DSN long baseline between Goldstone and Madrid was making monthly observations to determine UT1 for space navigation.[10]

Laying a foundation. In the summer of 1971, because ARIES relied on quasars, JPL initiated the development of a catalog of extragalactic S-band radio sources (mostly quasars) for use in geodesy using the Goldstone-Madrid long baseline interferometer. The first runs resulted in a preliminary catalog of about ten radio sources whose relative

---

Interferometry," 116-120 in GSFC, *Significant Accomplishments*; J. W. Ryan and Chopo Ma, "NASA-GSFC's Geodetic VLBI Program: A Twenty-Year Retrospective," *Physics and Chemistry of The Earth* 23 (1998): 1041-1052; Rogers, interview, 2-5; Shapiro, interview, October 1, 1993, 4-5; Alan R. Whitney, Rogers, Hans F. Hinteregger, Curtis A. Knight, James I. Levine, S. Lippincott, Thomas A. Clark, Irwin I. Shapiro, and Douglas S. Robertson, "A Very-Long-Baseline Interferometer System for Geodetic Applications," *Radio Science* 11 (May 1976): 421-432; Shapiro, Robertson, Knight, Counselman, Rogers, Hinteregger, Lippincott, Whitney, Clark, Arthur E. Niell, and Donovan J. Spitzmesser, "Transcontinental Baselines and the Rotation of the Earth Measured by Radio Interferometry," *Science* 186 (December 6, 1974): 920-922; Counselman, "Very-Long Baseline Interferometry Techniques Applied to Problems of Geodesy, Geophysics, Planetary Science, Astronomy, and General Relativity," *Proceedings of the IEEE* 61 (September 1973): 1225-1230; Counselman, "Geodesy by Very Long Baseline Interferometry," *Reviews of Geophysics and Space Physics* 13 (1975): 270-271; Robertson, "Special Relativity and Very Long Baseline Interferometers," *Nature* 257 (October 9, 1975): 467-468; Shapiro, "Very Long Baseline Interferometry, the Impact on Astronomy and Geophysics," 133 in George W. Morgenthaler and Howard D. Greyber, ed., *Astronomy from a Space Platform* (Tarzana, CA: AAS, 1972); Benjauthrit, 149.

[9] John L. Fanselow, Peter F. MacDoran, J. Brooks Thomas, James G. Williams, Clifford J. Finnie, Takeshi Sato, Lyle Skjerve, and Donovan J. Spitzmesser, "The Goldstone Interferometer for Earth Physics," 45-57 in DSNPR:V; MacDoran, "Radio Interferometry for International Study of the Earthquake Mechanism," *Acta Astronautica* 1 (1974): 1427-1444; Kwok Maw Ong, Peter F. MacDoran, J. Brooks Thomas, Henry F. Fliegel, Lyle J. Skjerve, Donovan J. Spitzmesser, Paul D. Batelaan, Stephen R. Paine, and M. G. Newsted, "A Demonstration of Radio Interferometric Surveying Using DSS 14 and the Project ARIES Transportable Antenna," 41-52 in DSNPR 42-26; MacDoran, "VLBI Earth Physics," 13-14; Benjauthrit, 151.

[10] Rogers, "Very Long Baseline Interferometry with Large Effective Band-Width for Phase-Delay Measurements," *Radio Science* 5 (October 1970): 1239-1247; MacDoran, "VLBI Earth Physics," 9 & 12; MacDoran, "VLBI," 69; J. Brooks Thomas and Henry F. Fliegel, "Time and Frequency Requirements for Radio Interferometric Earth Physics," 15 & 20 in *Fifth PTTI Meeting*.

positions were known to accuracies between 0.1 and 0.01 arc seconds. The Goldstone-Madrid interferometer also observed simultaneously at both S and X band in order to remove the effects of charged particles (dual-frequency calibration). The immediate goal was to develop a catalog of 20 or more sources with accuracies of 0.01 arc seconds.[11]

By 1974, the DSN had found only 60 radio sources, an amount deemed to be woefully inadequate for future VLBI needs by ARIES, radio astronomers, and navigation. That summer, the DSN undertook a literature search to identify potential S-band VLBI sources from among those already found by astronomers. The network then observed them to determine if they had the desired characteristics and incorporated the useful ones into the source catalog. The search went slowly.[12]

At the same time, similar campaigns were underway elsewhere, such as those led by Clark at Goddard, Shapiro at MIT, and Marshall Cohen at the Owens Valley Radio Observatory. These and other VLBI radio source searches soon joined with JPL to establish the so-called Quasar Patrol. The effort became one of the priority radio astronomy projects conducted by the DSN. The Goldstone 64-meter dish made monthly observations jointly with the MIT Haystack antenna and the NRAO 140-foot radio telescope. Participants included faculty and researchers from Caltech, Cornell University, the University of Maryland, MIT, the NRAO, and the Naval Observatory as well as Goddard and JPL. The quest became international with the participation of researchers in Argentina at the Observatorio de la Plata and in France at the Paris Observatory and the Observatory of Bordeaux. By 1978, the patrol had detected 564 VLBI sources, 166 of which lay within 10° of the ecliptic plane. Jim Williams wrote a program to cull candidate navigation sources from the list of known quasars, and by the fall of 1975, navigators had a list of 162 usable radio sources. That number continued to grow during the 1980s in particular to support navigators' needs.[13]

---

[11] MacDoran, "VLBI Earth Physics," 11 & 12.

[12] Claude E. Hildebrand, James S. Border, Frank F. Donivan, Susan G. Finley, Benjamin Moultrie, X X Newhall, Lyle Skjerve, Thomas P. Yunck, Frank R. Bletzacker, and Cory B. Smith, "Progress in the Application of VLBI to Interplanetary Navigation," 68 in VLBI Techniques; "Celestial VLBI Sources," 47 & 49; Robert M. Taylor and Richard N. Manchester, "Radio Astronomy," 2 in TDAPR 42-60; J. Brooks Thomas, Ojars J. Sovers, John L. Fanselow, Eri J. Cohen, George H. Purcell, Jr., David H. Rogstad, Lyle J. Skjerve and Donovan J. Spitzmesser, "Radio Interferometric Determination of Source Positions, Intercontinental Baselines, and Earth Orientation With Deep Space Network Antennas, 1971 to 1980," 128-155 in TDAPR 42-73.

[13] Curkendall, interview, 36; "Quasar Experiment: Part I," 32 & 33; Neil A. Mottinger, "Candidate Extragalactic Radio Sources for Differenced VLBI Tracking With Deep Space Probes," 65-68 in DSNPR:XVI; Robert A. Preston, Alan W. Harris, Martin A. Slade, James G. Williams, John L. Fanselow, J. Brooks Thomas, David D. Morabito, Donovan J. Spitzmesser, Lyle J. Skjerve, Benjamin Johnson, and David Jauncey, "JPL Catalog of VLBI Radio Sources," Bulletin of the American Astronomical Society 7 (September 1975): 517; Richard J. Fahnestock and Robert M. Taylor, "Radio Astronomy," 22 in DSNR 42-53; Kenneth I. Kellermann, "Structure and Apparent Motions in Compact Radio Sources: The Quasar Patrol," Bulletin of the American Astronomical Society 5 (March 1973): 285; Linnes, "Radio Science Support," 52-58 in DSNPR:X; Chopo Ma, Elisa F. Arias, T. Marshall Eubanks, Alan L. Fey, Anne-Marie Gontier, Christopher S. Jacobs, Ojars J. Sovers, Brent A. Archinal, and Patrick Charlot, "The International Celestial Reference Frame as Realized by Very Long Baseline Interferometry," The Astronomical Journal 116 (July 1998): 516-546; "Celestial VLBI Sources," 46; John L. Fanselow, Ojars J. Sovers, J. Brooks Thomas, George H. Purcell, Jr., "An Improved Celestial Radio Reference Frame: JPL 1982-4," 183-197 in VLBI Techniques.

The adaptation of VLBI to navigation proceeded gradually. In 1970, Jim Williams undertook a mathematical analysis of VLBI as a potential navigation tool, specifically to determine station locations and polar motion more precisely. He proceeded from the known (Doppler) to the unknown (VLBI). Williams showed the mathematical similarity between the fringe rate equation and the conventional Doppler equation. He concluded that VLBI had the potential to measure Universal Time, polar motion, precession and nutation constants, and relative station locations to a higher degree of accuracy than conventional Doppler or range. In turn, these improved values would translate into better estimates of spacecraft flight paths.[14]

The developer of the mathematics behind the adaptation of VLBI for navigation was J. Brooks Thomas. His study appeared in three parts between November 1971 and June 1973. Thomas' lengthy and detailed analyses focused on cross-correlation procedures for handling both natural radio sources and spacecraft tracking data. He wrote formulas that enabled navigators to analyze signal time delay to obtain the locations of radio sources (whether quasars or spacecraft), baseline information, Universal Time, and corrections for polar and other Earth motions. In the final installment, Thomas offered "practical data reduction steps" needed to obtain the necessary fringe and delay information.[15] The mathematical foundation for navigating by VLBI was now laid.

A critical question for navigators was whether VLBI could calibrate for media effects. Oldwig Von Roos led an effort to test dual-frequency and DRVID calibration methods for VLBI. The idea was to create a so-called wide-band very long baseline interferometer for the purpose of tracking spacecraft. "Wide band" referred to using a modulated signal, as opposed to a monochromatic, single-tone signal (narrow band). Von Roos quickly came to realize the crucial role of synchronizing station clocks. "Synchronization is essential for this system to work," he wrote. "Without it, all advantages of the spacecraft VLBI will be utterly lost. It is therefore vital to implement a viable system allowing for the best accuracy available." Hydrogen masers were very stable as time standards, but their drift rate was unpredictable: how long would they stay in synch? Von Roos proposed a clock synchronization technique that used laser ranging. Two VLBI antennas would point at a geosynchronous satellite at the same time that two lasers, one at each VLBI site, performed ranging on the satellite. A test of this approach took place on August 30, 1972, at the Pioneer and Echo dishes. But, instead of lasers, which Goldstone lacked, the trial used selected quasars.[16]

The Mariner 9 Quasar Experiment. A laser-ranging test did precede the undertaking of a full-scale demonstration of VLBI navigation with Mariner 9. In September 1971, Brooks

---

[14] James G. Williams, "Very Long Baseline Interferometry and its Sensitivity to Geophysical and Astronomical Effects," 50-51 in JPLSPS 37-62:II.

[15] J. Brooks Thomas, "An Analysis of Long Baseline Radio Interferometry," 37-50 in DSNPR:VII; Thomas, "An Analysis of Long Baseline Radio Interferometry, Part II," 29-38 in DSNPR:VIII.

[16] "Analysis for VLBI," 48 & 54; William J. Hurd, "Demonstration of Intercontinental DSN Clock Synchronization by VLBI," 59-66 in *Fifth PTTI Meeting*; Hurd, "A Demonstration of DSN Clock Synchronization by VLBI," 149-160 in DSNPR:XII; Hurd, "DSN Station Clock Synchronization by Maximum Likelihood VLBI," 82-95 in DSNPR:X.

Thomas, Marty Slade, and Peter MacDoran took advantage of transmissions from the Apollo Lunar Surface Experiment Packages (ALSEP), sets of scientific instruments that returned telemetry with an S-band transmitter.[17] The ALSEP tests also provided ersatz practice in tracking two probes simultaneously. Dual-spacecraft tracking—following two Mariners simultaneously within the same beam-width of a single antenna—had been incorporated in the Mariner Mars 1964 mission design, but the demise of Mariner 3 foiled that plan.[18]

The ALSEP tests used a long (8,500-km) baseline interferometer consisting of Goldstone's 64-meter Mars and Madrid's 26-meter antennas. Goldstone already had a hydrogen maser, but Madrid did not, so it relied on one furnished by the Smithsonian Astrophysical Observatory. The results did not answer navigators' questions. The ALSEP signals traversed different paths in the atmosphere (both the troposphere and ionosphere), so differencing them did not remove the media effects entirely. Moreover, unanticipated systematic error sources plagued the experiment.[19] This unpropitious demonstration did not bode well for the Mariner 9 Quasar Experiment.

The Quasar Experiment was the first full-scale attempt to use VLBI for navigation. The original plan had been to switch between a planet and a quasar. Irwin Shapiro, though, suggested that a craft in orbit around another planet would be a better target. Because the probe and quasar subtended a very small angle as viewed from Earth, the media effects would tend to cancel themselves, at least in theory.[20] Aside from being a first, the Quasar Experiment was noteworthy for at least two other reasons. One was its international breadth. The team included, from the United States, JPL's Marty Slade and Peter MacDoran and MIT's Shapiro, and three radio astronomy pioneers from Australia: Jack Gubbay, a professor at the University of Adelaide, and Anthony Legg and David S. Robertson, both researchers at the Space Research Group of the Weapons Research Establishment (now the Defence, Science, and Technology Organisation, DSTO) in Salisbury.[21]

The experiment equally was noteworthy as a first step in linking the star-based and quasar-based reference frames. The ephemerides of navigators and astronomers alike relied on a standard star catalog. Quasars, pulsars, and other extragalactic radio sources were located within their own radio reference frame. The tapes that the VLBI experimenters received from the navigation team related objects to the stellar reference frame. So, one of the goals of the Quasar Experiment was to locate Mars and Earth within the same radio reference frame as the quasars. Ultimately, to make VLBI practical, navigators needed a combined radio and optical reference frame.[22]

---

[17] James R. Bates, W. W. Lauderdale, and Harold Kernaghan, *ALSEP Termination Report* (Washington: NASA, April 1979); William D. Compton, *Where No Man Has Gone Before: A History of Apollo Lunar Exploration Missions* (Washington: NASA, 1989), 84-87, 95-96, 115, 134, 162, 166-167, 346 & 358.

[18] Slade, MacDoran, and Thomas, 35 in DSNPR:XII; *MM64 Final Report*, v, 2, 6, 22, 131, 192 & 325; Corliss, *History of the DSN*, 96 & 104; *MM64 TDA Report*, 3, 4, 6, 8, 9, 31, 32, 49 & 51.

[19] Slade, MacDoran, and Thomas, 36 & 37; Martin W. Levine and Robert F. C. Vessot, "Hydrogen-Maser Time and Frequency Standard at Agassiz Observatory," *Radio Science* 5 (October 1970): 1287-1292.

[20] "Quasar Experiment: Part I," 31 & 32.

[21] Raymond Haynes, *Explorers of the Southern Sky: A History of Australian Astronomy* (New York: Cambridge University Press, 1996), 277.

[22] "Quasar Experiment: Part I," 31-35.

A turf war won. Navigators hoped to apply VLBI not to more Mars flights, but to the Voyager and Galileo expeditions to the outer planets. But, developing VLBI for Voyager navigation would require approval and funding. Already, Voyager navigators had included conventional Doppler and range plus on-board optical navigation. Why add VLBI? Navigators tried to sell project managers and NASA Headquarters on the desirability of having all three observation types. "Under the canopy of Voyager," Curkendall recalled, they "developed all the aspects of the JPL navigation system and sharpened all the arrows in the quiver. Well, for whatever reason, if the cameras failed or degraded, you've still got a pretty capable system for almost all the experiments, right? So, as a backup and redundancy, you could really sell both. In those days, when the budgets weren't as constrained as they are now, it made a lot of sense. So we brought them all along." Nick Renzetti, who managed the engineering side of the DSN, promoted VLBI and other DSN uses "not so much because of their demonstrable critical importance to mission requirements," Curkendall explained, "but as a way to motivate and retain a vibrant technical and scientific staff for the laboratory."[23]

VLBI faced resistance from NASA Headquarters. "There were people at Headquarters," Frank Jordan explained, who opposed VLBI, because they thought it was unnecessary. Instead, they wanted to spend agency funds on "bigger dishes and higher telemetry [rates]." "Navigation doesn't need these crazy guys out there with VLBI. They're a bunch of academics anyway," Jordan added. Some Voyager scientists also saw the arrival of VLBI as a rationale for removing optical navigation from the project. "We had people in powerful places with budget streams either for or against either one of these things," Jordan recalled.[24]

Tom Hamilton, as Deputy Manager of the Systems Division, oversaw groups dedicated to all forms of navigation: radiometric, optical, and VLBI. He asked Jordan to evaluate the various data types for Voyager. Hamilton, Jordan explained, "was hoping that I would conclude that the development of optical [navigation]—and its utility on Voyager and Galileo—and the development of VLBI and its addition to the data set of Doppler [and] range—which we already had in the back—that it was a smart thing to do both of these. And I did it." As a result, though, "everybody [was] mad at me" for not endorsing VLBI or optical data as capable of providing the entirety of the mission's navigational needs. "I was unpopular in the other section, because I was questioning whether they could do everything. I was unpopular with my own people, because I didn't say optical NAV [navigation] is the whole thing."[25]

Jordan "carefully constructed a garbage-in, garbage-out probability analysis for Voyager and Galileo. I took into account the accuracies of the systems. It's no secret that optical was the defining accuracy. I also took into account the probabilities that maybe the camera could malfunction, and you couldn't rely on it. So, I put together a story. I don't know how I did it." He was able to convince the JPL Office of Flight

---

[23] Curkendall, interview, 49.
[24] Jordan, interview, 47.
[25] Jordan, interview, 47-48.

Projects, whose Director was Bob Parks, to agree that both VLBI and optical navigation were needed to assure the success of Voyager and Galileo.[26]

After winning over Parks' office, Jordan had to persuade NASA Headquarters. "Okay, I've been a section manager less than a year, and at the time I'm only 35," Jordan related. "I'm feeling like I'm back getting the Ph.D., back in the qualifying exam." He and Tom Hamilton had to make a presentation to the three Headquarters people in the Office of Advanced Research and Technology who were funding projects including the development of optical navigation as well as DSN research and development. They explained: "Everybody has a part to play, and they're synergistic."[27]

"It was a very high-risk endeavor," Jordan explained, "because one of the guys was a little antagonistic." The person overseeing funding for DSN research and development probably was the one who threatened to terminate funding, "if you can do all the missions with optical." "It was antagonistic, because you had these guys—their own egos—who were invested in what they were doing." Jordan and Hamilton tried to argue that navigation needed both optical and VLBI for the long-term. "We sold it." Headquarters agreed to underwrite development of both types of navigation.[28]

Preparing for Voyager. On January 2, 1976, NASA Headquarters gave its approval to start the VLBI Validation Project Task the following November. The idea was to compare VLBI and laser ranging using the four Goldstone antennas, MIT's Haystack dish, and the McDonald Observatory. Frequency and time were tied to the station's hydrogen maser frequency standard where available; otherwise, they relied on a rubidium standard.[29] Navigators also wanted to test drive VLBI on an actual spacecraft before attempting Voyager and Galileo. Viking was a promising possibility.

The Viking Orbiter-Quasar VLBI Experiment transpired over the months of July, August, and September in 1976. The Mars and Tidbinbilla dishes alternately recorded signals from the two Viking craft and quasars OL 064.5, PI 148-00, and 3C279. The goal was to determine the probes' location with precision and to link Mars and Earth to the quasar reference frame. The experiment shed a harsh light on the DSN's lack of preparedness for VLBI navigation. At the Mars and Tidbinbilla facilities, the DSN had installed equipment dedicated to acquiring the VLBI data, including a special VLBI S-band receiver, a hydrogen maser, a VLBI recording system, and a VLBI frequency synthesizer. But, the DSN still lacked the hardware necessary to process the VLBI measurements, so navigators had to run the data through Caltech's VLBI correlator.[30]

The DSN upgraded its navigation VLBI capabilities incrementally with the highest priority being assigned to preparation for the Voyager flights to Jupiter. Additional improvements would follow as the probes flew to Saturn. Ultimately, the network was

---

[26] Jordan, interview, 48.
[27] Jordan, interview, 48 & 49.
[28] Jordan, interview, 49.
[29] J. Brooks Thomas, John L. Fanselow, Peter F. MacDoran, Donovan J. Spitzmesser, and Lyle J. Skjerve, "Radio Interferometry Measurements of a 16-km Baseline with 4-cm Precision," 36-54 in DSNPR:XIX; Walter J. Ross, "VLBI Validation Project," 181 & 182 in DSNR 42-39.
[30] D. W. Johnston, Thorl W. Howe, and Gerald M. Rockwell, "Viking Mission Support," 12 in DSNPR 42-37.

not capable of handling navigational VLBI until 1979 with the completion of the so-called Very Long Baseline Interferometer (VLBI) System installed specifically to support Voyager. Among other features, it included station time synchronization to within 10 microseconds of the Goldstone master clock, polar motion measurements to within 50 centimeters, and Universal Time (UT1) to within 1.25 milliseconds.[31]

Those achievements, however, were in the future. In 1977, in preparation for Voyager and Galileo, navigators Jim Miller and Ken Rourke analyzed the VLBI system. Their analysis was encouraging for flights to both Jupiter and Saturn. They calculated that as early as 61 days before Jupiter closest approach, the VLBI would be able to detect small errors in the planetary ephemeris and that as early as 10 days before encounter, it would reduce orbit-determination errors to within 200 km. Their results for Uranus and Neptune were less than encouraging, though.[32] While VLBI measurements added precision to calculations, they also revealed certain inadequacies in navigators' geophysical and astrometric quantities and inaccuracies in station clock synchronization. As navigators achieved higher precision, they encountered again what Tom Hamilton had called the demons of navigation.

Failure. In 1979, navigators made their first VLBI-based attempt at determining the position of a spacecraft exploring the outer planets. The results were discouraging and again brought to light the essential shortcomings of the DSN's VLBI system. The trials entailed making a series of interrelated measurements of Jupiter, a quasar, and a Voyager probe in the hope of linking Jupiter and the craft in the stellar coordinate system with the quasar in the radio reference frame. The experimenters, Robert Preston, Lee Brunn, and others, chose the extragalactic radio source OJ287 because of its proximity to the Voyager flight paths.[33]

They began by constructing an improved Jupiter ephemeris using standard range and Doppler taken from Voyager 1. Armed with this improved ephemeris plus VLBI observations of Voyager 2 and the quasar, Brunn and Preston predicted encounter conditions for Voyager 2 about 30 days before encounter. They additionally acquired VLBI data from Voyager 1 before and after its Jupiter encounter to determine that probe's motion relative to OJ287. They combined these readings with the range and Doppler to make an accurate estimate of the motion of Jupiter with respect to OJ287. They used Voyager 2 pre-encounter VLBI to determine the probe's motion with respect to the radio source. Finally, they predicted the motion of Voyager 2 with respect to Jupiter.[34]

---

[31] Mudgway, *Uplink-Downlink*, 151, 153 & 155; Nicholas A. Renzetti, J. Frank Jordan, Allen L. Berman, Joseph A. Wackley, and Thomas P. Yunck, *The Deep Space Network: An Instrument for Radio Navigation of Deep Space Probes*, 82-102 (Pasadena: JPL, December 15, 1982), 33; Brendan D. Mulhall, "The DSN VLBI System, Mark III-79," 5-15 in DSNPR 42-46; Walter D. Chaney and North C. Ham, "DSN VLBI System MK 1-80," 26-34 in DSNPR 42-56; Gary L. Spradlin, "DSN Tracking System, Mark III-1979," 7-25 in DSNPR 42-56; Mudgway, *Uplink-Downlink*, 155.
[32] James K. Miller and Kenneth H. Rourke, "The Application of Differential VLBI to Planetary Approach Orbit Determination," 84 in DSNPR 42-40.
[33] "ΔVLBI Demo Part I," 111 & 114.
[34] "ΔVLBI Demo Part I," 112.

The experimenters used the same navigational computer (the Univac 1108) and software (a specially-adapted version of the ODP) as the Voyager navigation team. They processed their data in parallel with Voyager operations in "near real time," but always in such a way as not to interfere with flight operations. They used the same Jupiter ephemeris as well as the same astrodynamic constants, timing and polar motion parameters, and other a priori information as those used by the navigation team. The only different inputs were the improved ephemeris of Jupiter relative to OJ287 and the VLBI observations.[35]

Collecting the VLBI was not uncomplicated. A serious drawback was the inability to schedule tracking during the full duration of the view period. Shorter passes provided less accuracy. Pass durations typically were 2.5 hours long instead of 4.5 hours. More losses of accuracy occurred because of the relatively large angular separation (10 to 12 degrees) between the Voyagers and the quasar. The test really needed a radio source closer to the probe. The experimenters had felt that the advantage of using a single radio source throughout the trial outweighed the disadvantage of the large separation. That decision, though, meant that errors from media effects might not cancel well in the differenced data.[36]

Despite the failure to meet its accuracy goal, the experiment did leave navigators with a much clearer understanding of the problems of VLBI navigation, especially the sort of system and data acquisition procedures needed to reach the desired angular accuracies of 0.05 microradians (50 nanoradians). At the very least, the DSN needed to develop the ability to do VLBI with shorter passes. The DSN needed to undertake sweeping changes in how it handled VLBI data, so that it would be available to flight navigators within 24 hours of the relevant observations. The current system took weeks to deliver data tapes from overseas. The results of the Voyager VLBI navigation experiment were disappointing. It was an attempted use of an unproven new data type to support flight project operations.[37] By the time the Voyagers reached Saturn, the installation of a new VLBI system was underway, and an entirely new approach to VLBI had emerged.

---

[35] "ΔVLBI Demo Part I," 113.
[36] "ΔVLBI Demo Part II," 61, 63 & 64; "Innovations in DDOR," 4.
[37] "ΔVLBI Demo Part II," 64; Thornton, interview, 13.

Chapter 15

Optical Navigation

Navigators' quest for higher quality data also led to the development of optical navigation (opnav), a specialized form of navigation that exploits images from spacecraft cameras. Optical navigation offered several advantages over range and Doppler, advantages that made it, like VLBI, ideal for exploring the outer planets. While radio measurements tend to lose accuracy with increasing distance from Earth, opnav improves as the distance from Earth grows and as the probe moves toward its target. After many years of development and testing, optical navigation matured on Voyager.[1] Voyager opnav also contributed to the continuing relationship between navigation and science, as optical navigators discovered volcanism on Io and multiple moons.

Despite what the term may imply, optical navigation differs significantly, but not entirely, from celestial navigation, both of which rely on visual observations of stars. The basic form of opnav generally used an onboard camera—traditionally the same television camera that conducted mission-related science—to take photographs of a target body, such as a moon or planet, against a background that contained reference stars. Unlike celestial navigation, optical navigation also can work from images without the aid of a starry backdrop, often times searching for the center of a planet or other body in an image. Analysis of these pictures yielded a measurement of the relative position (apparent right ascension and declination) of the target body as seen from the craft.[2]

The opnav demo at Mars. The first test of optical navigation occurred during the Mariner Mars 1969 mission. The chief experimenters, Tom Duxbury, Chuck Acton, and Bill Breckenridge, were not in navigation, but in Richard Morris' Guidance and Control Analysis and Integration Section (Section 343). Later, a 1971 laboratory reorganization moved Duxbury to Melbourne's Tracking and Orbit Determination Section (housed in the Crest Building), while Acton remained in Section 343.[3] The separation of Duxbury and Acton into navigation and guidance reflected the unsettled question of where optical navigation should be within the JPL organizational structure. This lack of resolution eventually led to a turf war.

Although this first opnav demonstration would be with a Mars probe, the idea from the start was to prepare for navigating to the outer planets.[4] Duxbury and Acton ran their experiment as a skunk works operation, Dave Curkendall recalled. They

---

[1] "Brief History of JPL OpNav," 333.
[2] Maize, interview, 2-3.
[3] Duxbury, "Data from MM69," 219; "Optical Observables," 5-5; Acton, "Processing Onboard Optical Data," 746; Jordan, interview, 20; [JPL Telephone] "Directory, June, July, August, 1969," x, 1, 5 & 11, NHRC; [JPL, Telephone Directory, 1971], x, 1, 5 & 11, NHRC.
[4] Duxbury, "A Spacecraft-Based Navigation Instrument for Outer Planet Missions," *Journal of Spacecraft and Rockets* 7 (August 1970): 928-933.

followed "the easiest path, the path of least resistance," by calling on the science camera. "This was engineers taking over the scientist's plaything and using it for some good," Frank Jordan explained. "That's the way we looked at it, and they didn't always see it that way, because that was their baby. It competed with other proposers to put that camera on. It was their birthright, and we were using it for engineering. So, there was a little edge there. But it all worked out."[5]

Turning the science camera into a navigation instrument was, as Jordan pointed out, the reverse of what had happened with the DSN. Created for communication and navigation, the DSN ended up serving the purposes of science. In contrast, the Mariner camera, designed for scientists, wound up serving navigation. "This is poetic," Jordan reflected. "It's a wonderful story, I think, of how science and engineering benefited from one another."[6]

Wresting useful positional information from television camera pictures for navigational purposes was a complicated, labor-intensive process. Their approach was to use a combination of photographs and telemetry. The Mariners sent the television images to Earth in a digital format, and ground-based processing recorded them on magnetic tape. Duxbury and Acton wrote a program that searched the tapes for the lit limb of the Mars image and determined the center of Mars from the lit limb. The program next converted the center location into the direction to the center of Mars within the camera's coordinate system. As with all navigational data, the optical observations were susceptible to errors. For example, the camera distorted the geometry of Mars images. The experimenters calibrated for those distortions by testing the camera before launch.[7]

Once the images were processed, the next step in preparing the navigation data utilized the spacecraft telemetry to calculate the attitude of the Mariner and its camera and to determine the direction to Mars expressed in terms of right ascension and declination. The telemetry contained a treasure trove of information about the probe's attitude from the far-encounter planet sensor, the scan platform, and the attitude-control sensors. The positional measurements of right ascension and declination then entered the navigation computer program, which added them to the latest Doppler-based trajectory estimate and updated the orbit.[8] This, of course, is a highly simplified description of the optical navigation process.

Because of the experimental nature of the opnav demonstration, Mariner Mars 1969 project management expected that it would interfere minimally with routine flight operations. Processing optical data on mission-assigned computers could take place only during periods when operations normally were idle. Duxbury, Acton, and Breckenridge made their trajectory computations available to the head of the navigation team for his use only. In turn, the mission provided flight path estimates as well as Doppler, telemetry, and television pictures at scheduled times prior to encounter. The time

---

[5] Duxbury, "Data from MM69," 219; "Optical Observables"; Acton, "Processing Onboard Optical Data," 746; Jordan, interview, 20; Curkendall, interview, 36-37.
[6] Jordan, interview, 20.
[7] Duxbury, "Data from MM69," 219, 220 & 225; "Optical Observables," 5-1; "Brief History of JPL OpNav," 330.
[8] "Optical Observables," 5-1 & 5-2; "Brief History of JPL OpNav," 330 & 331.

required to estimate a trajectory from the raw data was around two hours. Most of that time (about 1.5 hours) went into finding the location of the center of Mars in the photographs. Duxbury and Acton measured the location of Mars in each picture by putting a clear plastic overlay on top of a hard copy of the image, matching a circle on the overlay to Mars and reading off the coordinates. Two individuals measured the center point of each picture, and the team averaged the two measurements. Because the idea was to demonstrate the feasibility of spacecraft-based data for flight operations, they processed the telemetry and television data as they became available and made trajectory estimates "almost" in real time.[9]

A key aspect of the test was the software for handling the visual information and integrating it into the flight navigation software. The first step was to convert the Mars image centers obtained by applying the clear plastic overlay to punched cards. At this point, the three different opnav computer programs ran on the IBM 7094 for about 30 minutes. These programs were prototypes of later, more sophisticated versions. Two programs computed corrections to the spacecraft orbit. One was the Radio Optical Orbit Determination Program (ROODP); the other was the Optical Observable Processor Program (OOPP). The Optical Observable Generator Program (OOGP) created data for use by the Optical Observable Processor Program. These programs necessarily were compatible with the standard navigation software.[10]

After encounter, the team assembled a complete set of data that filled in the gaps and thereby extended the data obtained during encounter. However, the degradation of the Mariner 7 telemetry channels, attributed inventively to the Great Galactic Ghoul, precluded analysis of that portion of the Mariner 7 flight. Still, the experiment met all of the objectives that Duxbury, Breckenridge, and Acton had set out to meet. They showed that one could apply spacecraft-based data to flight navigation. All the same, it was clear that the technique required improvement in certain areas.[11]

Turning to the stars. Mariner 9 (Mariner Mars 1971) was the occasion for a second test of optical navigation. The experimenters still had their sights set on the "stringent navigation accuracy requirements" for outer-planet flights. The Mariner 9 demo was an organizationally larger and more formal experiment funded by the NASA Headquarters Office of Aeronautics and Space Technology. The Optical Navigation Demonstration Team, to use its formal name, came into existence in early 1970, with members from both guidance and control and navigation sections. The Guidance and Control Section (343) contributed Chuck Acton, Bill Breckenridge, Marvin Bantell, Jr., and Hiroshi Ohtakay. Duxbury, George Born, Navin Jerath, Mohan Ananda, Srinivas Mohan, and Gary Ransford came from navigation. Later, several of them would be on the Voyager opnav team. A suggestion made at the end of the previous trial was to incorporate staff from JPL's Space Sciences Division, which included the Space Photography and Science Data Analysis Sections. Adding Space Science staff would have brought to bear the

---

[9] "Optical Observables," 5-8, 5-9, 5-11 & 5-13.
[10] "Brief History of JPL OpNav," 330; "Optical Observables," 5-11 & 5-20; *MM69 Final Report*, Vol. I, 394
[11] "Optical Observables," 5-13, 5-15, 5-17, 5-19 & 5-24; "Brief History of JPL OpNav," 330.

Division's broad expertise in such areas as imaging instrument performance and processing and displaying imaging data. The opnav team agreed that future missions should consider special image display products supplied potentially by JPL's own Image Processing Laboratory. The Optical Navigation Demonstration Team functioned as part of Mariner 9 operations, but again without interfering with normal flight activities. The opnav programs became part of the official suite of project software. The only obligation that the mission placed on the opnav team was to supply their results on a "best-effort basis."[12]

The Mariner 9 opnav demonstration tried out a new approach, but one that still relied on onboard vidicon images and telemetry. Navigators acquired images of the planet's moons, Deimos and Phobos, against a starry background and reconstructed the celestially-referenced directions to Phobos and Deimos. Also, they processed the images in terms of pixels and scan line numbers along with spacecraft attitude information from the telemetry.[13]

The optical navigators visually identified the positions of the Martian moons and background stars and measured them from computer printouts. They lacked the software to produce or the authority to request hardcopy photographs of the digital images in real time. Therefore, they depended on those produced normally by the Mission Test Computer, which created hardcopy image displays. This visual method proved effective in detecting most star images including some of the dimmer stars. Nonetheless, in order to use these display copies effectively, the optical team had to generate overlays using experimental software to remove television pointing, distortion, and other errors from the pictures.[14] This visual (eyeball) approach gave a new meaning to optical navigation.

The demonstration did not go off without a hitch, as one might expect with any test run. Several of the glitches threatened the application of optical navigation in near real time. For example, there was a delay of 4 to 6 hours in obtaining the digital imaging data on tape and another one hour delay in copying the data from tape into the navigation mainframe. The need for better image-display technology and software was evident. Dealing with Phobos and Deimos posed special problems of their own. Optical navigators repeatedly had to update their ephemerides during the flight, yet ephemeris errors continued to haunt the endeavor's success. Most picture attempts succeeded, but a few Phobos images were lost because of a software fault coupled with errors in the moon's ephemeris. Some ephemeris errors arose from an inaccurate model of the Martian gravity field.[15]

The dedication of the onboard camera to scientific research first and foremost also was no benefit to optical navigation. The pictures of Deimos and Phobos taken

---

[12] *MM71 OpNav Demo*, iv, 1-2, 4 & 5; Duxbury and Acton, 296.
[13] Breckenridge and Acton, "Detailed Analysis," F-2 & F-3; Duxbury and George H. Born, "Tracking Phobos and Deimos Aboard an Orbiting Spacecraft," D-5 in Appendix D, *MM71 OpNav Demo*; *MM71 OpNav Demo*, 1; Duxbury and Hiroshi Ohtakay, "In-Flight Calibration of a Navigation Instrument," Appendix B, *MM71 OpNav Demo*; Acton, "Processing Onboard Optical Data," 749-750.
[14] *MM71 OpNav Demo*, 4, 5, 7, 8 & 12; Duxbury and Acton, 295-307; Acton, "Processing Onboard Optical Data," 749.
[15] *MM71 OpNav Demo*, 5, 16, 20 & 21; Acton, "Processing Onboard Optical Data," 748-750.

during the Mariner 9 orbital phase had exposure times that were too short to capture star images in the background. However, the demonstration did show the feasibility of overexposing moon images to detect reference stars. The last picture from Mariner 9, taken at the request of the Voyager project office, was an overexposed image of Phobos. Ten stars were visible; picture distortions from overexposure were small. Consequently, Voyager resolved to acquire overexposed pictures of the Galilean moons of Jupiter (Callisto, Europa, Ganymede, and Io) to detect background stars.[16]

Tom Duxbury (left) and Chuck Acton receiving Burka Award (Courtesy of JPL)

The Optical Navigation Demonstration also tried out another new technique that held promise for probes nearing a target body. It determined the positions of Martian landmarks in topocentric coordinates. This landmark method gave optical navigators good estimates of the Mariner's position and velocity during both approach and orbital phases.[17] In their post-flight analysis, the optical navigators also tested an updated version of their initial scheme involving the lit limbs and center of Mars.[18]

---

[16] *MM71 OpNav Demo*, 16 & 17.
[17] Srinivas N. Mohan and Gary A. Ransford, "Cartography and Orbit Determination from a Mars Orbiting Spacecraft," E-2 in Appendix E, *MM71 OpNav Demo*; Mohan, "Results of Orbit Phase Navigation Study Using Landmark Data," Appendix Q, *MM71 OpNav Demo*; *MM71 OpNav Demo*, 22.
[18] *MM71 OpNav Demo*, 23 & 24; Duxbury, "Data from MM69," 219-225; Duxbury and Acton, 296.

The Mariner 9 demonstration laid the basis for future optical navigation in a number of ways. It developed camera models, data processing techniques, and team procedures that, in one way or another, remained in use. The basic software persisted, although some program names changed and, of course, the code underwent several rewrites. The demonstration's success motivated flight projects to commit to its use, including Viking, Voyager, and the Jupiter Orbiter Satellite Tour (known today as Galileo).[19]

For their efforts in pioneering optical navigation, NASA singled out Duxbury and Acton to receive the agency's Exceptional Scientific Achievement Medal. They also received wider recognition from the navigation community. In 1973, the Institute of Navigation gave them its Samuel M. Burka award for "outstanding achievement in the preparation of papers contributing to the advancement of navigation and space guidance." The award, presented in a ceremony held in San Diego in 1974, included the "princely sum of $175 each."[20]

Voyager dress rehearsal. The success of the Mariner 9 demonstration was a watershed moment. Mariner Venus-Mercury 1973 (Mariner 10) and Viking became dress rehearsals for operational optical navigation on Voyager for several reasons. Navigators utilized images of Mercury against a starry backdrop in the belief that these images would be "similar in nature" to those expected on a flight to Jupiter and Saturn. The camera was "essentially the same design" as the one that the Voyager mission planned for science and navigation.[21]

A critical goal of the Mariner 10 test was to demonstrate the ability to acquire and process optical data during the Mercury encounters in "near-real time," thereby "simulating a mission critical environment." Optical navigators returned to their original limb-fitting methods with updated software that could search automatically for stars. In most cases, though, detecting stars was still a manual (eyeballing) procedure, because the data frequently were too noisy, or the star images were too weak, for automatic detection.[22]

The number of opportunities to perform optical navigation during the mission was highly limited. The entire basis for the optical navigation experiment consisted of just 14 pictures taken during the probe's first departure from Mercury and another 30 from its second approach to the planet. A certain degree of inefficiency dogged the demonstration. Delays of several days in obtaining videotapes marred the first Mercury encounter. Those delays were just four hours during the second encounter, permitting a stab at "near-real-time" optical navigation. Still, the television pictures were extremely noisy and made star detection difficult and time consuming. Further aggravating factors were significant uncertainties in predicted star locations and the loss of two hours of

---

[19] *MM71 OpNav Demo*, 3.
[20] "Brief History of JPL OpNav," 332; Institute of Navigation, "Burka Award," http://www.ion.org/awards/burka.cfm (accessed November 3, 2009).
[21] Acton and Ohtakay, 2; "M10 OpNav Demo," 1; Dunne and Burgess, 94.
[22] Acton and Ohtakay, 3 & 8; "M10 OpNav Demo," 1 & 3.

computing time, when computer maintenance took place in the middle of the experiment.[23]

If optical navigation were going to succeed in abetting the exploration of the outer planets and their satellites, it had to be capable of detecting faint stars. Voyager estimated that its cameras needed to be able to discern magnitude 9 stars, when the probe photographed Titan during its Saturn approach between 30 and 5 days before encounter. Mariner Venus-Mercury 1973 did not detect magnitude 9 stars in any of its pictures, and magnitude 8 stars were only marginally visible, although in principle the camera could see magnitude 8.5 stars. Failures in the vidicon heaters had degraded camera sensitivity and had required longer exposures to detect dimmer stars, which inevitably caused images to smear and reduced navigators' ability to locate stars.[24]

File size was yet another limiting factor. By that day's standards, optical data files were relatively mammoth in size, about 6×10^6 bits per picture. Today, handling 6-megabit images is not a challenge. The state of the art in computer technology posed additional problems. Navigators characterized the video-processing software as being "I/O bound," meaning highly limited by input and output formats. These limitations made the software "nearly inoperable" in an environment of time-shared computing.[25]

Turf battle. The only two participants in the Mariner 10 demonstration who were not from Guidance and Control were Duxbury and Acton, both of whom were in navigation. Guidance and Control wanted optical navigation to be its bailiwick. That Section basically wanted to build flight equipment, and optical navigation gave it an opportunity to promote the design and use of special-purpose cameras. As Frank Jordan explained, "That's workforce, that's money, that's all those good things, building their stuff." "Because so much power went with" putting hardware on a spacecraft, Curkendall added, "the guidance and control people wanted their own sensor. Then they could control the whole thing." Flight projects, however, had been turning them down. In contrast, navigators were not interested in making equipment. They piggybacked on the science camera or whatever other camera might be available. "In our organization we were interested in results, not putting hardware on a spacecraft," Jordan clarified. "It's a different culture."[26]

The turf clash over optical navigation soon came to a head. Dick Stanton represented Guidance and Control in its push for onboard optical sensors. "But right away, we quelched that," Curkendall related. "Then there was a big political fight." The navigators argued for using available equipment—the science camera—rather than designing and building a specialized instrument. Guidance and Control wanted to design, build, and control its own instrument, and they wanted to do navigation with it. The resolution came during what Curkendall termed "a big come-to-Jesus meeting" between Guidance and Control and navigation. Bud Schurmeier, then the Assistant Laboratory Director, adjudicated the imbroglio. In the end, optical navigation went to navigation.

---

[23] Acton and Ohtakay, 2; "M10 OpNav Demo," 7-9.
[24] Acton and Ohtakay, 4; "M10 OpNav Demo," 3, 4, 7 & 10.
[25] Acton and Ohtakay, 8-9.
[26] Acton and Ohtakay, 1; Jordan, interview, 21; Curkendall, interview, 36.

Henceforth optical navigation was the purview of JPL navigators, who would adapt the science camera to their purposes. The decision also led to the creation of a separate group within the navigation area devoted exclusively to optical navigation.[27]

Viking: more rehearsing. The composition of the Viking opnav team reflected the definitive assignment of optical navigation to navigation. Ken Rourke, Navin Jerath, Acton, Breckenridge, Jim Campbell, Carl Christensen, Joe Donegan, Howard Koble, Neil Mottinger, George Rinker, and Bryant Winn all, with the exception of Breckenridge, were in Frank Jordan's Navigation Systems Section.[28]

Viking would be a rehearsal for outer-planet exploration. Viking project managers added opnav as a relatively inexpensive data complement with the potential to improve navigational accuracy. Ultimately, the resulting increased accuracy simplified mission operations, reduced fuel requirements, and benefited scientific goals. Viking optical navigation used the vidicon cameras—capable of sensing stars as dim as magnitude 9.5—to photograph Mars or one of its moons against a starry background. Navigators acquired two types of pictures. One consisted of stars-Mars-stars triads, used when the image of Mars was smaller than the field of view. The other type, for when Mars was larger than the field of view, consisted of Deimos against a starry backdrop.[29]

One of the keys to successful Viking optical navigation was the management of operations. Several organizations within the Viking flight team were involved in scheduling and handling navigation pictures. The processing of a large quantity of images occurred relatively easily as a result of two important decisions: 1) to allow those handling the optical data to manage the acquisition of the data; and 2) to give priority to navigation pictures over all other concurrent requests for image processing.[30]

Viking optical navigation was an undeniable success. The high accuracy of the resulting approach orbit estimates and the orbital insertion maneuver allowed the Viking 1 orbiter to fly over the proposed landing site on its first orbit, thereby rendering unnecessary two weeks of contingency operations. A small fuel leak became evident on Viking 1 sometime before Mars orbit insertion. The concern was that pressure in the propellant tanks might build up to a dangerous level prior to the long burn for orbit insertion. Therefore, the probe performed two separate maneuvers before orbit insertion. The ability to carry out both burns successfully without large errors appearing in the computed trajectory resulted from the application of optical data. Additionally,

---

[27] Curkendall, interview, 37; Cangahuala, interview, 8 & 33; "Brief History of JPL OpNav," 333; "Voyager Nav Strategy," 8.

[28] "Brief History of JPL OpNav," 332 & 333; Jordan, interview, 32-33; Kenneth H. Rourke, Charles H. Acton, Jr., William G. Breckenridge, James K. Campbell, Carl S. Christensen, Arthur J. Donegan, Howard M. Koble, Neil A. Mottinger, George C. Rinker, and F. Bryant Winn, "The Determination of the Interplanetary Orbits of Vikings 1 and 2," AIAA 77-71, 15th AIAA Aerospace Sciences Meeting, January 24-26, 1977, Los Angeles, CA.

[29] "Viking Interplanetary OD," 51, 55, 58, 73, 74 & 75; "Brief History of JPL OpNav," 332 & 333.

[30] "Viking Interplanetary OD," 74.

optical navigation eventually enabled close encounters with both Martian moons. The Deimos flyby at a distance of just 20 km was off by less than 2 km.[31]

MODCOMP mania. Voyager was a turning point in the evolution of optical navigation, when opnav went from an experimental or supplementary activity to a mission requirement. Optical navigation would show its mettle in helping to guide the Voyagers along their elaborate trajectories among the satellites of Jupiter and Saturn. However, despite all the preparation, optical navigation at Jupiter was disappointing. The limb and terminator of the gas giant were difficult to measure accurately. The thick atmosphere failed to produce a sharp boundary. "It was pretty much a wash at Jupiter," Frank Jordan reflected. Consequently, optical navigators focused only on satellites as targets. That is where the optical data really showed its value. As the Voyager probe flew past Io in 1979 at a distance of about 20,000 km, optical navigator Steve Synnott recalled: "you really needed to have the optical data pin down the ephemeris of Io in its orbit, as well as to get the spacecraft trajectory estimated with respect to the center of mass of Jupiter."[32]

Optical navigation on Voyager differed from previous practices in several ways. One was the shift of image processing to the optical navigation team from the JPL Space Sciences Division's Image Processing Laboratory that had provided printouts for navigators to analyze. The latter's software eliminated blurs, smears, and other errors. Such computer programs are rather common today, but back then, Curkendall explained, "it was pretty exotic." Moreover, the software ran on PDP (Programmed Data Processor) computers. "But there wasn't anything interactive about it," Curkendall elaborated, "it was all batch-processed."[33]

Curkendall and Acton sold the Voyager project on the idea of developing interactive image processing, "so that we could do this fast, right? That was what sold it." However, Curkendall added, "what we wanted to do was to use all the routines that had been actually developed over there" in the Space Science Division. It would be a relatively simple matter of transferring the software from one PDP to another PDP computer. The problem was that the DSN and the flight operations facility were now running MODCOMP computers.[34]

As a result, Curkendall recalled, "there was a big meeting" in which he and Acton insisted that the software had to be on PDPs: "We don't need no stinking MODCOMP computers." "So we laid the gauntlet," Curkendall continued, and told them how much it would cost. Their response was: "That's not much." "He just put a pinprick in our balloon. It was for good reasons really. To bring a foreign computer into the situation where they had probably 150 MODCOMPS chomping away would have been a mess. So we had to develop this bloody thing on a MODCOMP computer." The optical navigation group acquired a pair of MODCOMP IV computers for image processing and rewrote their image-processing programs. Now they could stretch

---

[31] "Brief History of JPL OpNav," 332-333; Euler, "Viking Mission Overview," 54 & 56.
[32] Jordan, interview, 45; "Brief History of JPL OpNav," 333; Synnott, interview, 5, 7 & 10.
[33] *MM71 OpNav Demo*, 5 & 8; Curkendall, interview, 38-39.
[34] Curkendall, interview, 39.

images with a joystick as part of limb fitting. "It was all interactive, all with joysticks," Curkendall explained.[35]

This huge bullpen area. As usual, opnav began with planning which pictures to take and when to acquire them, an activity that occupied a third of navigators' time. The critical issue was picking a time when a camera could see both a moon and enough stars to enable navigators to remove pointing errors. Once Voyager acquired and relayed images to the DSN, they made their way to the optical navigation team, who initiated their image-processing procedures.[36]

The workload stepped up dramatically in February 1979, as Voyager was closing in on Jupiter. Linda Morabito, a member of the optical navigation team, recalled "working a minimum of 14 hours a day.... I would get up with the Sun and would cross the laboratory to Building 264, feeling the excitement building as I got to watch Jupiter itself as the spacecraft was approaching it.... By February 1979, the data was falling down on us like rainfall and the images were coming in at all hours of the day and night." The gray-scale digitized images of Jupiter's moons against starry backgrounds arrived at the optical navigation area on large data tapes having been carried there by hand. After finding the image centers, Morabito sent them to Jim Campbell, who was in charge of the radio tracking data, and to Steve Synnott, who oversaw the optical data.[37]

The optical navigators, Synnott recalled, looked at digital images "in what at the time was a room-filling computer." It had "big TV screens that were the display devices. But yes, oh, we would actually see the raw pictures.... It almost always turned out that there were things very close to where they should be." Still, navigators could never anticipate the slight bobbing and weaving of the probe that occurred in three-dimensional space. As a result, they would have to search visually for the location of stars and other objects in the pictures. The improved software for automatically locating image centers allowed navigators to "press a button, and the computer would go off and find the little lump of light that was the star and center it as best it could."[38]

Although integrated organizationally into the flight navigation team, physically the optical navigators were in a different building. Nonetheless, Synnott described Voyager navigation as "one big team . . . very intimately connected and run under one navigation team chief, one head."[39]

Linda Morabito described the Voyager optical navigation area as being "this huge bullpen area." The imaging-processing MODCOMP "had its own corner, partitioned off with glass walls. Remember, this was in the era when a mini-computer like this could fill an entire room and had to be cooled. So when I would walk into the ONIPS area there

---

[35] Curkendall, interview, 39 & 40; "Brief History of JPL OpNav," 333.
[36] Synnott, interview, 5 & 10-11.
[37] "Linda Morabito Kelly."
[38] Synnott, interview, 12; Jean J. Lorre, "Automatic Technique for Accurately Locating Planet Centers in Voyager Images," 175-179 in Carol Clark, ed., *Image Understanding Systems II* (Bellingham, WA: Society of Photo-Optical Instrumentation Engineers, 1980).
[39] Synnott, interview, 9 & 10; "Linda Morabito Kelly."

would be this blast of these air conditioners from below the floor. It was cold, noisy, and an interesting environment in which to work."[40]

Over the years, many people participated in the Voyager optical navigation team. At one time or another, in addition to Synnott and Morabito, they included Chuck Acton, Joe Donegan, Adriana Ocampo, Bill Owen, Jr., Ed Riedel, Juli Stuve, Homa Taraji, and Robin Vaughan. All in all, Synnott reflected, "it was a small group." A departure from past practice in deep-space navigation was the addition of a team member, Joe Donegan, who was primarily responsible for the software.[41]

Seeing volcanism on Io. Voyager marked the entry of optical navigators into scientific discovery. Optical navigators already had made contributions to celestial mechanics. The opnav team on Mariner Mars 1971, for instance, had calculated values for the mass of Mars and the spherical harmonic coefficients of its gravity field and had contributed to an improved estimation of the direction of Mars' spin axis that Earth-based astronomy soon confirmed.[42] Perhaps the best-known (and certainly the most publicized) example of optical navigation's contribution to planetary science is the discovery of volcanism on Io by Linda Morabito, a member of the image-processing team whose sole purpose was to find the centers of pictures.[43]

On one of the brief occasions when she was able to go home, Morabito recalled watching the Science Imaging Team present the first close-up images of Io at a broadcast press conference. "What I was seeing in those images was so unlike anything we had ever seen in the Solar System before. I remember gazing at some of the images and weeping from the sense of what was being discovered." One color image of Io caught her attention. It "showed this unusual heart-shaped feature on the surface. It was so remarkable it was almost shocking. I had expected yet another dead, cratered moon, and yet this seemed alive. It moved me deeply and it's something I remember vividly."[44]

Morabito arrived at work on the morning of March 9, 1979, four days after Voyager 1 began its departure from Jupiter, and began processing several of the recent Voyager 1 images. She manipulated one of the images to look for a dim star and:[45]

> suddenly noticed an anomaly to the left of Io, just off the rim of that world. It was extremely large with respect to the overall size of Io and crescent-shaped. It seemed unbelievable that something that big had not been visible before, but my linear stretch popped it into view. It was a moment that every astronomer, every planetary scientist lives for. When you see something like that it evokes the deepest questions of your scientific interest. I have absolutely no recollection even to this day of the

---

[40] "Linda Morabito Kelly."
[41] Synnott, interview, 10; "Brief History of JPL OpNav," 333.
[42] *MM71 OpNav Demo*, 26-27; Andrew T. Sinclair, "The Motions of the Satellites of Mars," *Monthly Notices of the Royal Astronomical Society* 155 (1972): 249-274.
[43] "Linda Morabito Kelly."
[44] "Linda Morabito Kelly."
[45] "Linda Morabito Kelly."

star that I was looking for at the time. . . . [Morabito sensed that she] was seeing something that no one else had seen before. Without verification, it was only a sense, but I knew what I was seeing was that it was extremely important. Those moments were the stuff of dreams. They passed quickly as I dug in to determine what this anomaly was.

Morabito, who had studied astronomy at the University of Southern California, related that her "instincts as a scientist took over." "I wanted to know what this was. I immediately began considering each and every possibility. Was it real or not?" She and others on the optical navigation team "considered all possibilities, including that of a newly discovered satellite."[46]

Over the next six hours, Morabito, Synnott, and others eliminated every single possible glitch "until only one possible explanation remained—the anomaly was correlated with the surface of Io." This hypothesis crystallized as Morabito, Synnott, Duxbury, and members of the JPL Imaging Systems Section—Andy Collins and Peter Kupferman—reviewed the possible anomaly sources. They concluded that it must have something to do with Io. But, because the glitch was so large, they found it difficult to accept the notion that it had anything to do with the surface, although the image quality provided no surface details.[47]

Among scientists, volcanism on Io was the subject of an article in the journal Science that had appeared a week earlier on March 2, 1979. Stanton J. Peale, a professor at the University of California, Santa Barbara, along with Patrick Cassen and Raymond T. Reynolds of NASA Ames, argued that volcanism would arise from the tidal heating caused by the combined pulls on the solid interior of Io by the nearby moon Europa and the planet Jupiter.[48]

Morabito noted that the appearance of the Io image anomaly was consistent with "the way sunlight might illuminate a gas shell or volcanic plume." She and Synnott began correlating the anomaly with the moon's surface features and determined that the anomaly's point of origin appeared to match "very nearly with the large heart-shaped feature on Io." Morabito and Kupferman continued working the puzzle, when she remembered seeing the heart-shaped feature in a color image that the Science Imaging Team had presented during a televised press conference. The meaning of the anomaly dawned on her.[49]

On Saturday morning, as Morabito was processing more images, Ed Stone, the Voyager Project Scientist and future JPL director, and Bob Parks, the Voyager Project Manager, asked to see the image. On Monday morning, verification of the discovery came, after scientists went back over the science data and "realized that the volcanic plumes had been everywhere but had not been recognized." The two volcanic plumes

---

[46] "Linda Morabito Kelly."
[47] "Linda Morabito Kelly."
[48] Peale, Cassen, and Reynolds, "Melting of Io by Tidal Dissipation," Science 203 (March 2, 1979): 892-894; Charles F. Yoder and Peale, "The Tides of Io," Icarus 47 (July 1981): 1-35; Carl Sagan, Cosmos (New York: Random House, 1980), 126.
[49] "Linda Morabito Kelly."

discovered in the picture of Io comprised the first evidence of active volcanism on any body in the solar system other than Earth.[50]

The "Morabito incident." The discovery of volcanism on Io by a navigator, not by one of the project scientists, did not follow the procedures that Ed Stone had laid out for the announcement of discoveries. Announcing discoveries almost as they occurred, not to mention the aggregation of scientists into working groups, raised important questions of intellectual property rights and priority of discovery. Stone believed that scientists should share their discoveries. The initial publication of results also followed this group approach. Accordingly, Stone explained, "everybody had equal priority, because everybody was there at the same time."[51]

These measures accomplished their goal of protecting intellectual property and discovery rights. However, Morabito's discovery of volcanism "apparently embarrassed the doctoral scientists who had overlooked it," according to Eric J. Chaisson, a professor at Tufts University and Director of its H. Dudley Wright Center for Innovative Science Education. The episode became known as the "Morabito incident," and, Chaisson argues, had a direct impact on the organization of science on the Hubble Space Telescope project.[52]

The "Morabito incident" also reflected some of the friction that had arisen between some navigators and project scientists. Curkendall reflected on how the camera imaging team kept the navigators at a distance. "I remember trying to worm closer into their hair. Nyet." Consequently, Curkendall "took some pleasure in us discovering the volcanoes instead of them, because they'd really not frankly been very cooperative." Voyager scientists, in contrast, were willing to accommodate picture-taking requests at specific times when the stars were favorable. "We were running circles around the science team," Curkendall declared. "We got some kudos at the time, but it raised the barriers."[53]

Finding rocks. Such sentiments did not reflect the experience of everyone on the optical navigation team. Steve Synnott worked closely with project scientists throughout the mission. Moreover, it was that very same collaboration that led to his discovery of new moons in orbit around Jupiter and Saturn. Synnott credited his first discovery to Bradford A. Smith, a University of Arizona professor who led the imaging team. As

---

[50] "Linda Morabito Kelly"; Linda A. Morabito, Stephen P. Synnott, Peter N. Kupferman, and Stewart A. Collins, "Discovery of Currently Active Extraterrestrial Volcanism," *Science* 204 (June 1, 1979): 972. For subsequent scientific discussion of volcanism on Io, see Michael H. Carr, Harold Masursky, Robert G. Strom, and Richard J. Terrile, "Volcanic Features of Io," *Nature* 280 (August 30, 1979): 729-733; Allan F. Cook, Eugene M. Shoemaker, and Bradford A. Smith, "Dynamics of Volcanic Plumes on Io," *Nature* 280 (August 30, 1979): 743-746; Robert G. Strom and Nicholas M. Schneider, "Volcanic Eruption Plumes on Io," 598-633 in David Morrison, ed., *Satellites of Jupiter* (Tucson: University of Arizona Press, 1982).
[51] Stone, interview, 21-24.
[52] Eric Chaisson, *The Hubble Wars: Astrophysics Meets Astropolitics in the Two-Billion-Dollar Struggle over the Hubble Space Telescope* (New York: Harper-Collins, 1994), 102.
[53] Curkendall, interview, 39 & 50; "Brief History of JPL OpNav," 333.

Voyager was approaching Jupiter, project scientists wanted to take a picture of Amalthea, a moon much smaller than Io orbiting quite close to Jupiter. Smith told the navigators: "You can turn around and look at that picture right there of Jupiter. It [Amalthea] will look like a little lump of charcoal, because it's so dark compared to [Jupiter].... You will see it against the cloud tops." So, navigators looked for a "big lump of charcoal, just a black hole, a dark depression" against Jupiter's cloud tops. That became a way to look for other small black or very dark satellites. They were not visible against a starry background, but were discernible when imaged against Jupiter's bright clouds.[54]

In another instance, scientists saw an anomalous object trailing in a direction different from the stars in the same image of Jupiter taken with a long, 11-minute exposure. What was it? David Jewitt and G. Edward Danielson, two Caltech professors on the imaging team, asked Synnott to compute the object's orbit. Synnott had two pictures of the object taken three or four minutes apart. After performing the orbit determination, Synnott and the scientists concluded that the "rock" was a satellite of Jupiter. Discovered July 8, 1979, Adrastea, initially known as Jupiter XV, is the second satellite from Jupiter and the smallest of its four inner moons. Jewitt and Danielson shared discovery credit with Synnott.[55]

Synnott did not expect to find other satellites. Looking at Jupiter images, he took advantage of the "huge number of frames" of the moving cloud patterns taken to see the Red Spot evolve to study the planet's meteorology. Optical navigation also underwent an upgrade that allowed users to recognize new bodies and to compute their orbit,[56] making it easier to discover new satellites. Synnott tried looking for Adrastea using Brad Smith's suggestion, but instead found two more: Metis, discovered March 4, 1979, and Thebe, discovered March 5, 1979. "You can imagine what you'd see would be a picture with a lot of the cloud tops and stuff. Against it would be this little black hole. Well, I couldn't find Adrastea, but something bigger popped up." He knew it was not Amalthea. "That was Thebe. That was the bigger outer one."[57]

At Saturn, Synnott continued to seek out and find new satellites. The prolongation of the Voyager mission to Uranus and Neptune furnished additional opportunities for discovery. "I did that quite a bit at Uranus and Neptune. I forget how many I wound up with at Uranus. Maybe a half a dozen. Two or three more at Neptune. So yes, that was fun," and it was certainly one of the high points of his career.[58]

---

[54] Synnott, interview, 13-14.
[55] Jewitt, Danielson, and Synnott, "Discovery of a New Jupiter Satellite," *Science* 206 (November 23, 1979): 951; Synnott, interview, 14.
[56] Synnott, interview, 15; "Brief History of JPL OpNav," 334.
[57] Synnott, "1979J3: Discovery of a Previously Unknown Satellite of Jupiter," *Science* 212 (June 19, 1981): 1392; Synnott, "1979J2: The Discovery of a Previously Unknown Jovian Satellite," *Science* 210 (November 14, 1980): 786-788; Synnott, interview, 15.
[58] Synnott, "Evidence for the Existence of Additional Small Satellites of Saturn," *Icarus* 67 (August 1986): 189-204; "Brief History of JPL OpNav," 334; Synnott, interview, 16. Synnott is credited with discovering Cressida, Desdemona, Juliet, Portia, Rosalind, Belinda, and Puck, the moons of Uranus, and the Neptunian satellites Despina, Galatea, and Proteus: "On this Moon and Others," http://www.inconstantmoon.com/cyc_moon.htm (accessed November 3, 2009).

Figure 5. Joe Donegan, Ed Travers, Linda Morabito and Steve Synnott in the ONIPS room with the Modcomp IV minicomputer, 1979.

Voyager optical navigators (from left to right) A. Joseph Donegan, Edwin S. Travers, Linda A. Morabito, and Stephen P. Synnott and their MODCOMP minicomputer (Courtesy of JPL)

Part Four

Back to Earth

Chapter 16

Multi-Mission, Multinational Navigation

The decade of the 1980s saw fundamental transformations in international politics, space policy, and technology. Changes in the national security crisis and space policy threatened to derail navigation and the planetary exploration that it served. Space became an increasingly militarized and commercialized place. Congress slashed solar-system exploration. The Space Shuttle embodied the period's shifting emphasis from exploration to the exploitation of space, specifically near-Earth space. Navigation was not impervious to the terrestrial turn.

JPL navigators now dealt with Earth-circling probes. They adopted a new data type created by the Global Positioning System (GPS), established explicitly to serve Cold War ends. Along the way, GPS brought about major fundamental changes in navigators' software. The consolidation of Earth and space led to navigators, like the DSN, being "for hire" by foreign space agencies. In addition to assimilating themselves into international scientific endeavors, JPL navigators more and more operated in international settings as members of multinational space endeavors. Those missions, moreover, reflected scientists' growing attention to asteroids and comets.

The decade of the 1980s also witnessed continuing efforts to realize the ideal "direct" measuring method. GPS measurements held out promise as a new self-calibrating data type, just like VLBI. Despite the invention of a new and better form of VLBI, and despite plans to create a subnet dedicated solely to VLBI, the DSN lacked the hardware and procedures to make VLBI navigation practical. Only after a prolonged campaign of upgrades did the DSN begin to deliver VLBI data timely and reliably. Meanwhile, differenced data types, especially differenced Doppler, competed with VLBI for DSN resources and the attention of mission managers.

The "revolution in computing" was equally as transformative of deep-space navigation as the terrestrial turn. The advent of the minicomputer and workstation heralded the dénouement of the mainframe era. For the first time, navigators were freed of the restraints imposed by the old machines. They could update models during flights, thereby increasing navigational accuracy. Computer time seemed to be free; there was no longer any need for the computer section to charge for time. On the down side, all of the navigation software code had to be rewritten.

Another navigation-transforming technology, but perhaps less loudly heralded, was the Charge-Coupled Detector (CCD) camera. Optical navigation, one of the rising stars of the "direct" method, relied on the features of the onboard science camera. The CCD camera had none of the failings of its analog predecessors and became standard spacecraft equipment. It enabled a quantum leap in optical navigation improvements and subsequently made possible the introduction of autonomous navigation.

National security crisis redux. The election of Ronald Reagan in 1980 marked a sea change in the country's conduct of the Cold War. Following a decade of entente cordiale and cooperation between the United States and the U.S.S.R, the Cold War heated up only to end with the collapse of the Soviet Union and its partial reorganization in 1991 as the Commonwealth of Independent States. The original bilateral space agreements between the United States and the Soviet Union expired in 1982, a victim of the revived Cold War hostilities.

Reagan's entry into the White House also led to an escalation in the country's space activities. Space policy and initiatives held a high place. Reagan's two presidential terms saw the United States undertake more new and larger space programs than any previous administration since that of President John Kennedy. Indeed, Reagan's speeches often self-consciously imitated those of Kennedy.[1] Like Kennedy, Reagan passed on a lasting legacy to the nation's space program, but one that militarized and commercialized space.

Reagan-era new projects included the space station, the National Aero-Space Plane (NASP)—better known as the Orient Express—the Strategic Defense Initiative—a space-based antiballistic missile defense system—and the formulation for the first time of a national policy on the commercial use of space. The Reagan Administration was the first to pursue the stimulation of space business with vigor and to make it the subject of national policy. As a result of these programs and policies, it is very likely that the 1980s will be remembered as a major turning point in the history of space as important as, or possibly even more important than, the 1960s.

At the heart of Reagan space policies and projects was a new form of space transport, NASA's Space Shuttle. Before it began to fly in 1981, the Shuttle appeared to herald a new era of inexpensive routine space travel. Thanks to the "space truck," business ventures would play a much larger role in space, and space-based defenses would protect both military assets and the civilian population. The Shuttle also embodied the period's shifting emphasis in the country's space efforts as space exploration gave way to the exploitation of space, specifically near-Earth space. The Shuttle was the emblem and embodiment of this new terrestrially-centered emphasis. It launched Earth-orbiting civilian, military, and commercial payloads and was a platform for conducting experiments in near space, concentrating more on (commercial and military) space applications than on space exploration. The decision to launch everything—whether military, NASA, or commercial—aboard the Shuttle immediately bore devastating results for the country's fledgling commercial launch industry and soon had disastrous consequences for deep-space exploration.[2]

Planetary exploration, even before Reagan took office, was in disarray. One had to wonder: what golden age of planetary exploration? To some observers, solar-system studies had achieved their goals with the Viking, Pioneer Venus, and Voyager probes and that "there was little left to do." Other factors militated against new space missions. Chief among them was the mushrooming cost of the Space Shuttle. To cover the overruns, NASA raided its science budgets, "creating such havoc that these took almost a decade to recover," according to Paolo Ulivi and David Harland.[3]

---

[1] See, for example, Reagan, "State of the Union Message, January 25, 1984," *Public Papers of the Presidents of the United States: Ronald Reagan, 1984* (Washington: Government Printing Office, 1985), 90.
[2] Butrica, *SSTO*, 29-44.
[3] Ulivi and Harland, 1.

The entente cordiale with the Soviets initiated by President Richard Nixon a decade earlier had led to more cooperation rather than competition in space, but with little, if any, benefit to solar-system science. Reagan's rekindling of the Cold War likewise provided little hope for a return to deep-space exploration, although an unofficial defrosting in those relations during Reagan's second term did allow the DSN to collaborate in two Soviet-sponsored space ventures in 1985. NASA for its part shifted the focus of its science efforts away from the solar system towards Earth and a single, large-scale, expensive project, the Large Space Telescope (later renamed the Hubble Space Telescope).

On hiatus. For its part, Congress was out to make major budgetary reductions. Legislators had no sympathy for interplanetary missions bearing price tags of $500 million or more, which nonetheless was less than the cost of procuring almost any major weapon system.[4] Cost slashing would be a key leitmotif of deep-space exploration and navigation throughout the 1980s and beyond. At the decade's onset, NASA had only three planetary programs approved for development, none of which was on a particularly solid financial footing. They were known eventually as Magellan, Galileo, and Ulysses. Magellan's objective was to map the surface of Venus with a synthetic aperture radar, while Galileo would explore Jupiter and its satellites via an array of instruments. Their names reflected NASA's general plan of designating major planetary missions after famous scientists and explorers.[5] Ulysses, previously the Out-of-Ecliptic Mission followed by the International Solar Polar Mission, began as a dual-spacecraft joint venture of NASA and ESA intended to explore the heliosphere including the properties of the interplanetary medium that are of high importance to deep-space navigators. Between the 1978 lift-offs of the two Pioneer Venus probes and the 1989 launches of Galileo and Magellan (Ulysses flew in 1990), NASA sent zero probes into deep space.[6] At the time, JPL was the only NASA center building planetary probes.[7]

The prospects for planetary exploration grew bleaker, when Reagan entered office in 1981. NASA scaled back or canceled programs. As a political signal that the new president was serious about reducing federal outlays—or at least the civilian portion—budget czar David Stockman pressured NASA to sacrifice a major project. The agency chose the Venus Radar Mapper (the future Magellan).[8] NASA also considered closing the DSN and ending the Voyager 2 mission at Saturn. NASA Administrator James Beggs noted that "elimination of the planetary exploration program [would] make the JPL in California surplus to our needs." George Keyworth, Reagan's science advisor, suggested the complete suspension of deep-space projects for 10 years, "so as to enable NASA to focus on getting the Shuttle into service and then using this to

---

[4] Ulivi and Harland, 1.
[5] Butrica, *See the Unseen*, 192.
[6] See the list of launches in Asif A. Siddiqi, *Deep Space Chronicle: A Chronology of Deep Space and Planetary Probes, 1958-2000*, SP-4524 (Washington: NASA, 2002), 123-148.
[7] Ulivi and Harland, 2.
[8] Butrica, *See the Unseen*, 187.

conduct a variety of more worthwhile missions."[9] Meanwhile, the efforts of European and Asian space agencies continued apace, and ESA was preparing to join the Soviet Union in sending scientific probes into deep space with or without the United States.

NASA's policy and budgetary support for the Shuttle at the expense of solar-system science quickly impacted JPL. The Shuttle was part and parcel of a larger vision that included building a space station, and that meant diverting even more funds away from exploration. Headquarters' views of planetary science became unhappily clear when Hans Mark, NASA Deputy Administrator since 1981, visited JPL in January 1982. According to JPL historian Peter Westwick, Mark personified the integration of the military and civilian space programs. He also viewed the Shuttle as the focus of the agency's space efforts and a necessary step toward the space station.[10]

Johnny Gates, recently promoted to Assistant Laboratory Director of Technical Divisions, wrote about Mark's review of JPL. Mark spoke to representatives of three laboratory divisions and Director Bruce Murray. He told the gathering that the planetary program was "an aberration that would not recur," and that NASA could not support JPL at its then-current level. He estimated that JPL had a year to transition from planetary exploration to defense work.[11]

"All he could talk about," navigator Catherine Thornton remembered, was the space station. "He told us essentially that our deep-space exploration—the exploration of the planets—was done. That we really were not going to be having a whole lot more to do. . . . Yes, things got tight in the early eighties." "Of course," she added, "that's when the lab started doing some defense work. About 30 percent of our work was defense work for a while just to stay alive."[12]

JPL embarked on two major defense projects for the Army and Air Force. The Arroyo Center, established in 1982, was an external independent organization that performed various kinds of analyses for the Army. With DARPA funding, JPL undertook the classified and controversial Talon Gold project to develop one component of a space-based chemical laser weapon for the Air Force.[13] Caltech, which managed JPL for NASA, had serious concerns about undertaking this or any military research. In January 1984, the Caltech faculty voted for Caltech to divest itself of the Arroyo Center. The RAND Corporation ultimately took it over, and the facility continues today as a Federally-Funded Research and Development Center. The Talon Gold work, even more controversial, led to Caltech discussions about creating a separate JPL organization to conduct military research or, alternatively, whether Caltech should sever ties with JPL totally over the issue.[14] From the perspective of deep-space navigation, more important than the militarization of JPL research was the drive to reduce spending on solar-system exploration.

NASA Headquarters attempted to address the crisis in deep-space exploration by initiating a study by the ad hoc Solar System Exploration Committee established in

---

[9] Quoted in Ulivi and Harland, 2.
[10] Westwick, 52 & 101.
[11] Finding Aid, JPL235, 2.
[12] Thornton, interview, 19 & 20.
[13] Westwick, 104-105, 126-127, 129-141, 143 & 167.
[14] Finding Aid, JPL173, 3-4. JPL110 contains records relating to the Center.

1980. Its 1983 report called for the creation of two classes of missions. The Planetary Observer class (initially called the Pioneer class) flew within the inner solar system, borrowed technology from Earth-orbiting satellites, and cost no more than $150 million each. The Mariner Mark II probes explored the outer solar system and cost $300 million each at most. The Committee advocated four specific projects: Magellan, already underway; the first and only Planetary Observer, the Mars Geoscience and Climatology Orbiter (Mars Observer); the Mariner Mark II class Comet Rendezvous-Asteroid Flyby (CRAF); and the Saturn orbiter named after astronomer Giovanni Domenico Cassini.[15]

Consolidation and reimbursables. One of the cost-cutting measures that emerged as an offshoot of cutbacks within the DSN was multi-mission navigation. It marked a dramatic new direction in deep-space navigation that lasted into the next century. It changed the organization of navigators within the institutional structure of JPL as well as the very hardware, software, and techniques that were the tools of the navigators' trade. Multi-mission meant navigating not just to planets, but also to asteroids and comets, and navigating not just for NASA, but for foreign space agencies as well.

Multi-mission navigation grew out of a decision to merge tracking systems. Goddard, which managed the Spaceflight Data Tracking Network, wanted to replace it with the satellite-based Tracking and Data Relay Satellite System (TDRSS) beginning in 1984. Ray Amorose, the head of the JPL Tracking Data Acquisition Division, recalled that: "some wise men back east decided they would close the Goddard network, because Goddard was going to have the Tracking and Data Relay Satellite to track the low Earth orbiters. They would transfer three twenty-six meter antenna sites, which happened to be co-located with our deep space three sites at Goldstone, Canberra, and Madrid."[16]

While from Amorose's perspective the decision to merge tracking networks may have stemmed from "some wise men back east," the actual work started with the Networks Planning Working Group. It consisted of representatives of NASA Headquarters, JPL, Goddard, and the Spanish and Australian partners of both the DSN and the Goddard network. Between May and October 1979, the panel studied the proposed merger. The two main considerations were cost-cutting and the 1986 encounter of Voyager with Uranus. The new DSN architecture would have to make the best use of existing facilities and would have to track all flights not handled by TDRSS.[17]

The resulting Networks Consolidation Program called for creating a combined system capable of tracking both deep-space and so-called high-Earth-orbit missions. The trajectories of High Earth Orbit (HEO) flights had apogees in excess of 12,000 km. The DSN would track them with the 9-meter and 26-meter antennas acquired through

---

[15] Ulivi and Harland, 96-98, 100 & 102; David Morrison and Noel W. Hinners, "A Program for Planetary Exploration," *Science* 220 (May 6, 1983): 561-567; William H. Blume, "The Planetary Observer Program," *Journal of the British Interplanetary Society* 37 (August 1984): 355-360; Marcia Neugebauer, "Mariner Mark II and the Exploration of the Solar System," *Science* 219 (February 4, 1983): 443-449; J. W. Moore, "Effective Planetary Exploration at Low Cost," *Astronautics and Aeronautics* 20 (October 1982): 28-38.
[16] Amorose, interview, 2.
[17] "Networks Consolidation (42-59)," 108.

consolidation. Each complex—Goldstone, Spain, Australia—now had 5 antennas, only one of which was an X-band 64-meter antenna suitable for optimal deep-space communication and navigation. Each location also sported three 34-meter dishes, an increase of two achieved by enlarging existing 26-meter antennas. The DSN assigned the Goddard 9-meter S-band antennas specifically to HEO flights.[18]

The Networks Consolidation Program thus established a tracking system capable of both deep-space and near-Earth missions. Here was an instance in which the terracentric applications-focused space program shaped the configuration and capabilities of the DSN. One of the first projects tracked by the consolidated network was the GOES (Geostationary Operational Environmental Satellite) system, operated by a branch of the National Oceanic and Atmospheric Administration known as the National Environmental Satellite, Data, and Information Service (NESDIS).[19] GOES satellites were in a geosynchronous orbit about 35,800 km (22,300 miles) above the Earth. Space Shuttle launches also fell within the capability of the new network.[20]

The antenna merger also brought about an excess of tracking dishes over and above NASA's needs. Amorose wanted to market the DSN tracking capability to agencies outside the United States, mainly to European and Japanese space organizations. The Japanese, for example, had but a single tracking station. The Europeans had several antennas, but could not perform global tracking. As a result, the DSN was involved in a whole new class of external missions. Foreign space agencies reimbursed NASA at a standard hourly rate for the antenna time. These so-called reimbursables allowed Amorose to keep the DSN busy, while supplying NASA with paying customers.[21] This practice was consistent with the overall Reagan Administration policy of commercializing space and space assets.

Multi-mission navigation. The consolidation of deep-space and Earth-orbiting missions and the decision to set up a DSN "for hire" had a direct and immediate impact on navigation. Len Efron, a navigator on the Networks Planning Working Group, found himself becoming the representative to the DSN for missions that lacked JPL project officers, because the mission managers were at another agency in another country. At this point, Efron recalled, the Multi-Mission Navigation Team, "MMNAV as we refer to it ourselves," emerged. Funding for multi-mission navigation came through the DSN. Navigators charged their time against a special navigation account, giving rise on occasion to the use of the term "DSN Navigation Team."[22]

Amorose wanted the DSN to have enough navigational know-how, so that he would never have to worry about where to point the antenna. "With the larger antenna," Neil Mottinger explained, "the pointing accuracies are small: a tenth of a

---

[18] "Networks Consolidation (42-63)," 151; Mudgway, *Uplink-Downlink*, 169.
[19] NOAA, Office of Satellite Operations, "GOES Satellites," http://www.oso.noaa.gov/goes/ (accessed June 15, 2010).
[20] "Networks Consolidation (42-69)," 11; Efron, interview, 14; Cangahuala, interview, 16 & 18.
[21] "Networks Consolidation (42-63)," 150; "DSN Mark IVA Description," 255; Ellis, 262; Mottinger, interview, March 9, 2009, 60; Mudgway, *Uplink-Downlink*, 169.
[22] Efron, interview, 11, 12 & 19; Mottinger, interview, March 9, 2009, 60.

degree S-band or smaller when you are up into X-band. . . . You have to rely on an agency on the other side of the world, who possibly is not even the first one to see that spacecraft. . . . But if you don't know where to point the antenna, you're stuck."[23]

Orbit determination for HEO missions was to be done on a dedicated computer, a VAX 11/780. Programmers optimized navigation software to fit HEO requirements. In putting together this computing facility, "The possibility of cost savings to JPL-managed deep space missions, which might be derived from the use of this HEO facility, has, of course, been considered," JPL Assistant Laboratory Director Peter Lyman wrote to NASA Headquarters. He added that the "design will include the capability to perform deep space mission cruise OD [orbit determination]." Thus, although tailored to the needs of HEO flights, the facility's hardware and software resources were available to support portions of deep-space missions. "The facility could be extended to deep space missions at no charge," Lyman pointed out, "or perhaps a suitable reimbursement policy, which would help defray the sustaining HEO costs, would be drawn up." Lyman emphasized: "we intend to invoke all cost efficiencies which are made possible by this new facility at JPL."[24] As the organizer and first director of the HEO Multi-Mission Navigation Facility, Jordan Ellis instituted many general cost-cutting measures.[25]

The addition of Earth-orbiting missions made new demands on navigators. Navigation now focused on the first 24 to 48 hours of a deep-space launch, after which the project-dedicated navigation team took over. Multi-mission navigators also furnished predicts—DSN pointing directions—for the first two days of a flight. While deep-space flights allowed navigators to gather data and compute trajectories over several months, at least during cruise, Earth probes required much shorter periods between orbit computations. They also needed extra models for navigating around the Earth, such as models of the Earth's atmosphere and tides.[26]

The Multi-Mission Navigation group (in the navigation section) provided launch and early orbit support to all JPL missions. Over a 15-year period, they aided perhaps as many as 120 NASA, Japanese (ISAS and NASDA), and European (ESA, GSOC, CNES) missions. In one particularly busy calendar year, the team supported 15 launches. In addition to Ellis, the group's navigators included Len Efron, Tim McElrath, Prem Menon, Neil Mottinger, Mark Ryne, Tung-Han You, Kevin Criddle, and Earl Higa. The role of "navigation chief" rotated among the team members.[27]

A subset of Ellis' Multi-Mission Navigation group was the Radiometric and Data Conditioning Team (also known as the Radiometric Data Processing Group or the RMD for short), a group of JPL employees who became contract workers paid by the DSN. The team analyzed range, Doppler, and VLBI observations, the orbit determinations performed during flight, spacecraft maneuvers, and flight ephemerides. The Tracking and Systems Analytical Calibrations Group, established a decade earlier, furnished timing, polar motion, and media calibrations (for tropospheric, ionospheric, and interplanetary plasma effects). The multi-mission navigators used this information to produce predicts

---

[23] Mottinger, interview, March 9, 2009, 60.
[24] Lyman to Charles T. Force, July 8, 1982, F20/B1, JPL516.
[25] Ellis, 261-262; Thornton, interview, 14-15.
[26] Efron, interview, 15; Ellis, 263.
[27] Personal communication, Leonard Efron, June 7, 2012.

for the DSN and a validated tracking data file for transfer to external agencies or to the JPL navigation team assigned to that particular mission.[28]

The Radiometric and Data Conditioning Team served every mission, not just those assigned to Multi-Mission Navigation. Over the years the team included Lu Ann Brown, Laura Campanelli, Ted R. Eastman, Kyong Jae Lee, Margaret Medina, Sheryl Rinker, Shirley Stoneberger, and Theresa Thomas. Medina, Campanelli, and Thomas, in that order, subsequently headed the team after the departure of Gene Goltz.[29] Goltz, Efron recalled, "was a very influential leader and head of that group for a long time." He "knew more about what was going on because he wrote the software, and a lot of it was written on the fly. If you had a problem with anything that was going on, you went to see Gene Goltz. He was a fount of information. He's one of these people with a memory that I don't think anything's ever erased. When we had problems, we'd go down to see Gene. 'Gene, something's happening. I don't understand what's going on.' Gene would look at it, and he'd say, 'Okay, I'll look at it.' Then you knew it was a race. You had to get back to your office quickly, because the phone would invariably be ringing and he would tell you, 'I fixed it.' If you didn't get back there in time, you'd have to listen to it on voice mail. That was the kind of individual Gene was."[30]

Although the data processing group removed bad data points, sometimes navigators wanted the data before they cleaned it up. "We wanted to see all the noisy data, all the bad data," Efron explained, "because sometimes there was some information that would be lost if you did too good a job of cleaning it up." This was life in the fitter's universe. "We'd see all kinds of things in there. It's the old story of the child: When the child is good, they're very good. But when they're bad, oh, boy! So there were times when we wanted to see good and bad. Other times it was just routine. Let them clean it up, and we'll just process it."[31]

Comets and ICE. The advent of multi-mission navigation meant that JPL navigators were now dealing with probes in deep space and high orbit, with NASA and foreign space agencies, and with missions to asteroids and comets, the latter reflecting the new focus in solar-system exploration. The Multi-Mission Navigation Team worked with French, German, Japanese, and other space agencies on a number of projects. The first two such projects were AMPTE (the Active Magnetospheric Particle Tracer Explorers) and the International Comet Explorer (ICE). AMPTE consisted of three spacecraft built by three countries: NASA's Charge Composition Explorer (CCE), West Germany's Ion Release Module (IRM), and the United Kingdom Subsatellite (UKS). Ray Frauenholz and Prem Menon made up the navigation team for the AMPTE launch.[32]

---

[28] Ellis, 261-262; Thornton, interview, 14-15.
[29] Efron, interview, 18; Ellis, 262; Mottinger, interview, March 9, 2009, 24 & 25; Personal communication, Leonard Efron, June 7, 2002.
[30] Efron, interview, 17-18.
[31] Martin-Mur, interview, 26; Efron, interview, 18.
[32] Duncan A. Bryant, Stamatios M. Krimigis, and Gerhard Haerendel, "Outline of the Active Magnetospheric Particle Tracer Explorers (AMPTE) Mission," *IEEE Transactions on Geoscience and Remote*

ICE, managed by Goddard, was a rather memorable navigation experience. The original objective was to launch three probes known as the International Sun Earth Explorers. The third, ISEE-3, went into an intricate orbit devised by Goddard's Bob Farquhar, a mission designer who had invented the term halo orbit in 1968 to denote them. In 1982, upon completing its solar science goals, the craft took on a new name, the International Comet Explorer, and a new science mission. Thanks to five lunar gravity assists, ICE flew through the tail of the comet Giacobini-Zinner, on September 11, 1985.[33]

The Giacobini-Zinner flight was memorable for the rivalry between Farquhar and JPL navigators, or at least memorable for Len Efron, who was responsible for the orbit determination and led the navigation team, while Don Yeomans did the comet ephemeris. Farquhar omitted the incident in his memoirs. The three jokingly kidded each other about navigating ICE through the comet's tail. Efron, for instance, remembered telling Farquhar at a 1986 gather after the flyby: "You know, Bob, if ICE didn't get to the comet, I was going to blame it on you for a bad maneuver." Yeomans, in so many words, said that the ephemeris was perfect. "Don and I had decided that we were going to blame Bob," Efron recounted. Farquhar turned around, looked at the two, and said, "You know, I was going to blame both of you."[34]

Efron characterized the cooperation between JPL and Goddard on ICE as being better than usual. Indeed, the two institutions did not necessarily cooperate. In this case, though, "the personalities involved on both sides led to very close cooperation between the people doing the trajectory correction, maneuver, and design back there, and the people doing the OD [orbit determination] here. That synergism was just amazing," Efron reflected.[35]

The passage through Giacobini-Zinner's tail reflected cultural differences between Goddard and JPL navigators, who monitored ICE's progress in the Goddard control room. Everybody was looking at the clock on the wall that night, as the time approached for the probe to pass through the comet's tail. At the predicted time, everybody cheered. Efron, though, asked: "Does anybody know where we really went?"[36]

He decided to consult Edward J. Smith, the JPL Principal Investigator on the magnetometer experiment. Smith was in the backroom with the teletype machines. "Ed had been doing this long enough that he could look at the stream of zeros and ones on the teletype machine to know what was going on with this instrument," Efron recounted. After the cheering abated, Smith entered the control room. Had ICE passed through the tail? "He looked up, and he had a big smile on his face, and he said, 'The magnetometer reversed sign.'" The instrument reading, not the passage of time, was a

---

Sensing, GE-23 (May 1985): 177-181; Ellis, 262; Efron, interview, 28; Mottinger, interview, March 9, 2009, 61.

[33] Efron, interview, 24; Ellis, 264; Keith W. Ogilvie, Tycho von Rosenvinge, and Alastair C. Durney, "International Sun-Earth Explorer: A Three-Spacecraft Program," *Science* 198 (October 14, 1977): 131-138; Farquhar, 85-111.

[34] Efron, interview, 23; Farquhar, passim.

[35] Efron, interview, 24.

[36] Efron, interview, 24.

true indication that ICE had flown through the very center, the so-called neutral sheath, of the tail. Now, Efron recalled, "Ed Smith and I are cheering, and everybody else was just looking, because we were the only two in the room who knew that we really had success that night."[37]

Pathfinders. Another comet offered a noteworthy watershed moment in international cooperation in deep-space navigation. The United States did not participate in the multinational cooperative endeavor to explore Halley's Comet during its 1986 perihelion by sending a probe. At one point, NASA was planning a Halley rendezvous mission that navigators considered to be rather challenging because of the comet's retrograde orbit. In 1977, the agency held a so-called shootout between two competing approaches: solar sailing and solar electric propulsion, the eventual winner. It was a Pyrrhic victory; NASA declined to fund a Halley mission. ICE, after venturing through the tail of Giacobini-Zinner, took on a new, and again originally unintended, scientific objective: NASA's participant in the Halley collaboration.[38]

The so-called Halley Armada consisted of the Soviet Union's Vega 1 and Vega 2, ESA's Giotto probe, which approached closest to the comet's nucleus, and two Japanese spacecraft, Suisei and Sakigake.[39] The multilateral framework for coordinating Halley exploration, the Inter-Agency Consultative Group, included NASA, ESA, the Intercosmos Council of the Russian Academy of Sciences (now the Russian Aviation and Space Agency), and Japan's Institute for Space and Astronautical Science (ISAS).[40] In order to improve navigational accuracy for the armada's spacecraft, the Inter-Agency Consultative Group established Halley Pathfinder. Pathfinder was an opportunity for international cooperation and achievement in navigation. Its success garnered praise at the time as an exceptional example of international cooperation in space. It also opened up expanded levels of international coordination that have continued to this day. Frank Jordan led the NASA team.[41]

In the spring of 1982, to prepare for the Halley encounters, JPL navigators Frank Jordan, Jim Campbell, Jordan Ellis, John Ekelund, and Lincoln Wood plus Joe Wackley, a DSN tracking system expert, spent a week at the European Space Operations Center in Darmstadt, Germany. Al Cangahuala characterized its Orbit and Attitude Division as having "a small but capable orbit determination capability."[42] JPL navigators gave lectures on deep-space navigation and explored some of the particular navigational problems that the Giotto mission would face. As mission-specific questions came up, Frank Jordan and Lincoln Wood would call Neil Mottinger back at JPL during the evening (Central

---

[37] Efron, interview, 24; Farquhar, 102-104.
[38] Wood, interview, 15.
[39] Brandt and Chapman, 114-128; Yeomans, interview, 18.
[40] Joan Johnson-Freese, "A Model for Multinational Space Cooperation: The Interagency Consulting Group," *Space Policy* 9,4 (1989): 288-300.
[41] Jordan.
[42] Cangahuala, interview, 42; Martin-Mur, interview, 4.

European Time) to ask him to make some more computer runs and to send the results to ESOC for discussion the next day.[43]

On behalf of Halley Pathfinder, the DSN provided Doppler tracking to all five missions and VLBI tracking of the two Soviet missions. This was one of the earliest practical interplanetary uses of VLBI. Charles Stelzreid coordinated acquisition of the VLBI data. Jordan Ellis calculated Soviet spacecraft flight paths and the position of the comet's nucleus. JPL navigators then brought to bear the information garnered from assisting the Russian craft on improving the accuracy of the ephemeris of the comet's nucleus generated by Don Yeomans. Consequently, the ESA Giotto probe targeted the nucleus with greater exactitude, and the craft successfully flew closer to the nucleus than the four preceding probes to some added scientific advantage.[44] The two Vega probes thus served as pathfinders for Giotto.

Sakigake and Suisei. Working with the Japanese posed certain cultural shocks, although not out of unfamiliarity with the people. For example, Toshimitsu Nishimura, a Japanese national, had been in the JPL navigation section from 1962 to 1974, having joined after earning a doctorate from the University of California at Berkeley. His intention, Neil Mottinger recalled, "always was to go back to Japan and help his native space agency." Nishimura eventually returned to Japan and joined the ISAS. He helped to design and build the Institute's new 64-meter deep-space tracking station at Usuda-cho, Nagano. The dish became operational in 1984 in time to track the Sakigake probe, the first ISAS interplanetary spacecraft, which launched on January 8, 1985. Both the ISAS and the National Space Development Agency (NASDA) also had orbit-determination capability.[45]

The Sakigake and Suisei spacecraft represented Japan in the Halley Armada. Sakigake launched January 7, 1985, and encountered the comet on March 11, 1986, while Suisei left Earth on August 18, 1985, and met the comet on March 8, 1986. JPL navigators provided ephemerides to the Japanese for their Halley encounter. The JPL Multi-Mission Navigation Team also played an important role in navigating both craft to their targets. For Sakigake, they helped the ISAS to validate its navigation software by providing Voyager tracking data for testing. Additionally, they supplied trajectory solutions and antenna-pointing predicts for the Usuda-cho station during the first 10 days after launch and consulting services for 6 months after launch.[46]

Len Efron, like Neil Mottinger, recalled that working with the Japanese space agencies "often exposed us to cultural surprise(s)." During the Sakigake launch, real-time communication between the multi-mission navigators and the ISAS control room took place over a commercial telephone line. Mottinger calculated trajectories. Efron sat

---

[43] Personal communication, Lincoln Wood, May 20, 2013.
[44] Jordan; James Wilcher, Charles Stelzried, and Susan Finley, "Pathfinder Operations," 263-267 in TDAPR 42-87; Drechsel, 29-30.
[45] Mottinger, interview, March 10, 2009, 2; Tamiya Nomuraa, Tomonao Hayashia, Toshimitsu Nishimura, Haruto Hirosawa and Mitsuru Ichikawa, "Usuda Deep Space Station with 64-Meter-Diameter Antenna," *Acta Astronautica* 14 (1986): 97-103; Cangahuala, interview, 42; Martin-Mur, interview, 4.
[46] Brandt and Chapman, 114-128; Yeomans, interview, 18; Ellis, 264.

beside the phone to listen for and relay verbally the announcement of each launch phase.[47]

The DSN's ability to acquire the probe's signal depended on the launch stages following a predetermined timeline. When Efron heard the ISAS control center voice announce the times for fourth-stage ignition and then fourth-stage burnout, he realized that their timing disagreed with the navigators' anticipated times. Consequently, the DSN might not acquire the downlink signal. Efron, speechless, wondered: "Were we about to have an international disaster?" He asked to speak to Nishimura, the ISAS lead navigator, who promptly came to the telephone. Efron explained his concerns. Nishimura revealed that they had provided JPL with a so-called pseudo state for the position and time of the craft's injection into a nominal interplanetary trajectory. The Japanese believed that the pseudo state would have allowed the DSN to acquire Sakigake if it were en route to or close to that injection point and time. The navigators could relax.[48]

"The DSN I don't think could find the signal," Mottinger recalled. "I don't know if the DSN ever did find it." The spacecraft was so far off course that it required an enormous corrective maneuver. "They were like maybe 100,000 kilometers from Halley's Comet or something and made a big observation. . . . They had a 110 meter-per-second correction to the velocity. Blew us away!"[49]

Moreover, they later learned that the launch announcements had been a bit of kabuki theater. What Efron had heard was nothing more than someone reading aloud from a script, not results from instrument telemetry. "Had there not been any phone communication link," Efron mused, "MMNAV would have had a less stressful experience. All is well that ends well as was the case, but it was nerve wracking at that moment."[50]

---

[47] Personal communication, Leonard Efron, June 7, 2002.
[48] Ibid.
[49] Mottinger, interview, March 10, 2009, 2 & 3.
[50] Personal communication, Leonard Efron, June 7, 2002.

Chapter 17

Earth and Space Joined

One of the most transformative and lasting influences on deep-space navigation came from Earth-orbiting satellites. The Space Shuttle embodied the country's shifting emphasis in space on Earth-orbit civilian, military, and commercial applications over solar-system exploration. NASA's consolidation of tracking systems brought about the Multi-Mission Navigation Team that served probes orbiting Earth and flying into deep space. The ongoing search for better ways to adjust data, to cancel out error sources, now joined Earth and space with the adoption of the Global Positioning System (GPS) to calibrate data as well as to navigate spacecraft circling Earth. Adjusting tracking data would be one of many civilian uses of the GPS, whose primary purpose was to support prosecution of the Cold War.[1] As with many of the factors bending navigation toward Earth, GPS would involve navigators in international networks and would provide a new connection with geophysicists.

GPS-based terracentric navigation also benefited from the graduate-school training received by new JPL navigators. The navigation curriculum at key schools in Colorado and Texas was oriented toward Earth-orbiting applications. Many of the graduates hired from the Colorado program, for example, also had written theses on some aspect of GPS navigation. Space navigation began with Earth-orbiting satellites of the 1950s, and now during the 1980s, it seemed to be returning to its roots in Earth-orbiting satellites. This return to Earth reflected the shifting sands of the country's space effort.

Terracentric navigation. The adoption of GPS by the DSN and navigators began when research by the DSN Advanced Systems Program suggested that GPS satellites could provide the basis for calibrating Doppler and range data. Because GPS satellites fly above the Earth's ionosphere in well-defined orbits, ground stations can measure the delays imparted by the ionosphere and troposphere in a number of directions—as many directions as the number of satellites visible by a ground station at a given time—and use those measurements to calibrate Doppler and range. Moreover, knowing the satellite positions relative to the Earth's center of mass, not its surface, provided another method for observing and measuring the impact of terrestrial motions on station locations. Like VLBI, then, GPS promised to help correct for errors arising from both media effects and station locations. Although the DSN eventually used the GPS to provide those calibration measurements, it was not a total solution. Moreover, GPS-

---

[1] George W. Bradley, III, "Origins of the Global Positioning System," 245-253 in Jacob Neufeld, George M. Watson, Jr., and David Chenoweth, ed., *Technology and the Air Force: A Retrospective Assessment* (Washington: Air Force History and Museums Program, 1997); Ivan A. Getting, "The Global Positioning System," *IEEE Spectrum* 30 (December 1993): 36-38; Mark Denny, *The Science of Navigation: From Dead Reckoning to GPS* (Baltimore: Johns Hopkins University Press, 2012).

based navigators later applied their precise GPS-based orbit-determination methods to the Earth-circling GRACE (Gravity Recovery and Climate Experiment) and Orbiting Carbon Observatory (OCO).[2]

Navigating with GPS began in the early 1980s, when a group of JPL navigators, looking for alternatives to conventional Doppler and range, conceived a way to test the use of GPS for navigation. TOPEX/Poseidon became the first mission to demonstrate GPS navigation. Its scientific objective was to study ocean circulation and its interaction with the atmosphere as part of an international oceanographic and meteorological effort sponsored by the World Climate Research Program. The designation TOPEX/Poseidon derived from the names of the two radar altimeters given by the cooperating agencies, NASA and the CNES (Centre National d'Études Spatiales), France's space agency. The altimeters measured the satellite's height above the ocean, which, when combined with a highly precise calculation of the satellite's position relative to the center of the Earth, yielded the geocentric height of the ocean, the quantity of real interest. The probe transmitted the signals on two frequency bands to allow removal of the effects of ionospheric free electrons in them.[3] This is the "direct" method called dual-frequency calibration discussed in Chapter 13.

TOPEX navigation was a complex affair involving several institutions and data types. The TOPEX Precision Orbit Determination (POD) Team was a joint effort of NASA Goddard and the University of Texas at Austin's Center for Space Research in collaboration with JPL and the University of Colorado. JPL made trajectory estimates using GPS readings, while Goddard and Texas computed orbits from Doppler and laser ranging, but not GPS. The CNES' DORIS all-weather, global tracking network supplied an advanced form of Doppler.[4] Improved data accuracy stemmed from linking ground stations to the International Terrestrial Reference Frame (ITRF), which served users in the geodesy and geophysics communities. The ITRF of 1993 was the successor to the Clarke spheroid of 1866 and the Kaula spheroid that JPL navigators began using in the 1960s to establish DSN station locations.[5]

---

[2] Mudgway, Uplink-Downlink, 489; Cangahuala, interview, 14-15 & 19; Wood, interview, 59; Martin-Mur, interview, 10; Maize, interview, 33.
[3] "GPS Flight Experiment"; "TOPEX/Poseidon OD"; Hamilton, interview, 2010; "Precision OD," 24,383–24,404; "T/P 'Quick-Look' Ops," 1162.
[4] In English, DORIS (Détermination d'Orbite et de Radiopositionnement Intégrés par Satellite) stands for Integrated Orbit Determination and Radio-Positioning by Satellite. CNES, DORIS, "A Multipurpose System," http://www.cnes.fr/web/CNES-en/1514-a-multipurpose-system.php (accessed February 24, 2011); CNES, "DORIS The Space Surveyor," http://www.cnes.fr/web/CNES-en/1513-doris.php (accessed February 24, 2011); CNES, DORIS, http://smsc.cnes.fr/DORIS/ (accessed February 24, 2011); "3.4.4 International DORIS Service (IDS)," http://www.iers.org/SharedDocs/Publikationen/EN/IERS/Publications/ar/ar2001/ar2001__029,templateId=raw,property=publicationFile.pdf/ar2001_029.pdf (accessed February 28, 2011).
[5] ITRS, "The International Terrestrial Reference System (ITRS)," http://www.iers.org/IERS/EN/Science/ITRS/ITRS.html (accessed March 1, 2011); International Earth Rotation and Reference Systems Service, "Organization," http://www.iers.org/nn_10964/sid_2AE6306ACB7C4A00217022A0D2CE3123/IERS/EN/Organization/organization.html (accessed March 1, 2011); "TOPEX/Poseidon OD"; "GPS Flight Experiment"; "Precision OD," 24,383–24,404.

An orbit calculation known as a Precision Orbit Ephemeris (POE) was the principal basis for deriving scientific data—known as the Interim Geophysical Data Records (IGDRs) and later as the Geophysical Data Records (GDRs)—for geoscientists. The Goddard Orbit Determination Team computed these from range and Doppler collected over a ten-day period (127 orbits). The POEs were available thirty days after acquisition of the tracking data.[6] This simply was not fast enough for many scientists.

The MIRAGE. The need for faster IGDRs became apparent shortly after TOPEX's launch. The climate pattern known as El Niño, the cause of extreme weather—such as floods and droughts—in many areas of the globe, was in the news. Scientists understandably wanted reports on a more frequent basis. In response, the two navigators who made up the Precision Orbit Determination and Verification Team devised the so-called Medium Precision Orbit Ephemeris (MOE). It yielded science data faster—within a day or two—by relying on a single data type (laser ranging) as well as orbit estimates with smaller errors.[7]

The MOE, however, had two weaknesses: There were not enough laser-ranging stations in the Southern Hemisphere, and weather—rain and winter storms—often precluded their use. To get around these shortcomings, JPL navigators Ed Christensen, Bobby Williams, Al Cangahuala, and Peter J. Wolff decided to generate orbits from the GPS. It was available every day around the clock. Scientists now had their medium-fidelity data faster and more regularly. "It was still pretty high fidelity," Cangahuala explained. "I think it was a nice merger of our quick-turnaround techniques and high-fidelity modeling."[8]

Navigating with the GPS required writing new software. The original GPS-based program bore the designation GIPSY-OASIS for GPS Inferred Positioning System (GIPSY) and Orbit Analysis SImulation Software (OASIS). Created in 1979, as JPL planning for TOPEX was getting underway, GIPSY-OASIS provided the mission with precise probe position information based on GPS data and pinpointed satellite locations to within just 2 cm (less than an inch).[9]

TOPEX had an even deeper impact on interplanetary navigation software. Navigators refashioned the standard orbit-determination program into the Multiple Interferometric Ranging Analysis using GPS Ensemble (MIRAGE). Bobby Williams played a pivotal role in its overall design and development. MIRAGE superseded GIPSY-OASIS. Unlike GIPSY-OASIS, MIRAGE handled all three data types: laser range, Doppler, and GPS. The computer files generated from each data type were identical, so that formats were not a problem when creating the IGDRs with MIRAGE solutions. MIRAGE also had a utility that converted any MIRAGE orbit file into the POE format.[10]

---

[6] "TOPEX/Poseidon OD"; "T/P 'Quick-Look' Ops," 1162.
[7] Cangahuala, interview, 19 & 20; "T/P 'Quick-Look' Ops," 1162 & 1164.
[8] Cangahuala, interview, 20.
[9] Thornton, interview, 15-16; "GPS Software Packages Deliver Positioning Solutions," *Spinoff* (2010): 126-127.
[10] Guinn and Wolff, 143, 155 & 158.

The creation of MIRAGE resulted in navigators having two variations of the Regres mathematics discussed in Chapter 12. The original Regres formulation was based on the solar system's barycenter within a relativistic reference frame; it applied to spacecraft flying anywhere in the solar system, including to the Moon. The new version calculated positions and velocities within a local geocentric relativistic reference frame for missions near Earth, such as TOPEX/Poseidon. The need to achieve highly accurate flight-path estimates for TOPEX led to emendations of the navigation software's mathematical models. Among the forces modeled were the Earth's gravitation, the oblateness of the Earth and Moon, the tidal motions of the oceans and solid land, and atmospheric drag.[11] MIRAGE thus embodied the dual Earth and space navigation that emerged in the early 1980s.

Ted Moyer, who started working on Regres when he joined JPL in 1963, undertook yet another fundamental revision of the Regres math that JPL published in 2000, as Moyer retired from the lab. The formulas literally represented his life's work. Pete Breckheimer, John Ekelund, James B. Collier, Mike Wang, and Dah-Ning Yuan wrote the code that converted Moyer's formulas into the Regres program. The intent of publishing Moyer's formulas through the DESCANSO series (see below) was to make the Regres mathematics more widely available to users outside NASA, to "any organization that is developing an ODP" [orbit determination program] to navigate "spacecraft anywhere in the Solar System."[12]

The terracentric focus that TOPEX demanded also extended to navigation time scales. On top of the several time scales that navigators already used, such as Ephemeris Time, International Atomic Time, Universal Time, Coordinated Universal Time, and Station Time, GPS navigation necessitated the addition of two new ones: GPS and TOPEX Master Time (TPX). GPS master time was based on International Atomic Time. Time measurements obtained from clocks onboard Earth satellites were referenced to International Atomic Time. The Satellite GPS atomic time scale replaced UTC as a reference time scale for both GPS receiving stations on Earth and for GPS satellites. TOPEX Master Time was the atomic time scale used specifically for the TOPEX satellite.[13]

The adoption of the GPS by navigators did not mean the end of the DSN. Rather, it was one of several sources for determining spacecraft positions. The DSN supplied Doppler, range, and VLBI data, while onboard cameras furnished information for optical navigators, and now there was the GPS. "The DSN isn't going to go away," Al Cangahuala emphasized, "but we're just going to have access to more data types, and we have to get good at reconciling those different data types with each other."[14] Deep-space navigation had come a long way from the Doppler-only days of the early 1960s.

---

[11] Guinn and Wolff, 147; Moyer, *Formulation*, 1-2.
[12] Ekelund, "History of ODP at JPL," 16; Moyer, *Formulation*, ix & x.
[13] Moyer, *Formulation*, 2-1, 2-3, 2-4, & 2-5.
[14] Cangahuala, interview, 20.

The revolution. The shift to GPS coincided with the "revolution in computing" that was equally as transformative of deep-space navigation as the terrestrial turn. Minicomputers and workstations brought the mainframe era to an end and freed navigators of the restraints imposed by the old machines. New possibilities opened up. The arrival of the first new computers was a heady experience.

Amorose bought the navigators Digital Equipment Corporation (DEC) minicomputers. "I can remember the first time a salesman was here and talked about virtual memory," said Mottinger. "I couldn't believe it! . . . this salesman shows up and says, 'You know you need half a megabyte of storage, just ask for it. Let the computer worry about it.'" For a while, "any loading dock you looked [at had] empty DEC boxes. People were buying these things like wildfire."[15]

Users accessed the DEC minicomputers through VT100 terminals, also known as VAX (Virtual Address Extension) terminals, equipped with a video interface and keyboard. They rapidly replaced the Univac mainframe. "Everybody owned one," according to Mottinger. "You weren't going to be charged by the computer section to use their computer. It was free. The thing was going to hum 24 hours a day. . . . It wasn't too much before that the [Univac] 1108s had been on their way out, and they were going to have to be replaced."[16]

The shift to the new machines meant that navigators had to rewrite their software, while significant navigational work still was taking place on the Univac 1180. New generations of navigation software came along. The programs were not particular to any specific project, but rather applicable to a wide range of projects. Some projects used Sun [Microsystems] workstations. Mottinger recalled that at one time the same source code was supporting "easily three or four different platforms." The transfer of the navigational software to the VAX also marked the move of the navigators from Building 264 to Building 301, where they are today.[17]

The new software could be daunting to a recent hire. For example, when Shyam Bhaskaran started at JPL, he discovered that there was "a big difference between classroom teaching and real life." "I mean you really learn the stuff when you actually do it for real," he added. Classroom examples were spacecraft in Earth orbits and the data was "very simple" range and Doppler. In real life, the "actual Doppler observables are different," because "all sorts of things" are happening to the spacecraft. "Thrust happens. Gas leaks happen. The data doesn't always work; it's not always good. There are constraints on various things." All things considered, Bhaskaran reckoned, "The mathematical background was fine; it was perfectly adequate for the job." The challenge was "All these real-world things that you don't really learn about in school. There's a lot of real-world training."[18]

---

[15] Mottinger, interview, March 9, 2009, 61.
[16] Mottinger, interview, March 9, 2009, 61.
[17] Nonetheless, navigators working on projects in flight are in Buildings 230 or 264, which offer more restricted access and more reliable backup power. Ellis, 263; Richard B. Miller, "Pioneer Venus 1978 Mission Support," 15 in DSNProgR 42-40; Bhaskaran, interview, 68; Mottinger, interview, March 9, 2009, 61; Ekelund, "History of ODP at JPL," 14.
[18] Bhaskaran, interview, 10 & 11.

The software itself was one of the real-world challenges. "It's a very complicated program, and requires literally thousands of inputs," Bhaskaran explained. "So, learning all the software really took several months at least." Navigators entered some inputs manually, such as the mass and area of the spacecraft. Other inputs dealt with the Earth's orientation and calibrations for media effects.[19]

Long gone were the days of the single-precision program capable of estimating 20 parameters at most. Today's orbit determination program can handle literally a thousand parameters thanks mainly to advances in computers. Once navigation software migrated from the Univac mainframe, programmers had to rewrite the code for VAX machines, Sun workstations, and whatever other computer might be assigned to flight navigation.[20] There were at least as many software variations as there were computing platforms.

The multiplicity of program languages and platforms required maintaining an archive of old tracking data to test new programs. Tomás Martin-Mur, whose career in navigation started in Europe, helped to set up the archive. "We keep all the data, so then we can always go back to this data and investigate any possible issue or process the data again with new software and make sure that things work." The archive was separate and different from the Planetary Data System sponsored by NASA's Science Mission Directorate. The latter is an archive of science data from past NASA planetary missions maintained as a resource for scientists around the world.[21]

Unlike multi-mission navigation, DSN funding for the navigation archive was not forthcoming. The DSN, Martin-Mur pointed out, was "not in the business of keeping tracking data." Rather, "it is the responsibility of us, of the projects, to keep this data." The DSN retained tracking data for the use of a particular mission, but, "After a period of time, six months or one year, once they are sure that the project has the data, then they don't need to keep it anymore." At which point, the DSN erased the data. The current navigation archive has data back to Pioneers 10 and 11.[22]

Keeping old data has certain inherent technological problems. As Martin-Mur explained, "The problem then is that sometimes we have it retained in old media, like in tapes, and tapes deteriorate with time. So we have a big effort to convert the data from tapes to our modern digital form, like DVD's, so the data will not be lost. The other problem is that sometimes these data are archaic, in very old formats, or new software cannot read them anymore. So then we have to resurrect, you know, old software and process the data."[23]

Future navigators. The education of future navigators further reflected navigation's terracentric turn. Deep-space navigation was not a discipline in the sense of having its

---

[19] Bhaskaran, interview, 11.
[20] Mottinger, interview, March 9, 2009, 13-14 & 61.
[21] Martin-Mur, interview, 26 & 31; NASA PDS: The Planetary Data System, "About the PDS," http://pds.jpl.nasa.gov/about/about.shtml (accessed August 18, 2010).
[22] Martin-Mur, interview, 31 & 32.
[23] Martin-Mur, interview, 32.

own schools, societies, or institutions (unless one counts the JPL navigation section), but rather multidisciplinary and diffused within several larger scientific and engineering communities. Universities offered coursework in navigation, usually with an emphasis on Earth satellites, and navigators read papers at meetings of the American Institute of Aeronautics and Astronautics, the American Astronautical Society, the Institute of Navigation, or the American Geophysical Union, among several others.

Critical to the growth of navigation was the training of future practitioners. As we saw in Chapter 2, Samuel Herrick played a major pioneering role by establishing the Institute of Navigation, teaching the very first courses in space navigation, and educating future navigators at UCLA until his death in 1974. Many of Herrick's students found positions as JPL navigators. Well before 1974, two other universities were turning out space navigators, and many of their graduates joined the ranks of JPL navigators. Their graduates, though, did not account for the recruitment of all JPL navigators, as many came from other universities and had degrees in mathematics or engineering outside the aerospace field.

The older of the two programs was that of Prof. Byron D. Tapley at the University of Texas in Austin. Tapley taught the first orbital mechanics courses and pioneered space-related research at Texas during the early 1960s. In 1964, he moved to the Aeronautical Engineering Department (later changed to the Department of Aerospace Engineering). Tapley carried out his early research through an organ of his own creation, the Institute for Advanced Studies in Orbital Mechanics, and in 1981, he established and ran the Center for Space Research. Tapley's and the Center's research focused on a range of navigation and other problems associated with remote-sensing satellites, oceanography, geodesy, and climate change. More recently, Tapley was the Principal Investigator for the Gravity Recovery and Climate Experiment (GRACE) Mission that measured small variations in the Earth's gravitational field caused by interactions between the atmosphere, ocean, and land surfaces.[24]

Among Tapley's Texas colleagues was Victor Szebehely, who left the Yale faculty in 1968 to take the position. From 1977 to 1981, he chaired the Department of Aerospace Engineering. Szebehely is perhaps best remembered for his eponymous equation for determining gravitational potential. Other distinguished faculty included Ray Duncombe, former director of the Naval Observatory's Nautical Almanac Office (1963-1975), Roger Broucke, a former JPL navigator from 1963 to 1975, and Hans Mark, who joined after leaving NASA.[25]

---

[24] Fowler, 89, 90 & 92; "Byron D. Tapley, Ph.D.," http://www.csr.utexas.edu/info/staff/tapley.html (accessed February 23, 2011); University of Texas at Austin, Faculty Profile, http://www.utexas.edu/opa/experts/profile.php?id=389 (accessed February 23, 2011).

[25] Fowler, 89; University of Texas at Austin, Faculty Council, "In Memoriam: Victor G. Szebehely," http://www.utexas.edu/faculty/council/1998-1999/memorials/Szebehely/szebehely.html (accessed March 4, 2011); CSR Personnel Directory, http://www.csr.utexas.edu/info/staff/ (accessed February 23, 2011); University of Texas at Austin, Faculty Council, "In Memoriam: Roger A. Broucke," http://www.utexas.edu/faculty/council/2006-2007/memorials/broucke/broucke.html (accessed March 4, 2011).

The other and somewhat more recent major navigator training center was the University of Colorado at Boulder. University President Arnold Weber launched his Space Initiative in 1984, determined to make Colorado the nation's foremost space school. To that end he hired Don Hearth, formerly the NASA Langley director, to be the university's "space czar." Weber's successor, E. Gordon Gee, continued to fund the enterprise. He engaged George Morgenthaler from Martin Marietta to head the Department of Aerospace Engineering, and, more importantly for navigation, he signed up former JPL navigator George Born (a Texas graduate) to establish the Colorado Center for Astrodynamics Research in the fall of 1985. The Center has focused on the development and application of GPS technology to meteorology, oceanography, geodesy, and other areas.[26] The JPL navigators who graduated from Colorado included Peter Antreasian, Michael Armatys, Shailen Desai, Dolan Highsmith, Dan Kubitschek, Angela Reicher, and Shyam Bhaskaran.[27]

DESCANSO. Deep-space navigation did have its own institutions, namely NASA and JPL, as arenas in which to conduct professional activities, such as holding seminars, organizing symposia, and publishing research. JPL created a new organization for navigators—a center of excellence—that, in a sense, prolonged the "campus" atmosphere of the 1960s intended to cultivate employee competence in preparation for future missions. However, the idea behind creating NASA centers of excellence was to reduce costs and agency size by focusing each NASA field center on a given specialization. These centers thus were part and parcel of the cost-cutting culture that marked NASA from the 1980s to the present. JPL became the designated Center of Excellence for deep-space systems.[28] At the same time, JPL management decided to create its own centers of excellence on top of those already in existence[29] in order to build up certain laboratory assets and to give them more visibility. Management invited proposals for several dozen areas to become centers of excellence. The approval process was lengthy and went through several cycles simply to reduce the number of proposals. Among them was a suggestion for a navigation center of excellence.[30]

---

[26] Brian Argrow and Robert D. Culp, "Aerospace Engineering Science at the University of Colorado at Boulder," 445 in Barnes McCormick, Conrad Newberry, and Eric Jumper, ed., *Aerospace Engineering Education during the First Century of Flight* (Reston, VA: AIAA, 2004).
[27] CCAR, "CCAR Alumni," http://ccar.colorado.edu/alumni.html (accessed February 23, 2011); Cangahuala, interview, 19.
[28] NASA Press Release, 95-73, "Review Team Proposes Sweeping Management, Organizational Changes at NASA," May 19, 1995, http://www.jpl.nasa.gov/news/releases/95/release_1995_0519.html (accessed February 22, 2011); Office of Science and Technology Policy, Federal Laboratory Review: NASA Implementation Report, August 23, 1996, http://www.fas.org/irp/offdocs/pdd5status-d.html (accessed February 13, 2011).
[29] JPL still has a number of centers of excellence, such as the Interferometry Center of Excellence and the Laboratory for Reliable Software. JPL, Interferometry Center of Excellence, http://ice.jpl.nasa.gov/ (accessed February 13, 2011); JPL, "Laboratory for Reliable Software," http://lars-lab.jpl.nasa.gov/ (accessed February 13, 2011).
[30] Wood, interview, 35.

Lincoln Wood was responsible for shepherding the navigation proposal through the process. Immediate approval was not forthcoming. "I think we were one of three in the second cycle," he recalled. "When we were approved, we were like one of five centers." JPL approved the navigation proposal only after it merged with the competing submission from the Telecommunications Science and Engineering Division and the Deep Space Network. "Somewhere along the way somebody decided that those two ought to be merged together," Wood explained. "I do remember one review where Bruce Murray was a person on the review board, and he made that suggestion. Now, whether that was the decisive moment right there, and the two of them came together, I don't know, but within six months to a year the planning from that point on considered the two of them together as a merged center."[31]

The communication and navigation center of excellence differed from most of those approved. The others were in areas in which JPL had been doing well for several decades or areas that JPL wanted to develop more fully. Wood believed that creating the communication and navigation center ratified "the fact that we'd been doing this stuff for several decades, that we are the best in the world in it." JPL funded the center at a level lower than the others, because, in effect, it already had been established and its operation did not necessitate a large budget.[32]

In 1997, Wood set about establishing the center along with William Rafferty and Les Deutsch from the Telecommunications Science and Engineering Division and the Deep Space Network. Its activities focused on both navigation and communication. The center needed a name. Deutsch came up with DESCANSO named after the Descanso Gardens, which were a short distance from JPL in La Cañada Flintridge. As for DESCANSO as an acronym, it stood for DEep Space Communications And Navigation Systems. "You had to make up the O," Wood explained.[33]

Wood and Rafferty organized a series of DESCANSO symposia and monthly seminars. The papers dealt with the state of the art in navigation and communication. "Well, generally, the subject matter was both navigation and telecommunications, and we didn't break the two apart," Wood recalled. "In fact, in some people's view, one of the purposes of the center was to bring those two areas closer together." The navigation seminars dealt with such topics as orbiting around and landing on an asteroid, the subject of a talk given by Bobby Williams and Jim Miller; and an overview of NEAR (Near Earth Asteroid Rendezvous) navigation given by Williams. The DESCANSO symposia started in 1999 in Pasadena and continued to 2003. DESCANSO maintained two websites, one strictly internal to JPL and the other accessible by the public at descanso.jpl.nasa.gov. The internal website had more resources. "Well, at least one of them is a bibliography that I made up of references on Deep Space Navigation and

---

[31] Wood, interview, 35, 36, 37 & 55.
[32] Wood, interview, 35 & 36.
[33] Joseph H. Yuen, "foreword," xvii in Hamid Hemmati, ed., *Deep Space Optical Communications* (Hoboken: Wiley-Interscience, 2006); Wood, interview, 36 & 37. Descanso Gardens are on the grounds of the Rancho del Descanso ("Ranch of Rest") created by E. Manchester Boddy, publisher of the Los Angeles *Daily News*. Descanso Gardens, http://www.descansogardens.org/ (accessed February 25, 2011).

Mission Design," Wood added. "It's on the internal version, and we never decided that we necessarily wanted to get that approved for general release."[34]

Wood deemed that "[o]ne of the more significant things" that DESCANSO did was to start a series of technical books on navigation and communication. Initially published by JPL, Wiley Interscience eventually began printing them and reissuing the JPL imprints. All authors were in JPL navigation or (mostly) the communication sections. The very first monograph in the series was the magnum opus of deep-space navigation written by navigators Catherine Thornton and Jim Border. Their intention was to provide an introductory work for neophytes, a reference for professionals in allied fields, and a description of the state of the art. The work described the various radiometric techniques developed over the previous four decades to navigate interplanetary spacecraft by their colleagues past and present in both the navigation and tracking sections.[35]

Lincoln Wood, after being involved in DESCANSO from its inception and after serving as its Acting Deputy Leader (1997-2000) and Deputy Leader (2000-2003), "effectively" ended his involvement in 2003. Leadership or acting leadership passed successively from Rafferty (1998-1999) to Catherine Thornton (1999-2000) and finally to Joseph Yuen (2000-2003), who also was editor in chief of the DSN progress reports. After 2003, though, JPL no longer funded the effort, having "decided that the centers [of excellence] should not have infinite lifetimes but should have finite lifetimes." Nonetheless, in a sense, it continues to live through its websites.[36]

---

[34] Wood, interview, 37 & 38; DESCANSO, "DESCANSO Seminars," http://descanso.jpl.nasa.gov/Seminars/seminars.cfm?force_external=0 (accessed January 14, 2011); DESCANSO, "DESCANSO Symposia & Workshops," http://descanso.jpl.nasa.gov/Symposia/symposia.cfm?force_external=0 (accessed January 14, 2011). The reader will find a public bibliography at: DESCANSO, "DESCANSO Publication List," http://descanso.jpl.nasa.gov/Publication/publist.cfm?force_external=0 (accessed December 11, 2008).
[35] Wood, interview, 36-37; Thornton and Border.
[36] Wood, interview, 37, 55 & 65; DESCANSO, "DESCANSO People," http://descanso.jpl.nasa.gov/people/people.cfm?force_external=0 (accessed January 11, 2011); Joseph H. Yuen, ed., IPNPR 42-146.

Chapter 18

Opening the ΔDOR to VLBI Navigation

In 1979, VLBI was central to DSN planning along with other "direct" methods using two dishes, but not for long. As discussed in Chapter 13, these "direct" techniques, by changing the way one made spacecraft observations, potentially were the holy grail of self-calibrating data. They also were future rivals as navigation data types. The DSN VLBI system was adapted admirably for the many functions it performed: clock synchronization, geodesy, quasar exploration, and radio astronomy. It really was not suitable for navigating spacecraft, as the Voyager tests at Jupiter demonstrated. What was to be done?

The need for a solution became all the more crucial with the invention of a new "direct" approach to multiple-antenna navigation known as Delta Differential One-way Range or simply ΔDOR (pronounced "delta door" and written alternatively Delta DOR and DDOR). The salient characteristic that distinguished ΔDOR from other VLBI—not to mention other navigation data types—was its reliance on the transmission of tones by a spacecraft. Flight testing the new interferometric method required some inventiveness in light of the paucity of new mission starts during the 1980s. Ultimately, to enable VLBI navigation, the DSN had to undertake a radical transformation of its hardware and operations over a period of several years. Meanwhile, although ΔDOR was ideally suited to outer-planet exploration and was a mission requirement for Galileo, most of its earliest applications and achievements took place among the inner planets of Venus and Mars.

The all American NAVNET. VLBI was at the heart of DSN planning for the 1980s. System planners mulled over "technology development and its implications for long-range planning" during the DSN Advisory Committee's annual meeting. Formed in 1978, the Committee consisted of a mix of outside advisers plus former and current DSN leaders. The Chairman was DSN architect Eb Rechtin, now President of The Aerospace Corporation. Frank Lehan, another key member, had helped Rechtin to create the DSN and distinguished himself in several areas including space reconnaissance. John Pierce, Caltech Professor of Electrical Engineering and JPL Chief Technologist, renowned for his pioneering work on satellite communications at Bell Laboratories, was one of the outsiders. The other outsider was Albert Wheelon, a Vice President and Group Executive at Hughes Aircraft who earlier had been the first director of the CIA's Directorate of Science and Technology.[1] The Advisory Committee's makeup typified the integration of NASA, academia, industry, and national security that characterized the military-industrial complex.

---

[1] William H. Bayley to Dr. Alvin E. Nashman, June 3, 1980, F12/B1, JPL516.

On August 1, 1979, Rechtin and Lehan met with Allen Peterson of Stanford, William Schneider and Walter LaFleur from NASA Headquarters, and various DSN senior staff to assess the network's long-range plan. A critical part of the plan was the Navigation Network Project (NAVNET), an expansion of the system's ability to perform VLBI that had started just months earlier in March 1979. Donald McClure was the NAVNET Project Manager.[2]

In tackling "Navigation Requirements for the 1980's," planners determined that navigators were going to need "a factor of ten improvement in accuracy." NAVNET was based on the belief that "VLBI technology [was] better than X-band 2-way range and Doppler" and that "VLBI technology [was] needed to meet high accuracy navigation requirements." NAVNET included such additional dual-antenna data types as "near simultaneous range" (differencing two-way range measurements made by two antennas) and "Differenced Doppler,"[3] discussed in Chapter 13 as QVLBI.

VLBI and differencing approaches increased demand for antenna time for the simple reason that they needed two dishes instead of one. The idea, therefore, was to establish a DSN subnet earmarked for VLBI, the NAVNET. McClure's group considered three options that included upgrading existing antennas and building a few new ones. The preferred solution was the All-American Navigation Network. It would have added three navigation-dedicated installations in Hawaii, Washington, and Florida. Rather than incur the expense of building three new dishes, the Navigation Network Project had two existing dishes in mind. The candidate for the Hawaii instrument was the Kokee Park Geophysical Observatory. The former DSN Cape Canaveral Spacecraft Monitoring Station (DSS-71), which had closed in 1974 as an economy measure, was the Florida possibility. The All-American Navigation Network would be low-cost and designed for inexpensive maintenance and operation. The antennas would receive both S-band and X-band signals. Data would go to a central correlator via domestic communication satellites in real time. The network also would meet another important goal: minimizing dependence on foreign stations. All facilities would be in "Entirely US national locations." As a result, "All navigation could be done from US: Range and Doppler from Goldstone."[4]

The Networks Consolidation Project quickly cast a shadow over—and then crushed—these plans. The DSN postponed then abandoned the NAVNET as focus shifted to determining which Goddard stations to close or to retain. Budgetary limitations further curtailed any VLBI hopes. The final recommendations of McClure's group, made on May 6, 1980, provided for implementing VLBI technology within the existing DSN. The group acknowledged, as would become clear soon enough, that much more critical work yet was to be done in developing VLBI technology.[5]

---

[2] Hunter to Distribution; McClure to Bayley; McClure to Distribution.
[3] McClure to Distribution; "Long-Range Plan," MFE-9; presentation, Donald H. McClure, Navigation Network Project, Report to TDA Executive Committee, May 6, 1980, DHM-13 & DHM-BU3, F70/B2, JPL516.
[4] Memorandum, Robertson Stevens to William Melbourne, "Steering for Nav Study Team," January 21, 1980, F70/B2, JPL516; "Long-Range Plan," MFE-9, MFE-10 & MFE-15; Hunter to Distribution.
[5] McClure to Bayley.

The DOR of invention. By the time the DSN was contemplating a dedicated VLBI network, JPL navigator Dave Curkendall had invented a new and better form of VLBI: ΔDOR. Unlike other VLBI varieties, it relied on the transmission of tones from a space probe. The basic difference between these so-called DOR tones and the radio waves emitted by quasars was that the latter were akin to white noise, while the DOR tones were two or more modulated sine waves or square waves[6] generated aboard the spacecraft at a specified interval.

"It all came together that night," Curkendall recalled. "I'd just become deputy of the section, and all of a sudden I should know something about VLBI, and I'd never paid any attention to it before." He read "all the Brooks Thomas tomes on the theory of VLBI" published in the DSN progress reports. He also thought about "bandwidth synthesis and how it worked and everything." Bandwidth synthesis was a technique developed by Allan Rogers at MIT in 1970 and involved using multiple channels of data.[7] Curkendall also "knew about the Goddard sidetone [also known as harmonic] ranging system. That had some interesting aspects to it. I forget exactly why. But anyway, I put those two [bandwidth synthesis and sidetone ranging] together."[8]

Curkendall arrived in this way at the idea of collecting not just a broad spectrum of radio waves, such as those coming from a quasar, but detecting multiple, discrete narrow frequency channels (the tones) separated "at interestingly spaced intervals." He realized that "you couldn't just put a noise generator on board the spacecraft, because there's not enough power to do that." Instead, ΔDOR would detect and analyze discrete tones—narrow bands of radio signals—broadcast steadily from the spacecraft. That way, Curkendall explained, "You get away from the real problem in VLBI, that in effect you have to take noises and multiply them together to get a correlation. That squares the noise that's in the receiver. If you can get around that with a sine wave and just do a coherent tracking, you can do it with just a tiny signal, voila, and that's what the signal looks like on ΔDOR signal."[9]

Receiving one-way signals from a spacecraft, however, had the potential to invite errors into the data from a host of sources, such as the lack of precise time synch among DSN stations. "You had to have an independent measurement of that time synch," Curkendall added. The hardware at each station was an additional source of errors. To address such issues, Curkendall decided to have ΔDOR imitate ΔVLBI by "moving those antennas in tandem over to the nearest quasar. Then you're doing lots of things that way. You were solving for the time sync, and you were really figuring out where the spacecraft was relative to the quasar." That knowledge came from the ongoing effort to develop the quasar source catalog. Thus, ΔDOR continued to

---

[6] The terms "sine" and "square" refer to the shape of the radio wave.
[7] Allan E. E. Rogers, "Very-Long-Baseline Interferometry with Large Effective Bandwidth for Phase-Delay Measurements," *Radio Science* 5 (October 1970): 1239-1247.
[8] Curkendall, interview, 31. For the Goddard sidetone ranging system: George C. Kronmiller, Jr., and Elie J. Baghdady, "The Goddard Range and Range Rate Tracking System: Concept, Design and Performance," *Space Science Reviews* 5 (March 1966): 2656-307; Robert J. Coates, "Tracking and Data Acquisition for Space Exploration," *Space Science Reviews* 9 (1969): 361-418.
[9] Curkendall, interview, 30 & 31.

stimulate the DSN's search for extragalactic radio sources. Differencing the signals received at the two antennas, of course, also canceled out the effects of charged particles and contributed to producing more accurate navigation results.[10] Here was the epitome of the "direct" method.

In announcing ΔDOR in 1977, Curkendall outlined its many benefits for navigation. Many advantages derived from the fact that Doppler and range traditionally were two-way data types—involving both the transmission and reception of signals—while ΔDOR was a one-way data type: antennas just listened for the beacon signals (the DOR tones). One-way data eliminated the roundtrip light-time necessary for two-way ranging, and roundtrip light-times for probes exploring the outer solar system were sizeable. However, one-way range (part of the ΔDOR measurement) required that the spacecraft and tracking station clocks be in close synchronization, because any lack of coordination translated into a measurement error. Differencing one-way data received concurrently at two widely-separated stations removed this concern, because the spacecraft clock (and frequency drift) errors would be the same for each measurement. The clock discrepancy that mattered was the difference in time at the two observing stations, and that was a manageable issue. Station clocks had to be in step with each other to less than one nanosecond of accuracy. Installing hydrogen masers and synchronizing DSN clocks with VLBI measurements achieved the requisite level of precision.[11]

A VLBI system for navigation? Whether navigators wanted ΔDOR or the usual ΔVLBI, the DSN VLBI system stood in the way. The Mark II VLBI system, derived from the NRAO's VLBI, recorded data on videotape. Overseas facilities shipped the tapes to JPL for processing on a special purpose correlator located at Caltech. This scheme took weeks to turn around orbit calculations because of delays in shipping the tapes. Then there was the long correlation process at Caltech. "The problem," explained Barry Geldzahler, "was that we had these analog tape-recording systems. It took forever to correlate them. And they were only successful about 25 percent of the time."[12] To say the least, flight projects needed navigation solutions much quicker.

To circumvent these shortcomings, Robert Preston, Lee Brunn, and other navigators came up with an alternative design to allow data transmission and processing in near-real time. They processed VLBI measurements in an idiosyncratic fashion that included a data compression stage. The data also went through the usual adjustments for errors in station locations, UT1, station clocks, and polar motion. The Caltech PHASOR program performed this data massaging on the school's IBM 370.[13]

At this stage in the evolution of VLBI navigation, trajectory estimates in "near real time" meant that ΔDOR deliverables would be available within one day (24 hours). Part of the solution to expediting the movement of data was for the DSN to replace the

---

[10] Curkendall, interview, 35; Melbourne and Curkendall, 5.
[11] Melbourne and Curkendall, 5 & 10-12.
[12] "ΔVLBI Demo Part I," 120; Geldzahler, interview, 22.
[13] "ΔVLBI Demo Part I," 120.

shipment of tapes from overseas with the transmission of data sampled at 500 kilobits per second to JPL over the Ground Communication Facility wideband data lines. In this way, one could transmit the recorded data to JPL within hours.[14]

Additional upgrades accelerated the correlation work. The first step, ready in July 1978, implemented the correlator and related functions on the IBM 360 computers located in JPL's Space Flight Operations Facility, not at Caltech. The next stage was a hybrid of analog and digital components, but by the summer of 1981, the correlator was on a dedicated JPL minicomputer. The minicomputer disk storage expedited data processing dramatically compared to its mainframe predecessor, which had to search along the length of a long ribbon of tape inch by inch in order to find a particular piece of data. The delivery of ΔDOR observables to navigators now happened within 24 hours instead of weeks. A further advantage of the high-capacity disk was that data playback could take place while observations were still underway. This feature was crucial for navigation with tight time constraints on the delivery of processed data.[15]

It was not until 1985, after five long years, that the DSN completed the final phase of its massive upgrade of the VLBI system. The upgrade gave the DSN in effect two versions of VLBI: the Narrow Channel Bandwidth (NCB) and the Wide Channel Bandwidth (WCB) systems. The channel bandwidth designation was based on the data sampling rate. The NCB system provided fast data throughput and served primarily for ΔDOR, time synchronization, and monitoring variations in Universal Time and polar motion. The WCB system basically was compatible with external systems used worldwide by radio astronomers. Each had its advantages that lent it to a specific use. In the DSN, the wide-channel bandwidth VLBI served primarily for producing the position catalog of quasars and other extragalactic radio sources required for analysis of VLBI navigation data. Because of the greater sensitivity and precision of wide versus narrow bandwidth versions, the DSN also used the wide-channel configuration for radio astronomy and geophysical studies of plate motion.[16]

Shootout at Saturn. The scarcity of new mission starts militated against having a probe capable of transmitting the requisite DOR tones. How, then, to try out ΔDOR? Curkendall and his colleagues improvised with craft already flying in space or—in the case of Viking—already sitting on a planet. The first opportunity was the Voyager project, whose managers agreed to a demonstration during the Saturn encounters in 1980 and 1981. The probes would be at near-zero declination. As the declination viewing angle fell toward zero, Doppler performance degraded. This so-called zero-declination problem plagued Doppler navigation. Voyager project managers elected to supplement the Doppler with differenced two-way range. The ΔDOR would help to validate the differenced range.[17]

---

[14] "ΔVLBI Demo Part I," 120; "VLBI Implementation," 24; "Innovations in DDOR," 7.
[15] "ΔVLBI Demo Part I," 120; "VLBI Implementation," 24; "Innovations in DDOR," 7; Liewer, 240.
[16] "Innovations in DDOR," 6; "VLBI Implementation," 25; Liewer, 240; Border, interview, 5-7.
[17] "Innovations in DDOR," 5-6; Melbourne and Curkendall, 1 & 2; "Progress," 56.

The accuracy of two-way range at Saturn was rather limited. The extreme roundtrip light-time between Earth and the Voyagers meant that the tracking stations had to execute the two range measurements sequentially, rather than simultaneously, with each station obtaining a measure of the roundtrip light-time between the probe and the tracking station. The navigation team then had to process the data carefully by setting it to a common time period. The predicted accuracy for the differenced range with the Goldstone-Canberra baseline was about 1 microradian, a level of performance that Frank Jordan showed was commensurate with Voyager requirements. The ΔDOR experimenters hoped for an accuracy of 100 nanoradians, that is, ten times greater (0.1 microradians).[18]

The DSN VLBI navigation system was still a "research and development" activity recording data on magnetic tape. The ΔDOR observations made prior to the Saturn encounters used only two quasars. On the whole, the DSN did not collect a large amount of ΔDOR observations mainly because of its experimental nature. After editing out bad data, 58 ΔDOR data points (baselines) remained. The Voyager measurements made beginning in January 1980 provided angular position accuracies that approached 100 nanoradians, a degree of precision that allowed ΔDOR to contribute to the successful Voyager flybys of Saturn.[19]

The real test of ΔDOR and differenced range, however, was the delivery of accurate orbit determinations at near-zero declinations. No previous deep-space flight had to contend with such long periods of low declination as Voyager, especially during critical mission phases. The same roughly north-south (Goldstone-Australia) baseline that aided differenced range in overcoming low declination also helped ΔDOR. The long roundtrip light-time impaired the differenced range, but not the ΔDOR, as Curkendall had predicted. As the Voyagers neared Saturn, the roundtrip light time exceeded the mutual visibility period of the Goldstone and Australia antennas. Earlier range measurements chosen for differencing had been 15 minutes apart, but now the closest possible points ran from 2 hours 15 minutes to 2 hours 45 minutes. This duration was long enough to allow solar plasma effects to become significant. As a result, the value of the differenced range became questionable. The Voyager mission requirement for differenced range was now an accuracy equivalent to about 600 nanoradians on the Goldstone-Australia baseline.[20]

The differenced range and ΔDOR flight-path solutions disagreed with each other as well as with those derived from Doppler. The navigation accuracies expected prior to Saturn encounter were about 4.0 meters for ΔDOR and 6.4 m for differenced range. Voyager navigators decided that it *appeared* that differenced range *probably* met expectations, and that ΔDOR might have performed more poorly than anticipated. The

---

[18] Melbourne and Curkendall, 3; "Innovations in DDOR," 6; Carl S. Christensen and Herbert L. Siegel, "On Achieving Sufficient Dual Station Range Accuracy for Deep Space Navigation at Zero Declination," 1977 AAS/AIAA Astrodynamics Specialist Conference, September 7-9, 1977, Jackson Hole, WY; J. Frank Jordan, "Future Challenges in Space Navigation," Institute of Navigation, National Aerospace Meeting, Denver, CO, April 13, 1977.

[19] "Innovations in DDOR," 6; "Performance," 41-43; "Progress," 65.

[20] "Performance," 40 & 42-43.

ΔDOR navigators pointed out, though, that low noise levels were "the most significant feature for navigational purposes," because they enabled good orbit determinations. "From this standpoint, ΔDOR excels," they concluded. Only later did they understand that the large offsets in the ΔDOR residuals were caused by inconsistencies in the reference frame. At the time, the outer planet positions were given according to the optical FK4 catalog, while the ephemerides of the inner planets were starting to be tied to the radio reference frame. In March 1982, Voyager retired differenced range as a navigation data type.[21]

The value of ΔDOR for navigating the Voyagers beyond Saturn was not its superiority over Doppler for near-zero angle declinations. Still, there were other good reasons—particularly long roundtrip light-times—to continue using ΔDOR. At Uranus, for example, the lengthy roundtrip light-time reduced significantly the amount of collectable two-way Doppler. At Madrid, in order to acquire 1.5 hours of two-way Doppler or range while the Voyagers were at Uranus, the station had to track the probe for the entire 7-hour period. At Goldstone, the margin was just 1 hour better with an 8-hour viewing period. Only Australia had a spacecraft visibility period—12 hours—that was significantly greater than the round-trip light time. In addition, as the distance to the spacecraft lengthened, noise in the transmitted signals returned to Earth with a vengeance. The transponders would multiply that noise by a factor of 11/3 and retransmit it to the DSN. The Voyagers actually would be amplifying noise and sending it back.[22]

Because ΔDOR was a one-way observation, it lacked this inherent drawback. A few good ΔDOR data points could replace large amounts of Doppler in determining probe flight paths. Voyager began scheduling three ΔDOR measurements per month along with conventional two-way Doppler and range. The new data type continued to deliver even higher levels of accuracy, as the DSN began collecting X-band ΔDOR for the first time in December 1981. The goal now was to validate ΔDOR for future use on Galileo. The estimated accuracy of the X-band ΔDOR was about 30 to 40 cm, the equivalent of 69 to 80 nanoradians. In August 1982, Voyager made ΔDOR an operational data type for the remainder of the cruise to Uranus.[23] Voyager continued to use ΔDOR during the Uranus and Neptune encounters in 1986 and 1989, respectively. Final targeting for these two encounters also relied on optical navigation from the probe's own science cameras. Optical navigators imaged the planet's moons relative to the planet. Thanks to ΔDOR during the cruise phase, Voyager needed just small maneuvers to adjust the craft's trajectory, once camera images determined the Voyager's position relative to the planet.[24]

---

[21] "Performance," 44-45; Personal communication, James Border, October 4, 2012.
[22] "Performance," 45.
[23] "Performance," 41 & 45; "Innovations in DDOR," 7; "Progress," 67.
[24] Donald L. Gray, Robert J. Cesarone, and Richard E. Van Allen, "Voyager 2 Uranus and Neptune Targeting," AIAA 82-1476, AIAA/AAS Astrodynamics Conference, August 9-11, 1982, San Diego, CA, 4 & 7-8; "Innovations in DDOR," 7.

ΔDOR v Doppler. Galileo would be the next outer-planet mission to give ΔDOR a tryout. In the meantime, Venus offered two opportunities for sending up a ΔDOR "trial balloon." The first literally was a balloon, the Venus Balloon experiment, a cooperative venture of the Soviet, French, and United States space agencies. Two Soviet Vega probes released instrumented balloons into the Cytherean atmosphere on June 11 and 15, 1985, to extend the knowledge gained by earlier Soviet Venera and NASA's Pioneer Venus probes. The French Centre National d'Etudes Spatiales organized a multinational collection of DSN, Soviet, and radio-astronomy antennas to acquire scientific data from the craft.[25]

The experience highlighted the fundamental differences between JPL and Soviet approaches to deep-space navigation. The Soviet Union relied on daily short passes—lasting 10 to 20 minutes—of range and Doppler. Their orbit estimates fit the data over relatively long tracking arcs that reflected a certain degree of change in the spacecraft geometry relative to Earth. Their use of range, especially for planetary encounter navigation, increased the sensitivity of their trajectory estimates to errors in the planetary ephemeris, station locations, and other sources. JPL, in contrast, used continuous horizon-to-horizon passes of two-way Doppler lasting 8 to 10 hours. Trajectory computations derived from comparatively shorter tracking spans. The nearly continuous Doppler sensed changes in the flight path brought about by the planet's gravitational pull on the spacecraft. The time of closest approach was visible directly in the Doppler residuals during the flyby. Unlike the Soviets, JPL generally did not use range for flyby orbits because of its sensitivity to planetary ephemeris and station location errors. For the Venus Balloon Experiment, however, two-way Doppler was not an option.[26]

With the Magellan radar mapper, ΔDOR and its close relative, Delta Differential One-way Doppler (ΔDOD), became mission requirements. The purpose of the ΔDOD was to measure angular velocity during the orbit phase, while the mission would call on ΔDOR to measure angular position during the cruise phase. As with Vega and Voyager, the Magellan craft did not generate DOR tones, so navigators improvised them from the X-band telemetry system that returned radar images to Earth. Because the DSN tracked Magellan with 34-meter, rather than 64-meter, antennas, navigators had to select stronger radio sources at greater angular separations from the craft.[27] They successfully used ΔDOR during the Earth to Venus cruise, but ΔDOD did not fare so well.

Catherine Thornton recalled:[28]

---

[25] Robert A. Preston, James H. Wilcher, and Charles T. Stelzreid, "The Venus Balloon Project," 195-201 in TDAPR 42-80; Stelzried, Preston, Claude E. Hildebrand, James H. Wilcher, and Jordan Ellis, "The Venus Balloon Project," 191-198 in TDAPR 42-85; Mudgway, *Uplink-Downlink*, 190 & 193; Roald Z. Sagdeev, Vechaslav M. Linkin, Jacques E. Blamont, and Robert A. Preston, "The VEGA Venus Balloon Experiment," *Science* 231 (March 21, 1986): 1407-1408.
[26] Jordan Ellis and Timothy P. McElrath, "Determination of the Venus Flyby Orbits of the Soviet Vega Probes Using VLBI Techniques," 274 in TDAPR 42-93.
[27] "Innovations in DDOR," 8; Thornton, interview, 17; "Doppler and Interferometric," 919-939; "Reference Sources," 152-163; "Higher Density Catalog," 274-300.
[28] Thornton, interview, 13-14.

The Delta DOR actually gave us great angle data. The narrowband [ΔDOD] provided something analogous to Doppler. It is an angular rate measurement. Unfortunately, the first VLBI data acquisition system which supported Magellan was implemented by the DSN a little hastily. The equipment had its problems, and the operators had their own problems, and the project got frustrated with all the scheduling of the two antennas. They had to go off the spacecraft to view the quasar, which was another thing that projects never have liked. The failure rate for those early VLBI observations certainly did not endear them to the navigators. But in spite of those early problems we did acquire some valuable measurements, primarily Delta DOR, and we were able to identify ways the system could be improved.

Magellan project managers initially believed that ΔDOD was necessary to achieve the precision positioning and pointing of the spacecraft during its mapping mission. The craft's orbital period was substantially shorter than that for any previous NASA planetary orbiter, just 3.1 hours. Moreover, various factors made data available during only 60 percent of each orbit. Doppler was not available for a substantial portion of an hour during each orbit's periapsis, because the radar was using the high-gain antenna to image Venus, and for 15 minutes after apoapsis, when the probe was scanning reference stars to calibrate the attitude-control gyros.[29]

Magellan eventually dropped ΔDOD for differenced Doppler. The decision was not entirely a surprise. The inclusion of ΔDOD on Magellan had been met with suspicion in some quarters. In May 1987, during the Second Final Mission Design Review, Allen Berman, who reported to Ray Amorose, pointed out that the DSN might not "fully meet" the mission's ΔDOD requirements. "[O]pportunities" would "be lost due to [scheduling] conflicts with other flight projects." Berman proposed augmenting ΔDOD with differenced Doppler, which he believed could "deliver comparable accuracy." The basis of his assessment was a study by the DSN's Advanced Systems unit on the accuracy of differenced Doppler specifically as a backup to ΔDOD. Berman advised Magellan to consider differenced Doppler "as a backup to VLBI" and to "watch the navigation and DSN studies to determine orbit-determination accuracies at Sun-Earth-probe (SEP) angles smaller than $10°$."[30]

Despite Berman's warnings, Magellan managers still believed that ΔDOD was worth the added DSN scheduling expense to achieve the desired level of navigation precision at Venus. Throughout the cruise phase, from July 1989 to August 1990, navigators used both two-way Doppler and ΔDOR with highly accurate results. The targeting for orbit insertion was so accurate that no orbit trim maneuvers were needed for the scientific mission to begin. Even so, project management soon came around to Berman's way of thinking and decided that differenced Doppler alone was sufficiently accurate for the mapping operation. The main reasons for their decision were the lack

---

[29] "Magellan Lessons Learned"; Wood, "Nav Evolution, 1989-1999," 878-879; Ellis to Stelzried.
[30] Presentation, Allen L. Berman, "Magellan: Final Mission Design Review #2: Tracking and Data Systems," May 6, 1987, F56, JPL4, ALB-1, ALB-20 & EC-94; Ellis to Stelzried.

of ΔDOD's tight antenna scheduling needs and its need to skew antennas from the probe to observe a quasar. The project also understood that shifting from S-band to X-band Doppler would improve navigation accuracy and, at the same time, the DSN was discovering how to calibrate tracking data better with the GPS. Before making a final decision, project managers ordered a comparison test of differenced Doppler and ΔDOD as Magellan journeyed toward Venus.[31]

Subsequently, on May 9, 1990, at a project management meeting, John McNamee announced the elimination of ΔDOD from the mapping mission. Differenced Doppler was operationally simpler and cheaper. Project managers had concluded that both data types provided essentially the same navigational information. Also, simulations had shown that differenced Doppler was capable of the accuracies that mission scientists needed for mapping.[32]

Differenced Doppler reduced antenna scheduling conflicts with other projects, conflicts that had been a point of contention between the Magellan and Galileo projects. Its smaller appetite for antenna time translated into manpower reductions (costs savings) within the DSN and the navigation team. Unlike ΔDOD, the DSN collected two-way Doppler as a normal part of its tracking operations during any given flight. It did not demand any special scheduling to take advantage of spacecraft and quasar mutual visibility or a sequence of observations between spacecraft and quasars. It simply entailed scheduling a second antenna at a distant complex to acquire an overlapping pass of three-way Doppler. Additionally, differencing did not call for elaborate data correlation or extraordinary effort to transport data over great distances. Just fifteen minutes of differenced Doppler produced by an unplanned or coincidental overlap was usable for navigation. It was available in near real time within the normal delivery schedule of conventional radiometric data. The only drawback, it would seem, was that it was not as accurate as ΔDOD under flight conditions.[33]

Galileo: a frustrated first. Galileo repeated Magellan's cycle of acceptance followed by reluctance, but for a different reason. Like Magellan, Galileo made ΔDOR a navigation requirement. But, for the first time, the probe had the ability to generate DOR tones—both S-band and X-band DOR tones—from the start. The ΔDOR requirement accordingly led to an expansion of the JPL catalog of radio sources particularly around the ecliptic.[34] The complex flight path also called for Galileo to use ΔDOR to navigate within both the inner and outer solar system.

The Venus-Earth-Earth Gravity Assist (VEEGA) trajectory took the craft from Earth to Jupiter along a distinctly indirect route. It ultimately called for Galileo to perform 16 encounters—four planets, two asteroids, and 11 flybys of the Galilean

---

[31] "Magellan Lessons Learned"; "Doppler and Interferometric," 919-939; "Innovations in DDOR," 8; Mohan and D'Souza; Wood, "Nav Evolution, 1989-1999," 879.
[32] McNamee to Distribution.
[33] McNamee to Distribution; memorandum, Theodore Tate to John McNamee, "MCT Impact Resulting from Delta VLBI Deletion," May 17, 1990, F508/B30, JPL508; Ellis to Stelzried.
[34] JPL, Functional Requirement, Galileo Orbiter Navigation, February 10, 1983, 4, F8/B1, JPL169; "Reference Sources," 152-163; "Higher Density Catalog," 274-300.

moons Io, Europa, Ganymede, and Callisto—and a total of 68 propulsive maneuvers, by far many more than any previous planetary mission. The series of gravity assists at Venus and Earth (twice) demanded navigation of the highest precision. For example, the first Earth maneuver occurred at an altitude of only 960 kilometers. While still approaching Jupiter, the orbiter passed Europa and Io. Over a period of just 7 hours, scheduled to take place on December 7, 1995, the probe portion entered the planet's atmosphere and the orbiter segment flew just 1,000 km above Io (to avoid possible volcanic ejecta). The orbiter received signals from the probe and relayed them to Earth before entering orbit around Jupiter. The Io encounter provided a gravitational assist and helped to establish the Jupiter-centered orbit.[35] A key to navigating Galileo through these tight gravity assists was believed to be ΔDOR.

Navigation at Venus in February 1990 relied on S-band and X-band Doppler and range only. However, when precision became a priority, Galileo navigators went to the ΔDOR. It proved especially valuable for improving the accuracy of course correcting maneuvers. As Shyam Bhaskaran explained, ΔDOR solutions were more accurate, because it was "able to sense the motion of the spacecraft in the plane-of-the-sky directly." In fact, the sixteenth and seventeenth trajectory adjusting maneuvers performed in 1992 were so accurate that project managers canceled the next one.[36]

This degree of precision also was essential for the first Earth gravity maneuver, which took place on December 8, 1990, at an altitude of 960 kilometers over the Caribbean Sea. Navigators relied on ΔDOR as well as range and Doppler. The outstanding performance of ΔDOR also allowed the Galileo probe to conserve propellant for the Jovian science mission. Nonetheless, the actual fuel saving was only half the anticipated amount, probably due to a combination of low spacecraft declination during the encounter and a lack of sufficient ΔDOR observations.[37]

The navigation team looked forward to ΔDOR performance at Jupiter. A hardware mishap, though, flattened their expectations. In April 1991, between the first and second Earth gravity assists, the Galileo craft failed to deploy its high-gain antenna. The malfunction effectively crippled its ability to send DOR tones. The antenna loss also meant planning the mission anew even as the craft was in flight. Project managers dropped the ΔDOR requirement for Jupiter approach. Even so, the DSN did make some S-band ΔDOR measurements during the orbital phase (July 1996 to September 1997) to tie the position of Jupiter to the inner planets within the radio reference frame.[38]

---

[35] Christopher L. Potts and Michael G. Wilson, "Maneuver Design for the Galileo VEEGA Trajectory," 1019-1047 in Arun K. Misra, Vinod J. Modi, Richard Holdaway, and Peter M. Bainum, ed., *Astrodynamics 1993*, Pt. II (San Diego: Univelt, 1994); D'Amario, 116-117 & 141; Gray, 335-352; William M. Folkner, "Navigational Utility of High-Precision Radio Interferometry for Galileo's Approach to Jupiter," 35 in TDAPR 42-102.

[36] "Innovations in DDOR," 8; "1989-1995: Galileo Navigation," 10; "Galileo OD Earth-2," 3.

[37] Mudgway, *Uplink-Downlink*, 307; Gray, 335-352; "1989-1995: Galileo Navigation," 10-12 & 22.

[38] "Innovations in DDOR," 8; "Galileo OD Earth-2," 1; Louis A. D'Amario, Dennis V. Byrnes, William E. Kirhofer, Frank T. Nicholson, and Michael G. Wilson, "Navigation Strategy for the Galileo Jupiter Encounter and Orbital Tour: An Overview," 2, http://trs-new.jpl.nasa.gov/dspace/bitstream/2014/28053/1/95-0095.pdf (accessed July 8, 2010); "1989-1995: Galileo Navigation," 8.

Mars Observer: Mars (not) observed. Because the Galileo project abandoned X-band ΔDOR at Jupiter, the Mars Observer became the first spacecraft to use X-band ΔDOR with actual DOR tones for navigation.[39] The ill-fated probe, the product of NASA cost-cutting, was the chronologically first of the so-called Planetary Observer class of probes. It launched in 1992, instead of 1990, because of the *Challenger* accident and NASA's desire to avoid excessively burdening its fiscal 1987 budget.[40]

On August 24, 1993, as Mars Observer approached its target, the craft disappeared. The flight had been uneventful; the navigation team was surprised. Sam Dallas, the Mars Observer Mission Manager, recalled that "you could tell things went well" by the tracking data. "That's why we were very surprised on Mars Observer that something went wrong. We don't have any idea what went wrong." "We never got any information back. No signals from it after that," he explained. "As far as we know everything was working correctly right up to that maneuver. All the data indicated we were on the right trajectory." The Mars Exploration Rover that later explored the planet failed to find spacecraft fragments on the surface. "So we don't know. If it flew by, we would have gotten information back. It's a mystery."[41] The mysterious and still unexplained disappearance of the probe has given rise to a myriad of conspiracy theories.[42]

Navigators wanted to understand what had happened and to find the lost spacecraft. Al Cangahuala, a member of the navigation team, remembered that "they had a little barbecue for the staff," when Mars Observer launched. NASA lost contact on a Saturday.[43]

> [On Sunday,] we were all in doing what-ifs. You know, run out the trajectory, and do the best fit, and study that last bit of tracking. Did you see anything in the last few passes that would have you think that

---

[39] Memorandum of Understanding between Mars Observer and TDA Technology Development for Delta DOR Observations using Mars Observer, May 1993, F "DSN" B9, JPL496; Cangahuala, interview, 9; Traxler, 118.

[40] McCurdy, *Low-Cost Innovation*, 5 & 6; GAO, *Space Exploration: Cost, Schedule, and Performance of NASA's Mars Observer Mission*, NSIAD-88-137FS (Washington: GPO, May 1988), 9, 13 & 15.

[41] Dallas, interview, 18 & 33.

[42] In general, the basis of the theories is a belief that NASA is concealing the existence of aliens. For example: Steve Connor, "Triumph of the Martian Triangle: A NASA Spaceprobe goes Missing and Conspiracy Theories Abound and Multiply, But Why are We so Fascinated by the Red Planet?" *The Independent*, August 29, 1993, http://www.independent.co.uk/news/uk/triumph-of-the-martian-triangle-a-nasa-spaceprobe-goes-missing-and-conspiracy-theories-abound-and-multiply-but-why-are-we-so-fascinated-by-the-red-planet-1464065.html (accessed May 19, 2010); "Mars Conspiracy Theory," *Free Press Release*, October 18, 2009, http://www.free-press-release.com/news-mars-conspiracy-theory-1255884672.html (accessed May 19, 2010); "Life on Mars: Why Now?" http://www.theforbiddenknowledge.com/hardtruth/life_on_mars_now.htm (accessed May 19, 2010). An independent investigation led by the NRL concluded that the most likely cause was a rupture in the probe's propulsion system. "HQ94-1/Mars Observer Report Released," http://www.msss.com/mars/observer/project/mo_loss/nasa_mo_loss.txt (accessed May 19, 2010).

[43] Cangahuala, interview, 13-14.

something was awry? Then fit that data, run the trajectory out. Run the trajectory out, and now let's add, say, that there was an explosion or a leak. Let's open up the non-gravitational accelerations and see what happens. Then we had to design alternate Mars orbit insertions. You know, we go by the planned burn time, and we still have contact. Let's say we get contact a day later, can we still salvage the mission? So, for a week there was just a furious rush of doing those kinds of activities. After a week or so, there wasn't anything we could do. We had hoped that it had gone ahead and done the burn-off on its own, but we never found it.

Mars Observer was a failure from the perspective of attempting to cap mission costs and achieving its main scientific objectives. Yet, it was a navigation success. For one, navigators successfully demonstrated the value of X-band ΔDOR. VLBI data turnaround time had been shrinking. The DSN had acquired VLBI data during the Vega Halley encounters, and the 36-hour turnaround time represented the most stringent demand placed on the VLBI system until that moment. The Magellan project collected ΔDOR baselines daily with a turnaround time of 12 hours or less.[44]

On July 8, 1993, during the Mars Observer readiness review for Mars Orbit Insertion, DSN manager Marvin Traxler expressed concerns about the ability to deliver ΔDOR measurements within 12 hours. His concern arose from the fact that the navigation team and the DSN had not practiced delivering ΔDOR under those conditions. So, a drill was to take place on Saturday, July 24, 1993, with the goal of demonstrating "VLBI data delivered to project within 7 hours." The drill apparently worked. Delivery of ΔDOR measurements during the flight was right on time and without any problems, according to mission manager Sam Dallas.[45]

The accuracy of ΔDOR again proved itself to be particularly useful for tweaking flight paths. The Mars Observer ΔDOR X-band measurements had an accuracy approaching 20 nanoradians in the quasar reference system. The probe did not need as many trajectory adjustments as Galileo. It executed only three flight-path corrections, for which navigators used a combination of Doppler, range, and ΔDOR. They also relied on ΔDOR throughout the 337 days of interplanetary cruise up until 30 days prior to Mars orbit insertion. As with the Magellan mission, the mapping phase required neither ΔDOR nor ΔDOD.[46]

---

[44] Ellis to Stelzried; Sam W. Thurman and Gilbert Badilla, "Using Connected-Element Interferometer Phase-Delay Data for Magellan Navigation in Venus Orbit," 48-54 in TDAPR 42-100.

[45] Presentation, Traxler, "Delta Readiness Review for Mars Observer, MOI Phase," July 8, 1993, F "DSN" B9, JPL496; Dallas, interview, 18.

[46] Peter M. Kroger, James S. Border, and Sumita Nandi, "The Mars Observer Differential One-Way Range Demonstration," 2 in TDAPR 42-117; Presentation, Pasquale B. Esposito and Robert A. Mase, Navigation Team, "Mars Observer TCM-3 Preliminary Assessment," March 22, 1993, F "TCM-3" B9, JPL496; Presentation, Eric J. Graat, "Preliminary Orbit Determination Results for TCM-1," October 12, 1992, F "TCM & NAV" B18, JPL496; Pasquale B. Esposito, Mars Observer Navigation Team Plan and Status, December 3, 1991, B45, JPL520; memorandum, Thomas Penn to Pat Esposito, "Mars Observer Spacecraft Capability to Meet Navigation Tracking Requirements," November 27, 1990, B67, JPL520.

Chapter 19

The New Scientists of Navigation

Navigators were multidisciplinary scientists, unlike the single-discipline scientists of, say, the nineteenth century. They incarnated the convergence of science and engineering typical of the last half of the twentieth century. During the 1980s and 1990s, they participated in the newest and fastest growing field of solar-system research: asteroids and comets. The burgeoning number of asteroid discoveries reflected the field's expansion. The count of just near-Earth asteroids—those whose orbits have perihelions of 1.3 astronomical units or less—grew from 86 in 1980 to 170 in 1990 to 1,241 in 2000 and 7,565 in 2010, as the annual rate of discovery reached about 400.[1]

Navigation to these small, but gravitationally complex bodies benefited from a new imaging technology, the Charge-Coupled Detector (CCD) camera. The exploration of asteroids saw navigators, in their role as project scientists, continue their research in celestial mechanics, radio occultation, and gravity. This exploration also engaged them in a new way with ground-based astronomers who, using telescopes also equipped with CCD cameras, provided crucial observations for navigators' development of asteroid ephemerides.

The development of ephemerides remained one of navigators' central scientific activities. The nature of those ephemerides, though, needed to change to reflect the directions being taken by scientists. Star catalogs remained in use by astronomers and navigators, but optical navigators increasingly found them to be out of date. At the same time, navigators were using the DSN to seek out extragalactic radio sources for VLBI and Delta DOR. Navigators thus straddled both stellar and radio reference frames, but they also called upon yet more ephemerides that they had constructed for the inner planets and the Moon. Navigation had become complicated.

In an effort to establish a single unified reference frame, navigators undertook numerous observations of spacecraft with the DSN. Bringing the outer planets into this unified reference frame also entailed calling upon radio astronomers and the Very Large Array (VLA), the facility featured in the movie *Contact*. Eventually, navigators pulled together the disparate radio frames along with a new star catalog issued by the European Space Agency into the latest Development Ephemerides (DE405).

The CCD Revolution. Galileo marked a significant advance in optical navigation. It exploited a new technology—the CCD camera—that overcame the disadvantages of vidicon technology while improving resolution quality dramatically. Steve Synnott deemed the introduction of the CCD to be "the biggest single step up to a different

---

[1] John F. Kross, "Asteroids and Elsewhere: Getting There by a Flexible Path," *Ad Astra* (Winter 2010): 18; JPL Near Earth Object Program, "Near-Earth Asteroid Discovery Statistics," http://neo.jpl.nasa.gov/stats/ (accessed January 17, 2011).

plateau" in optical navigation.² The advantages of switching to CCD cameras were obvious in principal, but would they deliver during an actual flight?

"Of course, anytime you're doing something the first time, there are issues," Lincoln Wood explained. "Suppose the charge-coupled device can't see the stars that you want as background objects for optical navigation images and so forth. A lot of issues needed to be worked there, and a lot of interaction with instrument designers on the part of the navigation team. And in actuality it worked terrifically. But we didn't know in 1977 or 1978 that that was necessarily going to be the case."³

The same electronic technology was transforming ground-based astronomy. Observatories began switching from film to CCD cameras in many cases with financial help from NASA.⁴ Observations made with these cameras gave positions that were far more precise. This enhanced accuracy would be vital to optical navigation because of its use of starry backgrounds. The more exactly navigators knew star positions, the more accurately they could estimate spacecraft trajectories. Galileo's pioneering encounters with the mainbelt asteroids 951 Gaspra and 243 Ida exemplified the key role of CCD imaging.

Those asteroid encounters made use of a ground-based campaign by observatories in the United States, Chile, Great Britain, Japan, and New Zealand using CCD cameras. Although astrometric information on Gaspra was available back to its discovery in 1913, optical navigators needed more exact positional details. Don Yeomans, Paul Chodas, and others made advance preparations to observe the asteroid during the oppositions of 1989-1990 and 1991 prior to the October 1991 Galileo encounter. Under the direction of Arnold R. Klemola, Lick Observatory prepared two special star catalogs for the two oppositions. Ida equally was the subject of an intense ground campaign by experienced astronomers using CCD cameras, automated measuring devices, and advanced data reduction methods plus special Lick Observatory star catalogs to determine its ephemeris with the highest exactitude possible during the opposition of 1992-1993. These ventures improved the asteroid ephemerides dramatically, in the case of Gaspra by a factor of about three and for Ida by a factor of about four.⁵

At least as important as the CCD observations were the star catalogs that gave the position for each object navigators saw in the starry backgrounds of their images. Mariner and Viking optical navigation had relied on the whole-sky star catalog compiled by the Smithsonian Astrophysical Observatory and published in 1966. Most of that catalog's material came from the Yale Zone Catalogs of the 1930s, making the star positions three decades out of date.⁶

---

² Synnott, interview, 12-13 & 21; "Brief History of JPL OpNav," 336 & 344-345; Wood, "Nav Evolution, 1989-1999," 887; "OpNav for Gaspra," 361-369.
³ Wood, interview, 13.
⁴ Standish, interview, 17-18 & 33.
⁵ Donald K. Yeomans, Paul W. Chodas, Michael S. Keesey, William M. Owen, Jr., and Ravenel N. Wimberly, "Targeting an Asteroid: The Galileo Spacecraft's Encounter with 951 Gaspra," *The Astronomical Journal* 105 (April 1993): 1547-1549 & 1551-1552; "Galileo OD for Ida," 1028; "Galileo OD for Gaspra," 370-380.
⁶ SAO, *Star Catalog: Positions and Proper Motions of 258,997 Stars for the Epoch and Equinox of 1950.0*, 4 vols. (Washington: Smithsonian Institution, 1966).

Starting with Voyager, JPL undertook the creation of its own ad hoc star catalogs under contract with Lick Observatory. Klemola and his colleagues pointed their 20-inch dual astrograph[7] at coordinates indicated by JPL navigators. From time to time, navigators Linda Morabito, Homa Taraji, Adriana Ocampo, and Bill Owen traveled to Santa Cruz to select star targets. Lick exposed glass photographic plates; CCD technology lay in the future. The process was tedious and involved a joystick and a screen with crosshairs. The Lick computer generated precise coordinates for each image, and Klemola sent the reduced data to JPL as a catalog on magnetic tape.[8]

When Galileo optical navigators turned to the Smithsonian catalog positions for the Gaspra and Ida encounters, they discovered systematic errors that, if left undetected, might have imperiled the mission. Therefore, they called upon the special Lick catalogs as well as the CCD ground-based observations.[9] The introduction of CCD technology into observational astronomy soon enabled the creation of a more exacting star catalog for optical navigators and astronomers. The Hipparcos (HIgh Precision PARallax COllecting Satellite) and Tycho catalogs, released by ESA in 1997,[10] changed the world of stellar astrometry overnight.

Astronomers and navigators now knew the positions and motions of 120,000 stars to the highest degree of precision yet. The lower accuracy Tycho Catalog contained over a million stars, while the 2000 version catalogued and described 2.5 million stars. A similar campaign to produce a more accurate measure of star positions was underway by Norbert Zacharias and his Naval Observatory colleagues. Starting in February 1998, they measured and reduced the Astrographic Catalogue material from the turn of the twentieth century once again, but with the latest observational technologies. They imaged the entire sky with CCD cameras and produced the first USNO CCD Astrograph Catalog ("UCAC"). Version 2 of the catalog covered 86 percent of the sky, omitting the northern polar region.[11]

Ad asteroids! By including optical navigation from the start, Galileo mission designers had hoped to repeat Voyager's resoundingly successful use of images to navigate the outer solar system, specifically while Galileo was approaching and orbiting Jupiter and its moons. Unfortunately, optical navigation played a far lesser role because of the loss of

---

[7] An astrograph is a telescope especially equipped to take photographs; the dual astrograph had one lens corrected for yellow and the other for blue light.
[8] "Brief History of JPL OpNav," 345.
[9] "Brief History of JPL OpNav," 345.
[10] Michael A. C. Perryman, ed., *The Hipparcos and Tycho Catalogues: Astrometric and Photometric Star Catalogues derived from the ESA Hipparcos Space Astrometry Mission*, 16 vols. (Noordwijk, Netherlands: ESA Publications Division, 1997).
[11] Norbert Zacharias, Sean E. Urban, Marion I. Zacharias, David M. Hall, Gary L. Wycoff, Theodore J. Rafferty, Marvin E. Germain, Ella R. Holdenried, John W. Pohlman, F. Stephen Gauss, David G. Monet, and Lars Winter, "The First U.S. Naval Observatory CCD Astrograph Catalog," *The Astronomical Journal* 120 (October 2000): 2131-2147. Version 3 was released in August 2009, the fourth in August 2012. Zacharias, Charlie T. Finch, Terry M. Girard, Arne Henden, Jennifer L. Bartlett, David G. Monet, and M. I. Zacharias, "The Fourth US Naval Observatory CCD Astrograph Catalog (UCAC4)," *The Astronomical Journal* 145,2 (1 February 2013), 44-57.

Galileo's high-gain antenna and a hardware anomaly detected in the probe's tape recorder.[12] Instead, the highlight of Galileo optical navigation was a first: the exploration of the asteroids Gaspra and Ida. This success is ironic in that the initial mission design had not included any asteroid research.

Steve Synnott, the optical navigation group's supervisor, recalled that the "trajectory dynamicists" wanted "to find interesting things to do and optimum ways to get to planets and things. Well, one of the things that they were able to do is to fly the Galileo trajectory vicariously in the computer through a database of asteroids." Don Yeomans' celestial mechanics group had a database of all the known asteroids. They flew Galileo through the asteroids, at least as a computer simulation, "and they realized that Gaspra was not that far off our trajectory, if we take a little dogleg with a couple of trajectory change maneuvers, we could go have a close encounter. That's how it came about." Yeomans characterized it as "bonus science." "[T]hey could fly by these two objects without expending too much fuel. It allowed them to get the cameras exercised and utilized and calibrated prior to the encounter with Jupiter itself. It didn't cost much. The science benefits were likely to be fairly good, which turned out to be the case."[13]

A series of trajectory corrections brought Galileo to within just 1,601 km of Gaspra (6 km from its aim point) on October 29, 1991. The Ida encounter reflected equally precise navigation. Good orbit determinations allowed navigators to cancel the last of the three Ida flight-path modifications. The Ida flyby also took place less than 6 km from the desired aim point and at a distance of 2,391 km on August 23, 1993, when camera images revealed the first known asteroid satellite later designated Dactyl.[14]

The loss of Galileo's high-gain antenna in April 1991 might have thwarted this asteroid science, if it had not been for several clever modifications. As it was, using the low-gain antennas meant taking 70 hours, instead of minutes, to transmit navigation images. The total number of pictures available to navigators fell harshly from 100 to just 4 for Gaspra and only 5 for the Ida encounter. More spacecraft mishaps further restricted the return of images. Each picture had to count, especially as navigation competed with science for antenna access. The solution was to conserve available bandwidth. Ed Riedel, Robin Vaughan, and Mike Wang developed two data-compression fixes. One solution discarded as much background as possible while retaining only the meaningful parts. The other had Galileo's flight computers process the CCD images in real time, thereby greatly reducing the number of pixels transmitted to the ground. This fix, though, necessitated a colossal reprogramming of the camera's onboard software. Optical navigators also devised a new technique—called a single-frame mosaic—that squeezed the greatest visual information into each image. The procedure built up a series of images of Gaspra against background stars into a single picture by moving the scan platform in steps, while the science camera's shutter remained open. Each image frame contained the information equivalent to multiple conventional images. The

---

[12] D'Amario, 127-128; "Brief History of JPL OpNav," 336; Wood, interview, 12-13; "Galileo OD for Ida," 1031-1035.
[13] Synnott, interview, 19-20; Yeomans, interview, 15.
[14] "Galileo OD Satellite Tour," 1494-1495; D'Amario, 118-119 & 130-121.

method worked remarkably well during both encounters[15] and, as we shall see, found use on later missions.

Waltzing by Mathilde.
The first NASA mission explicitly intended to undertake asteroid studies was the Near Earth Asteroid Rendezvous (NEAR) probe. Later renamed NEAR Shoemaker, the spacecraft achieved a number of "firsts." It was the first spacecraft to orbit an asteroid and the first to land on an asteroid. The landing was all the more extraordinary because landing ability was not part of the probe's design, and it remained in communication with the DSN after alighting. From a navigational perspective, NEAR's orbit around the asteroid was a particularly intricate case of the standard two-body problem that arose from its peculiar shape and gravitational field not to mention its center and direction of rotation.[16] Managed by APL, NEAR's ultimate trajectory took it to Eros, the second largest near-Earth asteroid. Along the way, it also flew past 253 Mathilde, a main-belt asteroid a mere 50 km in diameter. Optical navigation was not just a mission highlight; it was a necessity.

Bobby Williams and the NEAR navigation team—Jim Miller, Bill Owen, Mike Wang, Clifford Helfrich, Peter Antreasian, Eric Carranza, John Bordi, and Steve Chesley—followed the probe's progress via X-band Doppler and range, reserving optical navigation for asteroid approaches. All in all, they achieved strikingly precise results. The accuracy of orbit estimates allowed project management to cancel a course correction scheduled for 13 hours before the flyby. The team benefited from CCD observations by astronomers during Mathilde's 1995 and 1996 oppositions as well as the new Hipparcos and Tycho star catalogs. As a result, navigators succeeded in flying NEAR past Mathilde at a distance of only 1,212 km.[17]

For the optical navigators, Steve Synnott, Bill Owen, and Mike Wang, Mathilde was a practice run for Eros. They acquired images during the last two days before reaching the asteroid. The pictures arrived in batches of sixteen on six separate occasions. Their acquisition was not straightforward. The position of the Sun relative to the spacecraft required pointing NEAR's solar arrays 50 degrees away from the Sun line. However, pointing the camera just 40 degrees from that direction let stray light into the navigation pictures and dimmed the background stars. Turning the probe in this direction also shifted the asteroid's center of brightness away from its center of mass.

---

[15] "Galileo OD Satellite Tour," 1494-1495; D'Amario, 119-120; Wood, "Nav Evolution, 1962-1989," 295-296; "Brief History of JPL OpNav," 336-338; Maize, interview, 10; "Galileo OD for Gaspra," 370-380. For a description of the single-frame mosaic technique, see "OpNav for Gaspra," 361-369.
[16] Peter Antreasian, Stephen R. Chesley, James K. Miller, John Bordi, and Bobby Williams, "The Design and Navigation of the NEAR Shoemaker Landing on Eros," AAS 01-372, 2001 AAS/AIAA Astrodynamics Specialist Conference, July 30-August 2, 2001, Quebec City, Quebec, http://trs-new.jpl.nasa.gov/dspace/bitstream/2014/12979/1/01-1395.pdf (accessed March 10, 2011); "NEAR Encounter Mathilde," 1157-1173.
[17] "Navigation for NEAR Shoemaker," 7; Bobby G. Williams, James K. Miller, Daniel J. Scheeres, Clifford E. Helfrich, Tseng-Chang M. Wang, William M. Owen, Jr., and George D. Lewis, "Navigation Results for NASA's Near Earth Asteroid Rendezvous Mission," 87-98 in *A Collection of Technical Papers: AIAA/AAS Astrodynamics Specialist Conference, July 29-31, 1996, San Diego, CA* (Reston, VA: AIAA, 1996); "NEAR Encounter Mathilde," 1157-1173; Wood, "Nav Evolution, 1989-1999," 886 & 892.

As a result, Mathilde was not visible in navigation pictures until the second imaging attempt, and only after navigators added several image frames together. In the end, the asteroid was visible in individual frames only on the fifth try.[18]

Do-si-do at Eros. The Mathilde exercise did not seem to presage a trouble-free flyby of Eros. The mission design called for four corrective maneuvers to slow NEAR's speed and to allow the asteroid to capture it into orbit. The first maneuver, commanded on December 20, 1998, was the first and largest of the four, but an onboard software problem forced its cancellation. The craft tumbled; communication was lost. Luckily, the craft corrected itself, but at the expense of a large amount of precious fuel. As a result, on December 23, 1998, NEAR zipped past Eros—too fast and too far away (at a distance of nearly 3,900 km)—for the asteroid to capture it.[19]

Nonetheless, controllers were able to command NEAR to photograph Eros as it sped by. The APL mission design team under Bob Farquhar came up with a recovery maneuver, executed on January 3, 1999, to match the spacecraft's speed to that of Eros. By then, the craft was nearly a million kilometers from Eros, so NEAR had to perform a long "U-turn" trek to return it to the asteroid on February 14, 2000. The rescue was possible only because NEAR had a generous fuel supply and adequate contingency plans.[20]

Eros optical navigation did not rely on pictures of Eros against a starry background for several reasons. The asteroid's irregular shape complicated finding its center with the opnav software. When NEAR was in low orbit, the camera saw only a portion of the surface, not the entire asteroid body. Additionally, the short exposures needed to avoid overexposing Eros were not long enough to capture any but the very brightest stars.[21]

Instead, navigators relied on a scheme based on the location of craters or "landmarks" in the images. In effect, the software deduced the position of the spacecraft relative to the center of Eros by referring to landmarks and from knowledge of the spacecraft's attitude contained in the telemetry. A rudimentary ability to process landmark observations had existed in the software for 15 years, but that program plotted landmarks as small plus signs. This clearly was inadequate. Navigators updated the code, so that the program could distinguish one crater from another on Eros' heavily cratered surface. The definition of a landmark went from a plus sign to a three-dimensional circle with a certain radius and orientation. Assuring that each observation of a given landmark always referred to the same feature out of the tens of thousands of craters required the use of surface modeling software, which lay in the future. So,

---

[18] "NEAR Encounter Mathilde," 1157-1173.
[19] "Maneuver History," 244; "Navigation for NEAR Shoemaker," 2.
[20] David W. Dunham, Robert W. Farquhar, James V. McAdams, Bobby G. Williams, James K. Miller, Clifford L. Helfrich, Peter G. Antreasian, and William Owen, Jr., "Recovery of NEAR's Mission to Eros," *Acta Astronautica* 47 (July-November 2000): 503-512; "Maneuver History," 244 & 258; "Navigation for NEAR Shoemaker," 2; Farquhar, 165-174.
[21] "NEAR OpNav at Eros," 1-2.

navigators modified their existing image display program ("xrover")²² and developed a landmark database and allied computer code. The writing of the software's initial code was complete before NEAR's December 1998 arrival at Eros. The missed encounter gave Owen and Wang time to continue improving the software.²³

Science for a new era. Asteroid encounters were particularly rich opportunities for navigators as project scientists to perform radio occultations and research in celestial mechanics and gravity. In particular, their peculiar shapes, gravitational fields, and other often bewildering attributes made them ripe for study. On the NEAR mission, Don Yeomans, Jon Giorgini, Alex Konopliv, and other navigators on the radio science team made an original contribution to asteroid science by deriving a detailed model of the shape, mass, bulk density, gravitational field, and polar rotary motion and direction. As usual, they obtained these measurements using Doppler because of its sensitivity to gravitational influences.²⁴

Navigators also participated on multinational science teams. For example, the Mars Observer science complement included eleven Russian scientists. Stanford's Len Tyler led the Radio Science Team and its contingent of navigator scientists. They did radio occultations and mapped the planet's gravity field more precisely. Their gravitational analysis increased understanding of Mars' internal structure and stresses and related Martian topography—as measured by the probe's laser altimeter—to variations in the gravitational field.²⁵ Bill Sjogren, a navigator on Tyler's team, pointed out that many researchers participated in their science experiments, but because they "were not on the initial proposals, they were not accepted as valid investigators, although they should have been." He described Alex Konopliv, one of the JPL navigators assisting him, as "a really sharp engineer." Konopliv, a University of Texas graduate,

---

[22] The picture display tool xrover in the Optical Navigation Image Processing System allowed one to measure features. A navigator dragged the cursor around the rim of a crater. The xrover program highlighted the screen pixels over which the cursor had traveled, fit an ellipse to those points, and displayed it. If the navigator judged the result to be good, xrover wrote it to a file for further processing. The process was tedious, but navigators gradually acquired some skill in its use. "Brief History of JPL OpNav," 339-340.

[23] "NEAR OpNav at Eros" 1-3; "Navigation for NEAR Shoemaker," 4; "Brief History of JPL OpNav," 339 & 340.

[24] Donald K. Yeomans, Peter G. Antreasian, Jean-Pierre Barriot, Stephen R. Chesley, David W. Dunham, Robert W. Farquhar, Jon D. Giorgini, Clifford E. Helfrich, Alexander S. Konopliv, Jim V. McAdams, James K. Miller, William M. Owen, Jr., Daniel J. Scheeres, Peter C. Thomas, Joseph Veverka, and Bobby G. Williams, "Radio Science Results during the NEAR-Shoemaker Spacecraft Rendezvous with Eros," *Science* 289 (September 22, 2000): 2085-2088.

[25] Mudgway, *Uplink-Downlink*, 327; Arden L. Albee and Frank D. Palluconi, "Mars Observer," 458 in James H. Shirley and Rhodes W. Fairbridge, ed., *Encyclopedia of Planetary Sciences* (New York: Chapman & Hall, 1997); Richard A. Simpson, "Mars Observer Radio Science (MORS) Observations in Polar Regions," 24-26 in Stephen M. Clifford, Alan D. Howard, and William S. B. Paterson, ed., *Workshop on the Polar Regions of Mars: Geology, Glaciology, and Climate History*, Pt. 1 (Washington: NASA, 1992); G. Leonard Tyler, Georges Balmino, David P. Hinson, William L. Sjogren, David E. Smith, Richard Woo, Sami W. Asmar, Michael J. Connally, Carole L. Hamilton, and Richard A. Simpson, "Radio Science Investigations with Mars Observer," *Journal of Geophysical Research* 97 (May 25, 1992): 7759-7779.

joined the Navigation Systems Section in 1991. He dramatically improved the gravity software, so that it characterized planetary gravitational fields with even greater precision.[26]

Another major scientific undertaking was in the field of radio astronomy to support navigating with the DSN's VLBI system. The search for extragalactic radio sources had woven radio astronomers, navigators, and the DSN into a web of mutually beneficial relationships. Over the course of 1984 and 1985, JPL divested itself of the geodetic research initiated as the ARIES study of crustal motions and transferred the project to its ultimate users, Goddard and the National Oceanic and Atmospheric Administration's National Geodetic Survey.[27] The quasar searches also served more practical purposes, such as the synchronization of DSN station clocks to Goldstone standard time. Navigators benefited from these synchronizing measurements, because they helped to maintain antenna frequency standards and to determine station locations with an accuracy of 5 to 10 centimeters,[28] a far cry from Neil Mottinger's quest two decades earlier to achieve an accuracy of just 1 meter. By 1988, the accuracy of DSN VLBI measurements of radio sources was better than 30 nanoradians.[29]

The resulting collection and catalog of extragalactic radio sources formed a new kind of reference frame within which navigators determined spacecraft positions. This radio reference frame was suited ideally for making ΔVLBI or ΔDOR observations of spacecraft and quasars. Because the quasars were so far away—literally outside our galaxy—any motion they might have was negligible. In effect, they provided what physicists call an inertial frame of reference, that is, a space-time coordinate system that neither rotates nor accelerates. A further benefit was that the positions of the radio objects were known to an accuracy that was 10 to 100 times better than for the star-based reference frame.[30]

The stellar reference frame used by navigators and astronomers was the fourth version of the fundamental catalog (FK4) and consisted of positions for about 1,500 stars obtained by optical meridian transit measurements. The next version, the FK5, contained some 4,500 stars positioned in the J2000 system (noon on January 1, 2000, in Julian years), whereas the FK4 had been based on the B1950.0 system (January 1950 in Besselian years). In order to avoid the nightmare of constantly changing coordinates,

---

[26] Sjogren, interview, 2008, 28-29; Lunar Prospector, "Alexander Konopliv," http://lunar.arc.nasa.gov/results/scientists/alkon.htm (accessed May 19, 2010).

[27] D. Lee Brunn, "ORION Mobile Unit Design," 6-32 in TDAPR 42-60; John M. Davidson and Donald W. Trask, "Utilization of Mobile VLBI for Geodetic Measurements," 248-266 in TDAPR 42-80; Mudgway, *Uplink-Downlink*, 542-543.

[28] John V. LuValle, Richard D. Shaffer, Mitchell G. Roth, Thomas M. Eubanks, and Philip S. Callahan, "Operational VLBI Clock Synchronization and Platform Parameter Determination," 307-318 in TDAPR 42-66; Thornton, interview, 15; Mitchell Roth and Thomas Yunck, "VLBI System for Weekly Measurement of UTI and Polar Motion: Preliminary Results," 15-20 in TDAPR 42-58; Liewer, 240.

[29] Chopo Ma, Elisa F. Arias, T. Marshall Eubanks, Alan L. Fey, Anne-Marie Gontier, Christopher S. Jacobs, Ojars J. Sovers, Brent A. Archinal, and Patrick Charlot, "The International Celestial Reference Frame as Realized by Very Long Baseline Interferometry," *The Astronomical Journal* 116 (July 1998): 516-546; Liewer, 240.

[30] "Relating Ephemerides and Frame," 1; Christian De Vegt, "Interrelation of Present Optical and Radio Reference Frames," 357-360 in Roberto Fanti, Kenneth Kellerman, and Giancarlo Setti, ed., *VLBI and Compact Radio Sources* (Boston: D. Reidel, 1984).

astronomers generally specified an object's location in terms of a particular reference date known as the equinox, such as B1950.0 or J2000. Both the radio reference frame and JPL's Development Ephemeris 200 (DE200)—the basis for the world's almanacs starting in 1984—shared the J2000 equinox, but DE200, like the FK4 and FK5, was based on stars, not radio sources.[31]

In addition to these stellar and radio reference frames, navigators routinely used a third one, one of their own making, that consisted of the ephemerides that they had generated for the inner planets and the Moon. The observational basis of these ephemerides was the wealth of precise range measurements of the Moon and planets made over the years with radar, lasers, and spacecraft ranging, especially to the Viking Lander. Because it was based on range—a direct measure of the distance to a spacecraft or target body—this reference system was sensitive to the motion of the planets, but not to background objects such as stars or quasars. The accuracy of probe positions calculated within this frame was better than accuracies obtained when estimating spacecraft positions within the star-based reference frame. Its use, though, was somewhat limited. It contained a small number of objects (four planets and their moons), and the objects themselves were large in terms of angular extent,[32] especially compared to the angular size of a probe.

All together now. If navigators were going to continue using VLBI (whether ΔVLBI or ΔDOR), they would have to link the radio and stellar reference frames. For example, they would have to know the position of Mars within both reference frames. Navigators approached the problem by obtaining positional measurements of probes relative to radio sources as well as relative to a planet. They began with the inner planets—Mars and Venus—where their planetary reference frame also served as their guide. The Viking Landers proved to be especially valuable in this effort.

The Landers became available for this purpose in June 1978, when NASA Headquarters transferred Viking program management from Langley to JPL for the Viking Continuation Mission. The Landers were ideal objects for linking the radio and planetary frames, because navigators knew their positions on the planet's surface with some precision, and both the Landers and Orbiters had radio "beacons" that located them (and Mars) within the radio frame.[33] Frank Donivan and Skip Newhall hoped to reduce the uncertainty in the Mars ephemeris from the current 40 km to just 20 km within the radio reference frame. However, they discovered that the Viking Lander's telemetry spectrum lacked sufficient bandwidth, and time ran out before they were able to take enough measurements to make the technique work. On the other hand, they did succeed in positioning the Viking Orbiter and the Pioneer Venus Orbiter in the radio reference frame and established ties between it and the stellar reference frame.[34]

---

[31] "Relating Ephemerides and Frame," 1; "Orientation of DE200/LE200," 297-302.
[32] "Relating Ephemerides and Frame," 2.
[33] Mudgway, *Uplink-Downlink*, 207; "Delta VLBI Observations," 555.
[34] "Delta VLBI Observations," 555; "Innovations in DDOR," 4 & 6; "Relating Ephemerides and Frame," 3; "Relating Frame and Ephemerides," 789-794; F. Bryant Winn, Mohan P. Ananda, Francis T. Nicholson, and

The Soviet Phobos landers gave them another opportunity to tie Mars to the radio frame. Phobos was the first international deep-space initiative of any extent following the successful Comet Halley flyby missions of Vega and Giotto in March 1986. The mission craft consisted of a "Frog" or "Hopping Lander" and an automated Lander (abbreviated DAS in Russian). Navigators hoped to improve both the existing frame-tie measurements and the accuracy of the Mars and Phobos ephemerides by making ΔVLBI observations of a lander and a nearby radio source. Despite the mission's premature termination, all was not lost. Even though the DSN garnered only the sparsest of data, analysis yielded satisfying results for the "frame-tie" experiment. The uncertainty of the alignment of the planetary and radio frames had fallen from about 250 to 40 nanoradians.[35]

The limit to these frame linking experiments was the inner solar system. Exploration of the outer planets required that navigators tie the radio and stellar reference frames to each other for both Jupiter and Saturn. Astronomers, of course, were working on the same problem. To start the task of bringing the outer planets into the united stellar-radio reference frame, JPL navigators began with Jupiter and requested help from the radio astronomers at the VLA, a radio astronomy observatory located about 80 km from Socorro, New Mexico, and managed by the NRAO. The VLA observed Jupiter and the Galilean moons relative to extragalactic radio sources and succeeded in situating Jupiter within the radio reference frame by measuring the positions of its satellites Ganymede, Europa, and Callisto relative to two extragalactic radio sources.[36] However, it soon became apparent that the VLA's determination of Jupiter's declination was not consistent with other known data. Fortunately, the Ulysses mission had scheduled a flyby of Jupiter in 1992 on its way to a polar orbit around the Sun. DSN ΔDOR observations of Ulysses during the Jupiter flyby enabled navigators to determine the planet's declination component with 10 times greater accuracy in the radio reference frame. These measurements also made their way into DE200 as well as DE400 and DE405.[37]

The effort to tie the radio and stellar reference frames culminated in the approval of new ephemerides for astronomers and navigators. The IAU had adopted the existing ephemerides, DE200 (and LE200), to be effective in 1984. Now, effective 2003, the IAU adopted JPL Planetary and Lunar Ephemerides DE405. Created in 1997, DE405 benefited from astrolabe, VLA, and ground-based CCD observations.[38] It featured the IAU (2000) System of Astronomical Constants and a major revision of the section on asteroids and minor planets. DE405 included some 300 asteroids, indicative of the

---

Edward W. Walsh, "Improvements in Galileo Mars Navigation Using the Viking Lander," 48 in TDAPR 42-60.

[35] Mudgway, Uplink-Downlink, 190 & 279-283.

[36] "Relating Ephemerides and Frame," 2-4; "Quasar Experiment: Part I," 31-35; Dick, Sky and Ocean, 444-449.

[37] Standish to Distribution; Moyer, Formulation, 3-3; Standish and Williams, 1; Memorandum, IOM 314.10-109, Standish and William M. Folkner to Distribution, "DE400/LE400 and DE140/LE140: Delivery and Covariances," January 20, 1995, copy furnished by James Border.

[38] Memorandum, E. Myles Standish to Distribution, "The Effect of Future VLBA Measurements upon the Covariance of Saturn: A Preliminary Assessment," May 1, 2006, F347, DSN Files; Geldzahler, interview, 30.

importance of asteroids to astronomical research and space exploration. The primary star catalog was ESA's Hipparcos catalog. The FK5 was no more. Instead, the International Celestial Reference System (ICRS), based on extragalactic radio sources rather than stars, replaced it. The ICRS retained the solar-system barycenter as its center, a convention of the FK5, and contained precise positions for 212 extragalactic sources (mainly quasars) instead of stars. As usual, DE405 was available from JPL on a CD or downloadable via the Internet.[39]

---

[39] "HIPPARCOS - Hipparcos Main Catalog," http://heasarc.gsfc.nasa.gov/W3Browse/all/hipparcos.html (accessed March 16, 2010); "The International Celestial Reference System," http://aa.usno.navy.mil/data/docs/ICRS_links (accessed March 16, 2010); USNO, "The International Celestial Reference Frame," http://rorf.usno.navy.mil/ICRF/ (accessed March 1, 2011); Standish, interview, 29-30; Standish to Distribution; Standish, "Observational Basis for DE200," 252-271; "Orientation of DE200/LE200," 297-302; Standish and Williams, 2, 22 & 28-29.

Part Five

Things Go Awry

Chapter 20

Navigating with a Bull in the China Shop

The collapse of the U.S.S.R. brought an end to the Cold War, but it did not stop the influence of the national security crisis on the country's space program. President Reagan had created, and his successors continued, the Strategic Defense Initiative (SDI). That organization's management style was replacing the Manhattan Project as a model for large organizations to imitate. NASA was no exception. The agency seemed in need of sweeping reorganization following the loss of the *Challenger* shuttle and the Hubble Telescope affair. With help from within the White House, reformers imposed a new Administrator on NASA, Dan Goldin, who instilled SDI management practices at NASA under the guise of Faster, Better, Cheaper. SDI's impact on NASA lasted throughout the 1990s and into the next decade.

The budget cuts of the 1980s continued into the 1990s under the rubric of "reinventing government." NASA funding and personnel shrank. The cuts and managerial changes had repercussions within navigation culture. The navigation section's new name reflected its growing size and diversity, especially with the arrival of the mission designers and their computer programmers. Navigation also reflected the emphasis on commercializing space that continued from the Reagan years. Many perceived an erosion of JPL's traditional line management in favor of project and program offices throughout JPL, but to a particularly greater degree in the Mars Program Office, the pride of Goldin's reforms. All in all, navigation seemed to have more and more discontinuities rather than continuities with its past. The same might be said of NASA and the country.

SDI takes over. The space program under President George H. Bush seemed to continue the Reagan Administration's legacy of large, expensive projects. SDI, the National AeroSpace Plane (NASP), and the Space Station lingered, as did the poster child of large and expensive, the Space Shuttle. To these Bush added yet another, the Space Exploration Initiative. Speaking on the steps of the Air and Space Museum on July 20, 1989, the twentieth anniversary of the first Apollo landing, Bush announced the Space Exploration Initiative. He proposed returning to the Moon, setting up a lunar base, and landing an astronaut on Mars. Speaking at Texas A&M University ten months later, Bush set a deadline for the Mars landing: the year 2019.[1]

Although it is tempting to view the Bush Administration as simply a continuation of that of President Reagan, and certainly most Reagan space initiatives persisted through the Bush years, the period was one more of change than of continuity, of turmoil rather than stability. SDI was under increased attack from a growing number of scientists, engineers, clergy, and members of the Left. The Iran-Contra affair came to

---

[1] Hogan, *Mars Wars*, 1-2; Ragsdale, 161.

light in the fall of 1986, and the stock market crash of October 1987 wiped out over 20 percent of the market's value amid revelations of insider trading and the failures of numerous savings and loan institutions.

More than any of these, one of the greatest influences on the fate of the nation's space program was the end of the Cold War. The Reagan Administration, to paraphrase Sir Winston Churchill, had fought the Cold War on the ground, under water, in the air, and in space. Now the Cold War was over. The Soviet Union dissolved definitively on December 8, 1991, as Russia, Belarus, and Ukraine established the Commonwealth of Independent States (CIS). The national security crisis, the longest-lasting influence on the evolution of deep-space navigation, appeared to be over.

One of the fundamental factors shaping the nation's space program took place nearly six years earlier, on January 28, 1986. The *Challenger* accident started a chain of events that would lead to a management revolution at NASA. In contrast to those who came along later, Ronald Reagan supported the Shuttle throughout the space agency's darkest hours. "We'll continue our quest in space. There will be more shuttle flights and more shuttle crews," Reagan pledged. "Nothing ends here."[2] A different, more critical assessment of NASA arose from the ashes of *Challenger* and led to a call to reform the agency. The official incident investigation revealed fundamental flaws in the way NASA operated. The malfunctioning Hubble Space Telescope seemed to signal furthermore a general incompetence within the agency. Criticism grew, as did calls for reform. Vice President Dan Quayle and the National Space Council (established in 1988) were first among those seeking to change NASA.

Quayle, the Council's titular head, wanted to "shake up" NASA, which he believed was "to a great extent, still living off the glory it had earned in the 1960s." He complained that NASA projects were "too unimaginative, too expensive, too big, and too slow."[3] A typical example that Quayle failed to point out would be the Space Exploration Initiative. Although imaginative, it certainly was, to use Quayle's words, "too expensive, too big, and too slow." He, like many other NASA reformers, wanted the agency to undertake projects that were "faster, cheaper, smaller." If NASA shifted from large, prolonged, expensive projects to smaller, faster, cheaper projects, critics argued, the agency would be able to accomplish more science for less money.

The real struggle between Quayle and NASA was over whether the White House or the space agency should determine space policy. NASA "wanted to keep making space policy themselves," Quayle complained.[4] What lay behind the charge was a personal conflict between Quayle and the NASA Administrator, Vice Admiral Dick Truly. President Bush had named Truly to the position on July 1, 1989. If any NASA Administrator ever stood for the Space Shuttle, it was Dick Truly. He had been a Shuttle astronaut and, before becoming Administrator, had overseen the heroic struggle to return the Shuttle to flight and to institute measures intended to safeguard future flights. But, for those pushing to reform NASA, the Space Shuttle symbolized what was wrong with the space agency.

---

[2] Ragsdale, 154.
[3] Quayle, 179 & 180.
[4] Quayle, 179 & 180.

Quayle became convinced that he wanted to replace Truly. NASA, he believed, needed not a "caretaker" (Truly) but "a bull in its china shop." Acting under pressure from President Bush, Quayle, and the Space Council, Truly eventually resigned. Quayle and the Council favored Dan Goldin for the role of "bull" in NASA's "china shop." Quayle characterized him as "just the person we needed, and after he went to NASA he started breaking some china."[5]

A former research scientist at NASA's Lewis Research Center (1962-1967), Goldin was Vice President and General Manager of TRW's Space and Technology Group (satellite systems), where he dealt in military reconnaissance, electronic intelligence, MILSTAR (Military Strategic and Tactical Relay) communications and other defense programs. He had firsthand knowledge of, and a high regard for, SDI and its "smaller, faster, cheaper" approach. Goldin had increased TRW work on smaller, lighter spacecraft, and he had introduced novel management approaches to cut program costs.[6]

The Goldin decade. Goldin became the new NASA Administrator on April 1, 1992. He remained in that position through the two terms of President Clinton and into the incumbency of President George W. Bush. No other NASA Administrator has served as many consecutive years, or under as many presidents. Goldin initiated the changes demanded by reformers and made "faster, cheaper, smaller" his own, though as "Faster, Better, Cheaper."[7] Quayle had used the expression "faster, safer, cheaper and better."[8] Goldin's long incumbency seemed to assure the triumph and continuing dominance of "faster, cheaper, smaller" within NASA.

The "faster, cheaper, smaller" (sometimes simply "rapid") project management approach stood in contrast to the typical government program. In the formula, both "faster" and "smaller" worked to keep costs down, a requisite for projects with modest budgets. Reducing project length also held costs down, because the longer a program runs, the more it costs just for overhead and labor. Similarly, keeping staff levels at a minimum decreases costs. The goal, therefore, was to produce results rapidly at relatively low cost.[9]

The Strategic Defense Initiative Organization (SDIO) had come to exemplify this approach, although it was actually a much older bureaucratic practice with its roots in the early years of the Cold War. Known as special projects offices, these featured a small staff, a modest budget, and a short time table for completion. A major advantage

---

[5] Quayle, 181, 184, 185 & 189-190.
[6] "Pre-Hearing Questions Submitted to Daniel S. Goldin Submitted by the Majority," nd, F4511, NHRC; Craig Covault, "Nominee for NASA Chief Fits Space Council Approach," *Aviation Week & Space Technology*, March 16, 1992, 21; "Daniel S. Goldin," data for news release, F4512, NHRC; "Newsmaker Forum: Daniel Goldin," *Space News*, 1991, np, F4512, NHRC; David C. Morrison, "NASA's Big Bang," *Government Executive* 25 (February 1993): 16-18, 39-41; Hogan, *Mars Wars*, 103 & 132; McCurdy, *Faster, Better, Cheaper*, 44.
[7] McCurdy, *Faster, Better, Cheaper*, 46, indicates that the expression "faster, better, cheaper" had been around since at least 1990, but Goldin made the saying his own.
[8] Quayle, "Prepared Remarks of the Vice President to the American Institute of Aeronautics and Astronautics," May 1, 1990, 10, F513, X-33 Files.
[9] Butrica, *SSTO*, 135-139.

of the approach was the avoidance of acquisition rules and regulations that applied to major defense systems, which were under constant attack and change during the Reagan years.[10]

This alternative management approach began migrating to NASA in 1989, well before Quayle began looking for a new NASA chief. The first attempted application was to the Space Exploration Initiative. The Council endeavored to alter NASA's culture by appointing leaders who would introduce SDI practices. To that end, in 1991, Michael Griffin left the SDIO, where he was the director of technology, and took charge of NASA's Space Exploration Initiative. In early 1992, as a replacement for Truly, before settling on Goldin, the Space Council chose Lt. Gen. James Abrahamson, the first SDI director. The selection encountered unbending opposition from Congressional Democrats who objected to the SDI in general and who worried about the waxing influence of "star warriors" such as Griffin and Abrahamson. Quayle and the Space Council found someone who shared their philosophy, but lacked the Star Wars taint, Dan Goldin.[11]

Although appointed by the Bush Administration, Goldin served most of his years during the two presidential terms of Bill Clinton. The policies and management principles that Goldin introduced during those eight years reflected the larger political and budgetary realities of the era. Among his achievements was a renewal of NASA's technical credibility through a dramatic 1993 mission to repair the Hubble telescope. Goldin directed a new Space Station design and helped to induce Russia to become a partner in the enterprise. In the name of cost-cutting, he oversaw the privatization of certain Space Shuttle operations by negotiating a sole-source contract with United Space Alliance, a partnership of Lockheed Martin and Rockwell formed specifically for that purpose.[12]

Once the Soviet Union collapsed, support for SDI faded. President Clinton repeatedly vetoed Republican attempts to revive it. But, when North Korea tested a ballistic missile, and it landed near Alaska, Clinton revived the SDI idea and signed the National Missile Defense Act of 1999. The impact of the Clinton Administration on navigation, though, would come mainly from its conservative fiscal policy.

President Clinton enacted into law the 1993 Deficit Reduction Plan—passed without a single Republican vote—and the bipartisan Balanced Budget Agreement of 1997. These bills cut federal spending, reduced interest payments on the national debt, and turned record budget deficits into record surpluses. In contrast, between 1981 and 1992 the national debt had quadrupled, and the annual budget deficit had grown to $290 billion in 1992, the largest ever until then.[13] These federal budget cuts resounded through NASA as the leitmotif of the period, which saw few peace dividends and scores of budget cutbacks.

---

[10] Butrica, "Overview of Acquisition," 199-223.
[11] Hogan, *Mars Wars*, 103; McCurdy, *Faster, Better, Cheaper*, 44, 46 & 47.
[12] *Report of the Space Shuttle Management Independent Review Team* (Washington: NASA, 1995); Lambright, 21.
[13] White House, fact sheet, "The Clinton Presidency: Historic Economic Growth," January 9, 2001, http://clinton5.nara.gov/WH/Accomplishments/eightyears-03.html (accessed February 18, 2011).

Reducing expenditures was all part of the new way of managing government known as New Public Management or, more familiarly in the United States, "reinventing government." New Public Management became popular in Reagan's America and Margaret Thatcher's United Kingdom during the 1980s as a way to renovate the public sector, but found fertile soil in other countries around the world as well. The main supposition behind the movement was that by increasing the market orientation of government operations, one could realize greater cost-efficiencies in the public sector without any negative side effects. In short, it was the embodiment of the notion that private enterprise can perform any function better than government.[14]

The Clinton Administration enthusiastically embraced New Public Management. On March 3, 1993, Clinton announced the National Performance Review, later known as the National Partnership for Reinventing Government. It was an attempt to make government more "businesslike" by making government less expensive and more efficient by cutting waste and red tape. "Reinventing government," "empowerment," and "customer service" were program buzzwords.[15]

Zero-base good times. By 1999, JPL navigators—and much of NASA—had little to celebrate, as Faster, Better, Cheaper and "reinventing government" took their toll in the form of a dramatic shedding of civil service and contractor jobs. The agency was beginning to resemble the theme song from the 1970s television show "Good Times":

> Keepin' your head above water,
> Making a wave when you can,
> Temporary lay offs
> Good Times . . .
> Scratchin' and surviving
> Good Times
> Ain't we lucky we got 'em
> Good Times.

The downsizing—a sugarcoated term for massive layoffs—began in 1992. A second round hit in early 1995, when JPL completed a so-called Zero-Base Review. The Review came at the behest of NASA acting in compliance with Clinton's National Performance Review and a Presidential directive to cut NASA's budget. Goldin appointed an internal review team, known as the Zero-Base Review, to find

---

[14] Wolfgang Drechsler, "The Rise and Demise of the New Public Management," *Post-Autistic Economics Review* 33 (September 14, 2005): 17-28.

[15] David Osborne and Ted Gaebler, *Reinventing Government: How the Entrepreneurial Spirit is Transforming the Public Sector* (Reading, Mass: Addison-Wesley, 1992); John M. Kamensky, "Role of the 'Reinventing Government' Movement in Federal Management Reform," *Public Administration Review* 56 (May-June 1996): 247-255.

management and organizational changes as well as budget cuts and cost saving measures intended to slash the agency's fiscal 1996 and 1997 budgets.[16]

The net result of the JPL dismissals was a loss of more than 1,400 employees from 1992 to 1998, plus a cut of on-site contractors from 7,600 work-years in 1992 to 5,344 by 1999, a 30% staff cut. The layoffs tended to affect many employees with the greatest amount of experience and institutional knowledge. The loss of expertise would haunt NASA and the federal government (including the Department of Defense) for years to come. Goldin said he wanted to take NASA back to the "Wonder Years" of the 1960s and symbolically replaced the so-called worm logo introduced in 1975 with the NASA "meatball,"[17] but the changes left many wondering whether they would keep their job.

When Faster, Better, Cheaper came to NASA in 1992, JPL had a new director, Ed Stone, who had replaced Lew Allen, Jr., in late 1990. On top of cost-cutting and downsizing, JPL introduced a major change in how it operated within its matrix organizational structure. The core of JPL's technical expertise was the line organization that consisted of the technical divisions, sections, and groups. Navigators working on a flight project had line organization bosses in their group, section, and division plus the project office management. This was the JPL matrix that had worked "pretty well over the years," Lincoln Wood judged.[18]

During the Surveyor program of the 1960s, George Null recalled, Hughes attempted to "paint green" all JPL employees paid by Surveyor and to move them to a dedicated area where Hughes project staff would supervise them. JPL management successfully fought this effort to subvert normal line management.[19] The conflict between line management—where people fit into the usual matrixed JPL hierarchy—and their participation in flight project—program management—came to the forefront during the 1990s.

Shyam Bhaskaran recalled that since 1992, when he joined JPL, "we've always been told at JPL to make your way up the chain. It always helps to work both the line and project side. You shouldn't stay in one exclusively."[20] Nonetheless, there was conflict between the line and project sides. Wood observed that some people "on the project side" believed that the line organization was "too powerful" and "was looking after its own interests and not after the interests of the projects." This attitude became "particularly acute" as projects became smaller and more cost-conscious, as Faster, Better, Cheaper became the space agency's management philosophy. "So what happened was a tremendous, I feel, erosion of the influence of the line side of the organization in favor of project and program offices." Instead of going to the line organization, authority went to "different people in projects" down to the "first-level technical management positions" known as the Project Element Managers (PEMs). They

---

[16] NASA, Press Release, "Review Team Proposes Sweeping Management, Organizational Changes at NASA," May 19, 1995, http://www.jpl.nasa.gov/news/releases/95/release_1995_0519.html (accessed February 22, 2011).
[17] Westwick, 17, 211 & 243.
[18] Wood, interview, 50 & 51.
[19] Personal communication, George Null, June 12, 2012.
[20] Bhaskaran, interview, 58.

were people in the line organization assigned to a project who had been told "basically to listen to what the project tells you and not what your section manager tells you. Do things our way, and just recognize who your boss really is," meaning the project.[21] What Hughes had failed to impose on Surveyor during the 1960s had become *de rigueur* during the 1990s, and not necessarily for the best.

The emphasis on the project side, according to Wood, extended "to a considerable extent across the laboratory, but to a particularly large degree in the Mars Program Office. The Mars 1998 mission [the Mars Climate Orbiter and the Mars Polar Lander] was, I would say, the ultimate in how far that was pushed. The line organization counted for very little in terms of influence. All the decisions were intended anyway to be made at the project level." As a result, the projects were making a lot of technical decisions "without much if any review at the line level." Wood cites the decision by the Mars Pathfinder project not to use ΔDOR or ΔVLBI navigation data. "Not only did they not need it, but future Mars missions didn't need it either," the project decided. "That's not something that was ever run by the line organization as something to agree with or disagree with."[22]

The Great Divide Redux. The management mantras of the era—whether Faster, Better, Cheaper or "reinventing government"—all impacted and transformed the institutional life of JPL navigators. NASA grew increasingly more bureaucratic and imposed a greater degree of formality, procedures, and management levels on project activities. The institutional milieu and culture of navigation reluctantly took on those features as well. Laboratory reorganizations, some fleeting, others permanent, at times did not appear to arise from any inherent need to reorder navigation, but rather issued from needs sensed at the highest levels of JPL management.

From the 1980s into the 1990s, the navigation Section grew larger and more diverse. In 1981, when Frank Jordan managed the Section (314), it included groups dedicated to orbit determination and conditioning tracking data. Optical navigation already was a separate group supervised by Tom Duxbury followed by Steve Synnott. In 1981, Bill Sjogren led the "solar system dynamics" team that developed ephemerides and conducted celestial mechanics, gravity, and other studies during flights. After Jim McDanell succeeded Jordan as Section leader, he began adding new groups, the first of which, Jordan Ellis' "multimission nav support" group, had arrived by 1989. By 1992, McDanell and Deputy Manager Wood had split "solar system dynamics" into two groups. Don Yeomans led the Solar System Dynamics Group, and Sjogren oversaw the specialized Planetary Gravity Analysis Group.[23]

In addition, from 1981, Jordan's navigation section had a new group, led by Lincoln Wood, for "future mission studies." Jordan set it up because he felt that future missions were not receiving enough attention. Wood characterized its efforts as being

---

[21] Wood, interview, 51.
[22] Wood, interview, 51.
[23] OrgChart, Section 314, November 1981, JPL556; OrgChart, Section 314, January 1985, JPL556; OrgChart, Section 314, July 1989, JPL556; OrgChart, Section 314, October 1992, JPL556; OrgChart, Section 314, August 9, 1993, JPL556; OrgChart, Section 314, April 25, 1995, JPL556.

"perhaps a little bit fragmented." Future Mission Studies interacted closely with mission designers, who were not in the navigation section. As soon as the latter began studying a new mission concept, Future Mission Studies became involved and provided them with navigational analysis ranging from likely navigational challenges to what accuracies could be realized and "what relative reliance we'd place on radio data and optical data and that sort of thing," Wood explained. Future Mission Studies involved itself in a wide range of projects, not all interplanetary in scope. For example, it considered space station navigation, air traffic surveillance, precise instrument pointing, and orbiting Very Long Baseline Interferometers.[24]

Future Mission Studies dissolved in 1989, and Wood became Jim McDanell's Deputy Manager. During Jordan's entire tenure as Section Manager from 1976 to 1984, he had no assistant. Jim McDanell also was without an assistant until Wood became the Deputy Manager in 1991. At the time, navigation, with about a hundred engineers, was the largest JPL section without a Deputy Manager. As Wood noted, "there's a certain amount of bureaucracy that has to be dealt with to have a new deputy position approved." Once JPL managers gave their approval, McDanell advertised for the job, and Wood "wound up being the person selected for that job."[25]

In 1995, the recently renamed Navigation and Flight Mechanics Section (312) was a much larger and more complex organization thanks to the addition of several mission design and analysis groups. Jordan, now at the division level, had decided that it was time for a few changes in the groups making up the navigation section, and that included bringing over the mission analysts and designers. At that point, the only similar team was Willard Bollman's trajectory maneuvers unit known in the 1990s as the Maneuver Analysis Group. The new section order added dozens of new design and analysis staff dedicated to "terrestrial planets," "outer planets," and "launch and Earth orbit." The merger with mission design also appended a group of programmers dedicated to "astrodynamics and mission design software" that stood alongside Pete Breckheimer's Navigation Software Development Group, set up in 1982.[26]

The section's 13 groups of navigators and mission designers varied quite a bit in size, so a reordering of 1996-1997 brought about a more balanced structure of just 10 groups. In late 1999, another mission design group transferred into the section. In early 2002, as a result of splitting two groups and making other adjustments, the total number of groups was again 13. A Division reorganization cut the number of sections from eight to just four making all sections larger than before. Meanwhile, the navigation section, now with 130 engineers and scientists, continued to house mission designers and their programmers. The 1995 Zero-Base Review required that line organization managers—such as the section managers and deputies—reapply for their jobs. McDanell and Wood remained the Section Manager and Deputy Manager.[27]

---

[24] Wood, interview, 19-20.
[25] Wood, interview, 35.
[26] Wood, interview, 19 & 40; OrgChart, Section 314, July 1982, JPL556; OrgChart, Section 314, October 1991, JPL 556; OrgChart, Section 314, August 1993, JPL556; OrgChart, Section 312, June 7, 1995, JPL556.
[27] OrgChart, Section 312, March 3, 1997, JPL556; OrgChart, Section 312, November 1, 1999, JPL556; OrgChart, Section 312, April 8, 2002, JPL556; Wood, interview, 40.

Navigation also reflected the emphasis on commercialization in the political and JPL institutional environment. Organization charts across NASA sprouted all kinds of "enterprises," even "space enterprise" and "Earth Science Enterprise." As JPL historian Peter Westwick observed, from the perspective of JPL, "More generally, faster-better-cheaper brought more of an industrial mindset, where cost became the driving factor for deep-space missions instead of success-at-any-price."[28]

Navigation was not immune to this "entrepreneurial" mindset. By January 1999, the Navigation Section had its own Business Operations office that merged the section's administrative and secretarial staffs. McDanell and Wood thought that gathering these employees into a single unit reporting to a single person who, in turn, reported directly to section management (themselves) seemed to be a good idea. The merger would boost efficiency, and neither McDanell nor Wood had expertise in hiring or supervising a similar grouping of employees.[29]

Actually, the navigators' Business Operations preceded the laboratory reorganization instituted by John Beckman as Manager of the Systems Division (1998-2001) that gave every JPL section a Business Enterprise Office. "It's almost like they don't work for the section," Neil Mottinger observed. Within the section, they reported more to the division than to the section. Mottinger further noted with a laugh: "I've heard a few hallway comments about that."[30]

The full MONTE. The organizational merger of navigation and mission design along with their programmers found its embodiment in the software used by both communities. The so-called Next Generation Navigation Software, known more formally as the Mission-analysis, Operations, and Navigation Toolkit Environment (MONTE),[31] replaced the orbit-determination program as well as the trajectory program utilized by both navigators and mission designers. MONTE users could plan and design trajectories for flights as well as calculate actual flight paths and generate corrective maneuvers.[32]

The earlier separation of mission design from navigation had led to software specialization and separate software libraries. The partition between navigation and mission design programs was costly and caused the software to lack interoperability,

---

[28] Westwick, 252.
[29] Personal communication, Lincoln Wood, May 22, 2012.
[30] Mottinger, interview, March 10, 1999, 11; John Beckman, "Biography," in "Agenda," June 2002, http://trs-new.jpl.nasa.gov/dspace/bitstream/2014/8870/1/02-1425.pdf (accessed February 21, 2011); OrgChart, Section 312, January 1, 1999, JPL556; OrgChart, Section 312, December 2001, JPL556; OrgChart, Section 312, January 2003, JPL556.
[31] Martin-Mur, interview, 32. For a description, see the internal publication, Steve Flanagan, *Architectural Design Document for MONTE—Mission Analysis and Operational Navigation Toolkit Environment*, draft, November 29, 2000; Steve Flanagan and Todd Ely, "Navigation and Mission Analysis Software for the Next Generation of JPL Missions," 16th International Symposium on Space Flight Dynamics, December 3-7, 2001, Pasadena, CA, http://trs-new.jpl.nasa.gov/dspace/bitstream/2014/42884/1/12-0510_A1b.pdf (accessed March 11, 2010); NASA Tech Brief.
[32] A new mission design software set, called MASAR, is underway and draws heavily on MONTE. Wallace S. Tai, *Deep Space Mission System (DSMS) Services Catalog, Version 7.5*, D-19002 (Pasadena: JPL, May 19, 2003), 10-7 & 14-19; "Nav Software," 1.

even though navigators and mission designers often collaborated. MONTE was a solution for this split world. It maintained a single software library for navigators and mission designers alike. The unified software library also promised both budgetary economies and operational efficiencies.[33]

MONTE had several advantages over previous navigation programs. Firstly, it was not an adaptation of old program code, but rather one that had been written completely new, from the ground up by Richard F. Sunseri, Hsi-Cheng Wu, Robert A. Hanna, Michael Mossey, Courtney Duncan, Scott Evans, James Evans, Ted Drain, Michelle Guevara, Tomás Martin-Mur, and Ahlam Attiyah. The old software had been expensive to maintain because of its origins in aging, mainframe computer technology and because of the mutually dependent way that the software and hardware had evolved over time. Moreover, finding new people with the necessary skills in FORTRAN, the navigation software language for the last five decades, was becoming increasingly difficult.[34]

Like MIRAGE, MONTE could handle multiple probes in contrast to the single-spacecraft capability of the former orbit determination programs. It could plot and display several flight scenarios at the same time using colors to differentiate among them, and it could permit side-by-side analysis of scenarios. These and other graphical user interfaces were a key characteristic of MONTE. A future MONTE will provide onboard spacecraft navigation. Additionally, the intent was to make the software more accessible to novices, but without hampering experts. In September 2009, JPL made MONTE Release 7.3 available to the public in exchange for a commercial licensing fee and licensed it to several universities free of charge.[35]

Discovering a program. Meanwhile, Dan Goldin was establishing programs that embodied his Faster, Better, Cheaper formula. He talked about launching one or two spacecraft per year, but soon upped the ante by proposing a launch rate of one a month within a decade. The idea was to send a succession of small and medium-sized spacecraft to all the planets and their major satellites as well as to asteroids and comets. Faster, Better, Cheaper would build and launch a lot of probes, but also would accept some failure as part of the cost. Failure was now an option at NASA. Indeed, Goldin encouraged project managers to take risks.[36]

The Discovery Program became the programmatic embodiment of Faster, Better, Cheaper. However, it already existed before Goldin's selection to run NASA. The same impulse to cut costs that had led to the creation of the Planetary Observer and Mariner Mark II mission classes had engendered a new generation of probes in NASA's long-running Explorer program. Explorer flew, and continues to fly, probes to investigate solar and astrophysical phenomena. The first one launched in 1958. The earliest Explorers were small and lightweight, but they became larger multipurpose (and

---

[33] "Nav Software," 1.
[34] "Nav Software," 1; NASA, Tech Brief.
[35] Guinn and Wolff, 143; "Nav Software," 1; NASA Tech Brief.
[36] McCurdy, *Faster, Better, Cheaper*, 50 & 51; Westwick, 214.

therefore more expensive) and flew less often, when the Space Shuttle came along in 1981 with the ability to carry up to 65,000 pounds.[37]

In May 1988, the Goddard scientists in charge of Explorer announced the Small Explorer (SMEX) program. The idea was to launch smaller, lighter, less expensive probes one to two times a year, depending on the availability of funding, launchers, and mission cost. Each mission was to take three years from selection to launch. The first such probe, SAMPEX (Solar Anomalous and Magnetospheric Explorer), was intended to study interplanetary energetic particles. It launched in June 1992 aboard a Scout rocket from Vandenberg Air Force Base. Upon Goldin's arrival at NASA, SMEX became an example of his Faster, Better, Cheaper philosophy.[38]

Just as Goddard scientists were creating the SMEX program, the NASA Headquarters Solar System Exploration Division began leaning in the same direction. In June of 1989, they organized a workshop at the University of New Hampshire to seek advice from the scientific community and to garner support for a broad range of solar system exploration activities. The workshop led to the formation of the Science Working Group run by Robert A. Brown, who had chaired one of the workshop panels. Brown's group met in 1989 and 1990, but did not recommend any specific missions. Nonetheless, they did propose a name for the overall concept of low-cost projects: the Discovery Program. In 1990, the new head of NASA Solar System Exploration, Wesley Huntress, asked the Science Working Group to select a mission to test the feasibility of the low-cost approach. At another workshop, held this time in Woods Hole, Massachusetts, in July 1991, some 60 NASA managers and space scientists agreed to shift programming from big missions (so-called flagship missions) to smaller, more frequent flights.[39]

Within days of Goldin's taking the oath as Administrator, NASA announced the creation of the Discovery program which Goldin referred to as "the world's best kept secret."[40] To qualify for Discovery funding, a project had to meet certain guidelines. They could not cost more than $150 million and had to take three years from approval

---

[37] McCurdy, *Low-Cost Innovation*, 9; GSFC, "History of the Explorers Program," http://explorers.gsfc.nasa.gov/history.html (accessed February 17, 2011). For a historical overview of Explorer I, see JPL, *Explorer I*, http://www.jpl.nasa.gov/explorer/downloads/Explorer1.pdf (accessed February 17, 2011).

[38] W. Vernon Jones and Nickolus O. Rasch, "NASA's Small Explorer Program," *Acta Astronautica* 22 (1990): 269-275; Daniel N. Baker, Gordon Chin, and Robert F. Pfaff, Jr., "NASA's Small Explorer Program," *Physics Today* 44 (December 1991): 44-51; George P. Newton, "Lessons Learned from, and the Future for, NASA's Small Explorer Program," *Acta Astronautica* 28 (August 1992): 307-318; Gilberto Colon and John Catena, "SAMPEX," in JPL, *Deep Space Network: Mission Support Requirements*, D-0787, Rev. AF (Pasadena: JPL, October 1991), 37-1 to 37-4; "SAMPEX," http://sunland.gsfc.nasa.gov/smex/sampex/ (accessed March 1, 2011); "Welcome to the Small Explorer's Web Site," http://sunland.gsfc.nasa.gov/smex/ (accessed February 17, 2011), which is an archived version of the program's website; GSFC, "NASA's Small Explorer Program: Faster, Better, Cheaper," January 1998, http://www.scribd.com/doc/49005357/NASA-Facts-NASA%E2%80%99s-Small-Explorer-Program-Faster-Better-Cheaper (accessed February 17, 2011).

[39] Stephanie A. Roy, "The Origin of the Smaller, Faster, Cheaper Approach in NASA's Solar System Exploration Program," *Space Policy* 14 (August 1998): 153-171; *SSB Annual Report*, 12-13; McCurdy, *Low-Cost Innovation in Spaceflight*, 8-9; Westwick, 210.

[40] Quoted in McCurdy, *Faster, Better, Cheaper*, 52.

to launch. Probes had to be small and light enough to be launched on a Delta II or smaller rocket. Candidate projects would use "mature instrument and spacecraft technology" and entailed "the acceptance of a modest increase in the level of risk." The initial call for Discovery applications met with enthusiasm: 89 scientists submitted proposals. The submissions from scientists reflected another Discovery novelty; instead of institutions, principal investigators proposed missions.[41]

Discovering a rival. Discovery started with Near-Earth Asteroid Rendezvous (NEAR), discussed in Chapter 19. In 1982, before the Discovery program's inception, JPL had proposed NEAR as a low-cost Pioneer-class mission that would have launched in 1994. NASA put the proposal on ice, then breathed new life into it for the Discovery series. The bid that JPL tendered was well beyond the requisite cost range, so NASA awarded NEAR to Johns Hopkins University's Applied Physics Laboratory (APL). Its proposal came in with a substantially lower budget. The loss of NEAR shocked JPL managers.[42]

Henceforth, APL was JPL's main competitor for deep-space projects and won frequently, the most recent success being the MESSENGER flight to Mercury. The two rivals were alike in many ways. Universities (Caltech and Johns Hopkins) managed them, their origins lay in undertaking research and development for the military, and they had been entrenched for a long time in the nation's space effort. Johns Hopkins created APL in 1942 primarily to perform defense work. APL's decades of space work included management of several Explorer missions and construction of flight instruments for JPL spacecraft, including Galileo. Before NEAR, though, its major space-related projects had included the Navy's TRANSIT satellite navigation (also known as NAVSAT) and its Earth observing GEOSAT (GEOdetic SATellite) systems.[43]

APL also had substantial involvement in SDI. For example, it built an experimental interceptor called Vector Sum or Delta 180 that launched in May 1986. APL built the demonstrator in fourteen months from approval to launch at a cost of $150 million. The speed and low-cost of the effort typified the SDI management approach. The project also exemplified what Goldin meant by Faster, Better, Cheaper. JPL's main rival thus had an edge by already having adopted NASA's new management style.[44]

The second Discovery project was an overdue flight to Mars. During the late 1980s, JPL and NASA revived plans for a flight to land a rover and bring a sample back to Earth. The 1991 Woods Hole meeting resulted in a reformulated Mars effort known as MESUR for Mars Environmental SURvey. The intent was to explore Mars in preparation for the Space Exploration Initiative astronauts. MESUR consisted of 16 flights to establish a network of small landers conducting research in meteorology and seismology in conjunction with Mars Observer. Planners saw it as a way to survey Mars

---

[41] McCurdy, *Faster, Better, Cheaper*, 56; Westwick, 211, 214 & 215.
[42] Westwick, 212.
[43] Greg Rienzi, "APL: Sixty Years in War and Peace," *The [Johns Hopkins] Gazette* 31 (March 25, 2002), http://www.jhu.edu/~gazette/2002/25mar02/25sixty.html (accessed December 28, 2010); APL, "Our Heritage," http://www.jhuapl.edu/aboutapl/heritage/default.asp (accessed December 28, 2010).
[44] Westwick, 217 & 224.

inexpensively—by sending multiple relatively inexpensive probes—and with low risk, because the loss of a spacecraft would not be fatal to the program.[45]

MESUR was to begin with a single demonstration flight. NASA's Ames Research Center and JPL vied for the contract, which NASA awarded to JPL. MESUR soon underwent fundamental modifications. In February 1992 Huntress informed JPL that it could do MESUR under one condition: it had to be a Discovery project and have a price tag under $150 million. With the disappearance of the Mars Observer in 1993, NASA deferred the remainder of the MESUR endeavor. The first would be the only flight.[46]

NASA called the Discovery version MESUR-Pathfinder, but soon went with the simpler Mars Pathfinder. The agency made Pathfinder a showcase of the Faster, Better, Cheaper approach to planetary missions. The cost of the project, $171 million, was substantially below the $1 billion price tag of the Mars Observer. Pathfinder took off on December 4, 1996, and landed on July 4, 1997, during a holiday weekend. Upon landing, it released a small robotic rover—Sojourner—that crawled over the nearby terrain. This was the first mobile probe to roam another planet. Its performance was extraordinary, and performance was the operational word. The craft transmitted pictures to Earth that people at home could see on television, or they could access the Pathfinder website on their computers and follow along as Sojourner studied such colorfully-named rocks in the Ares Vallis as "Barnacle Bill" and "Scooby Doo." The Pathfinder subsequently became a token of remembrance as the Carl Sagan Memorial Station.[47]

The Mars Program. Goldin also reconstructed Mars exploration along entirely new lines. He had inherited the expensive, mammoth Space Exploration Initiative. Its estimated cost in 1993, when Congress and President Clinton terminated it, was $400 billion. Also in 1993, the DSN lost communication with the Mars Observer, which had a price tag of $1 billion. The loss caused NASA great chagrin, to say the least. Goldin exploited the moment to restructure Mars exploration in a way that was congruent with Faster, Better, Cheaper. More importantly, he turned the Mars program into the de facto flag carrier for his managerial reforms. He hoped that this policy decision would enable NASA to do more exploration for less money. Mars exploration now became a sequence of probes that were "faster, better, cheaper" than anything conceived previously.[48]

A second key moment further shaped Mars exploration. In August 1996, NASA announced that a primitive form of single-cell life may have existed on Mars some 4 billion years ago based on an examination of a Martian meteorite found on Earth.[49] The

---

[45] *SSB Annual Report*, 12-13; Westwick, 212; Wilford, "16 Landers"; Wilford, "NASA Return Trip."
[46] Westwick, 212 & 213; Wilford, "16 Landers"; Wilford, "NASA Return Trip."
[47] Westwick, 213; Lambright, 4-5 & 23.
[48] Hogan, *Mars Wars*, 62; Lambright, 8 & 19-20.
[49] David S. McKay, Everett K. Gibson Jr., Kathie L. Thomas-Keprta, Hojatollah Vali, Christopher S. Romanek, Simon J. Clemett, Xavier D. F. Chillier, Claude R. Maechling, and Richard N. Zare, "Search for Past Life on Mars: Possible Relic Biogenic Activity in Martian Meteorite ALH84001," *Science* 273 (August,

claim did not go undisputed. Nonetheless, it gave Goldin the rationale for focusing the entire space program on the search for extraterrestrial life. Ironically, only a few years earlier, Congress had canceled funding for the SETI (Search for ExtraTerrestrial Intelligence) program.

Goldin established a new effort, known as Origins, to seek out the beginnings and future of the cosmos and of life itself (but not to find intelligent life) and charged the Mars program with concentrating on the search for life (present or past). NASA would launch Mars probes about every two years, when the Earth and Mars were closest to each other (that is, in opposition), culminating in a mission to bring back Martian soil to Earth. Each flight would push Faster, Better, Cheaper farther than ever before.[50]

Goldin also seized on a new defining characteristic for his management mantra: high technology. Rather than rely on available hardware, as had been the original intent, Goldin decided that Faster, Better, Cheaper would push technological development. Peter Westwick considers this decision to be Goldin's "main contribution to the faster-better-cheaper philosophy and one that meshed with the Clinton administration's emphasis on high-tech industry." The policy, though, did not dovetail with accepted (and reasonable) engineering practices. Engineers did not want to risk spacecraft failure by flying unproven technology; without flight-testing, all new technologies remain unproven. It seemed that, at least from a technological perspective, mission failure became a choice.[51]

Pathfinder already had been a "proof-of-concept" for various technologies such as the deployment of an airbag to help to land the craft and its rover safely on the surface. Now NASA created a separate mission class specifically to flight-test new technologies. In 1994, NASA's Office of Space Science and its Office of Earth Science jointly established the New Millennium program to "identify and flight-test new technologies that will enable science missions of the early 21st century."[52] The first New Millennium flight, Deep Space 1, validated various experimental technologies, including a novel propulsion system consisting of a xenon ion engine and autonomous navigation of a planetary spacecraft, as we shall see in Chapter 22.

How successful was Faster, Better, Cheaper? Pathfinder clearly marked the high point of Goldin's personal reputation as NASA Administrator. The Pathfinder success swayed many doubters, albeit begrudgingly, to admit that Faster, Better, Cheaper could succeed. Howard E. McCurdy, a professor of public administration and policy at American University, studied Faster, Better, Cheaper missions that flew between 1992 and 1998. He concluded that the approach worked well during those years. Out of 10 probes launched, the only failure was the Lewis Earth-observing satellite, which spun out of control shortly after taking off in 1997. Cost overruns led to the cancellation of an

---

16, 1996): 924-930; NSSDC, "Evidence of Ancient Martian Life in Meteorite ALH84001?" http://nssdc.gsfc.nasa.gov/planetary/marslife.html (accessed February 18, 2011).

[50] Westwick, 256; NASA, "Origins," http://origins.stsci.edu/ (accessed February 18, 2011); Lambright, 22.
[51] Westwick, 214-215.
[52] Westwick, 215; McCurdy, *Faster, Better, Cheaper*, 57; Bhaskaran, interview, 17.

eleventh project, the Clark satellite, which with Lewis was part of the Small Satellite Technology Initiative to demonstrate smaller, cheaper, advanced technology satellites.[53]

The NASA History Office lauds the success of Goldin's "new approach to mission design [that] resulted in an extraordinary increase in the number of space missions launched by NASA." In praising the sheer number of flights, the Office states: "The past decade had more than triple the number of flights attempted during the 1980s." The comparison is unfair. After 1978, NASA did not fly any planetary probes until Galileo and Magellan, whose launches the *Challenger* disaster postponed. The History Office further states that NASA was able to "undertake over 170 missions worth some $23 billion during his tenure. His acceptance of the potential for failure on lower cost missions was critical to this accomplishment. Thus, while the number of mission failures increased in absolute terms, the eleven failures cost the taxpayers only $0.5 billion."[54] This assessment fails to consider the losses to science and the agency's prestige produced by those failures.

In contrast, McCurdy discusses the utter failure of Faster, Better, Cheaper missions in 1999. During that *annus horribilis* four out of five flights failed. The Widefield Infrared Explorer (WIRE), launched on March 5, 1999, into Earth orbit, was to make a four-month infrared survey of the entire sky. A malfunction shortly after takeoff depleted the cryogenic hydrogen required by its science instruments. The Mars Climate Orbiter, launched on December 11, 1998, went missing on September 23, 1999. Its companion probe, the Mars Polar Lander, disappeared on December 3, 1999. That mishap took with it Deep Space 2, the second New Millennium project, which it was carrying aboard. Deep Space 2 was actually a pair of probes, named "Scott" and "Amundsen" in honor of the first human explorers to reach the Earth's south pole, Robert Falcon Scott and Roald Amundsen. They would have been the first probes to penetrate below the surface of another planet.[55]

---

[53] Lambright, 23; McCurdy, *Faster, Better, Cheaper*, 2 & 5, 57 & 58; Robert J. Hayduk, Walter S. Scott, Gerald D. Walberg, James J. Butts, and Richard D. Starr, "NASA's Small Spacecraft Technology Initiative Clark Spacecraft," *Acta Astronautica* 39 (November-December 1996): 677-686.
[54] NASA, History Office, "Administrator Daniel S. Goldin's Accomplishments," http://history.nasa.gov/goldin-accomplishments.pdf (accessed January 17, 2011).
[55] McCurdy, *Faster, Better, Cheaper*, 5 & 58; Westwick, 276.

Chapter 21

The Angry Red Planet

NASA's exploration of Mars changed dramatically after Dan Goldin became Administrator in 1992. He turned the Mars program into the de facto flag carrier for his broad managerial reforms encapsulated in the mantra Faster, Better, Cheaper. The Discovery class was the mission model, and Goldin had a ready-made pioneer for that program, Mars Pathfinder. The Mars Global Surveyor—in many ways the heir to the Mars Observer's science mission, preceded Pathfinder to the red planet. Then came Mars Surveyor '98—the Mars Climate Orbiter and the Mars Polar Lander—to study Martian weather and climate change. These flights would have rounded out the string of successes—scientific and navigational—of the preceding years, but taking risks and making cost the chief metric sooner or later leads to disaster.

Voyages to Mars historically have been fraught with failure. The abnormally high number of disappearances and malfunctions has become the stuff of myth. The so-called Mars Curse, once attributed to the Galactic Ghoul, reflected this reality of failure. Of 45 probes launched from Earth to Mars, only 20—less than half—succeeded. Twelve of the missions included attempts to land on the surface, but only seven transmitted data after landing. The majority of failures took place during the early years of space exploration, and the Soviet Union suffered most of them. The United States' record stands at 14 successes out of 20 flights. That success rate of 70% has grown in recent years. Nonetheless, failures seem to be inevitable.

Despite the planet's relative nearness, navigating probes to Mars is not easy. The greatest challenges, as illustrated by the Mars Global Surveyor and Mars Pathfinder, arise as spacecraft near the planet. The Global Surveyor had to perform a maneuver that would put it into an appropriate elliptical orbit for carrying out science imaging. In contrast, Pathfinder landed and released a rover. The trick was to guide the craft on a unique path that took it directly from its cruise trajectory into the Martian atmosphere and to a determined landing area. The vehicle decelerated using its heat shield, then parachutes, small rockets, and finally airbags, each transition requiring a new adjustment of navigation parameters. Navigating craft to Mars was not as tricky as putting them in orbit or landing them on the planet.

Mars Global Surveyor. The Mars Global Surveyor[1] was the first NASA probe to visit Mars since the Viking mission two decades earlier. Touted as among the first of Goldin's Faster, Better, Cheaper flights, it was the scientific heir to the missing Mars Observer including the radio occultation and gravity experiments conducted by navigators. The

---

[1] For background: Arden L. Albee, Frank D. Palluconi, and Raymond E. Arvidson, "Mars Global Surveyor Mission: Overview and Status," *Science* 279 (March 13, 1998): 1671-1672; Albee, Palluconi, Arvidson, and Thomas Thorpe, "Overview of the Mars Global Surveyor Mission," *Journal of Geophysical Research* 106 (2001): 23291–23316.

innovative, but thorny, flight leg was the orbit insertion maneuver. A 22-minute engine burn slowed the craft just enough to allow the planet's gravity well to capture it into an elliptical orbit.

"Well, the big challenge was getting into orbit. The correct orbit," Sam Dallas recalled. The specter of Mars Observer hung over the navigation team. "After having the Mars Observer fail, that was such a big concern that we did a computation of a maneuver to be done, and we had to redo it to make sure that it was the right maneuver.... So Mars Observer influenced the operations from that perspective. We were much more careful to ensure that we were going to get a good insertion and get into the right orbit."[2]

Surveyor used a technique called aerobraking over a period of four months to lower the high point of its orbit from 54,000 km (33,554 mi) to an altitude near 450 km (280 mi). Once it achieved an apoapsis altitude of 450 km, the spacecraft maneuvered to establish its final mapping orbit. Instead of slowing the probe by firing its thrusters in a direction opposite to its motion, the aerobrake saved fuel by using the atmosphere to slow down the spacecraft. The fuel saving also translated into a smaller launch mass that allowed more weight for scientific instruments. In short, as Tomás Martin-Mur noted, "it allows you to have more science for the same buck." Aerobraking was not new, having been demonstrated in 1991 by the Japanese MUSES-A (Hiten) spacecraft while coursing through the Earth's atmosphere. The Mars Surveyor relied on the experience gained in 1993 from Magellan which aerobraked to circularize its orbit in order to increase the precision of its Venusian gravity field measurements. Aerobraking also enabled the Mars project to squeeze into its very tight budget by reducing the launcher cost by at least $100 million.[3]

The price for all the project benefits of aerobraking was increased navigational precision. Navigators had to predict exactly the time and location of entry into the atmosphere to ensure that Surveyor's angle-of-attack was maintained throughout the drag pass. Additionally, the altitude at periapsis (closest orbital approach to Mars) had to be maintained within a particular range: flying too low could damage the probe. The navigation team continually updated the craft's estimated trajectory as well as key models, the most critical of which was the one for atmospheric density. They also were on high alert for any sign of an impending dust storm, because Surveyor was arriving at the height of the dust storm season.[4]

The event that ended up challenging the entire mission occurred right after launch. One of the two solar wings failed to latch. These wings controlled the probe's passage through the tenuous upper atmosphere and, through aerobraking, lowered the apoapsis of its orbit over the course of time. Mission designers recalculated the orbit insertion and aerobraking to accommodate the wing failure. The reconfiguration worked as planned. The actual periapsis altitude of 262.9 km was only 12.9 km higher than the

---

[2] Dallas, interview, 19.
[3] "MGS Aerobraking," 208; Martin-Mur, interview, 23 & 24; "MGS Aerobraking Broken Wing," 276; "MGS Aerobraking Overview," 313.
[4] "Nav and MGS Mission," 375; "MGS Aerobraking Broken Wing," 284-285; "MGS Aerobraking," 208.

250 km target, while the actual 44.993 hour orbital period was within 25 seconds of the 45 hour target. This was the definition of precise navigation.[5]

It was now time to begin smoothing out the probe's elliptical orbit into a circular orbit suitable for achieving scientists' objectives. Navigators made a flight-path estimate for each orbit to determine the atmospheric density at periapsis for the benefit of aerobrake operations. Aerobraking continued until October 7, 1997 (orbit number 15), when pressure on the craft decreased and certain inconsistencies in the telemetry came to light. Project management came to believe that the solar array failure had been far worse than originally believed. They suspended aerobraking temporarily on October 12, 1997, raised the periapsis altitude to 172 km, and changed the orbital period to slightly less than 35 hours.[6]

During the 25-day aerobraking hiatus, project managers and scientists considered various options. Glenn Cunningham, the Project Manager, concluded that achieving a circular orbit was essential for meeting mission science goals and that only aerobraking could realize an adequate science orbit. In November 1997, a new regimen of aerobraking began: two periods of aerobraking separated by a pause of several months called the science phasing orbit. The latter allowed low-altitude high-resolution imaging and certain scientific experiments. A second period of aerobraking began in September 1998 and lasted until early February 1999, during which time the Surveyor would make nearly 700 orbits around Mars.[7]

Monitoring the aerobraking was a great challenge. "The first passes are like two or three days apart," Al Cangahuala explained. "By the end, when you have a pass every two and a half or three hours, that requires around-the-clock staffing."[8] Because the craft flew at such low altitudes, as had been the case with the Apollo astronauts, a key aid to aerobrake navigation was a highly precise description of the Martian gravity field developed by navigators on the project's radio science team.[9]

Alex Konopliv and Bill Sjogren constructed the initial gravity model from Mariner 9 and Viking readings using new techniques and an HP755 workstation. The revolution in computing—providing greater computer capacity and speed at lower cost—enabled them to build harmonic models of greater and greater exactitude to show detailed topographic structures and to allow scientists to analyze individual features. Using one-way, two-way, and three-way Doppler and range acquired during various Surveyor orbits, Konopliv and Sjogren refined their gravity model to one of 75th degree and order, the highest level of precision ever attained.[10]

---

[5] "MGS OD Uncertainties," 1636; "MGS Aerobraking Overview," 307-309. "MGS Aerobraking Broken Wing," 275-284, details the solar wing problem and its repair.

[6] "MGS Aerobraking," 215; "MGS Aerobraking Overview," 309-311; "MGS Nav and Aerobraking," 1013.

[7] "MGS Aerobraking Overview," 311-313; NASA, *Mars Global Surveyor Arrival, Press Kit September 1997* (Washington: NASA, 1997), 34; "MGS Aerobraking," 216-218; "MGS Nav and Aerobraking," 1013 & 1020; "MGS Orbit Evolution," 1617.

[8] Cangahuala, interview, 29.

[9] "MGS Aerobraking Overview," 322.

[10] Alexander S. Konopliv and William L. Sjogren, *The JPL Mars Gravity Field, Mars50c, Based Upon Viking and Mariner 9 Doppler Tracking Data*, P 95-5 (Pasadena: JPL, February 1995), 1; Sjogren, interview, 2008, 26; "MGS Orbit Evolution," 1629; "MGS Mapping OD," 4 & 6; "MGS OD Uncertainties," 1638; "Nav and MGS Mission," 371.

Mars Pathfinder. Mars Pathfinder was the second Discovery series mission. When it landed on July 4, 1997, Pathfinder released Sojourner, the first robotic rover on another planet. Pathfinder was the first NASA flight to enter the Martian atmosphere directly from its interplanetary course. The craft decelerated by successive deployment of its heat shield, rockets, and airbags.[11]

Pathfinder navigation software reflected the cost-cutting philosophy of the Discovery program. Software development represented a large and complex part of flight navigation costs, so navigators took advantage of existing programs, but still found ways to innovate. As a result, Pathfinder became the first interplanetary flight to use the TOPEX/GPS MIRAGE program written for near-Earth navigation. They successfully tested MIRAGE with both Magellan and Ulysses tracking data, and to be on the safe side, they ran the standard orbit determination program as a backup during the Pathfinder flight.[12]

For the most complex flight phase, the descent through the atmosphere, navigators cobbled together specialized software. The Atmospheric Entry Program modeled the craft's trajectory dynamics as it descended and landed. They integrated that program into MIRAGE along with POST (Program to Optimize Simulated Trajectories) developed for NASA Langley by Martin Marietta. POST dated back to 1970, when it was the Space Shuttle Trajectory Optimization Program. Written in FORTRAN, POST underwent several revisions that added capabilities such as trajectory simulation. The combination of POST and the Atmospheric Entry Program helped to target Pathfinder to the desired surface location with great accuracy. Calculations showed that Pathfinder had arrived roughly southwest of, and less than 30 km from, the target site, well within the goal of a 200 x 100 km footprint around the target.[13]

The upgraded MIRAGE software featured an improved filter that provided more exact estimates of the spacecraft's flight path. Filters historically described non-gravitational forces acting on the spacecraft—solar pressure, velocity changes from course corrections, the impact of attitude control jets—as a mixture of parameters assigned constant values. This filter, in contrast, estimated and, as necessary, altered these values during the mission, thereby creating a better fit between the physical world and the "fitter's universe." Other filters dealt with DSN station locations as well as up to ten parameters in the Earth and Mars ephemeris prior to the probe's arrival at Mars. The intent of this novel approach was to achieve higher navigational precision using only Doppler and range and no other data type (such as ΔDOR). Once Pathfinder validated it, Pieter Kallemeyn and his team looked forward to applying these filters to future Mars flights. Before using the filter on Pathfinder, though, Sam Thurman, Jeff Estefan, and

---

[11] "Nav Flight Ops for Mars Pathfinder," 644.
[12] "Nav Flight Ops for Mars Pathfinder," 643; "Guidance and Nav for Pathfinder," 546 & 551; Thurman, *Pathfinder Project*, 10, 12, 15 & 16; Kallemeyn, "Nav Ops," 4, 7, 9, 11, 26 & 28; Spencer, 1-2.
[13] "Nav Flight Ops for Mars Pathfinder," 653, 658-659 & 661; Robert D. Braun, David A. Spencer, Pieter H. Kallemeyn, and Robin M. Vaughan, "Mars Pathfinder Atmospheric Entry Navigation Operations," 2 & 4-7, AIAA 97-3663, http://www.cs.odu.edu/~mln/ltrs-pdfs/NASA-aiaa-97-3663.pdf (accessed July 5, 2008); G. L. Brauer, Douglas E. Cornick, and R. Stevenson, *Capabilities and Applications of the Program to Optimize Simulated Trajectories (POST)*, CR-2770 (Washington: NASA, February 1977), 1-2.

Shyam Bhaskaran tested it during Galileo's second Earth encounter. They also looked at Mars Observer data and found a performance improvement of a "factor of 2 to 4" over the standard filter. They further claimed that a "50 nrad navigation accuracy or better (10 to 15 km at Mars) [was] possible with X-band counted Doppler."[14]

Because the filter handled such a large number of variables, it would not have run with the usual orbit determination program. But, the availability of relatively inexpensive, powerful computers, such as the Hewlett-Packard J5600 workstation, and graphic-oriented software, such as ARDVARC (see Chapter 25), made it possible to use such sophisticated sequential filters routinely.[15] Still, despite all the testing and hopes, the filter's use was not uncomplicated.

During the seven-month cruise, Pieter Kallemeyn, Robin Vaughan, and Robert Braun, among others, estimated Pathfinder's position and velocity using the filter. Although the filter was supposed to accommodate changes in the probe's course resulting from solar pressure, its model of solar pressure was not up to the task. At the heart of the matter was the craft's complex shape. The model treated it as a collection of various standard shapes, such as flat plates and cylinders. The backshell was the hardest part to model; it did not have a standard shape. Ultimately, Pathfinder used four models of different component groupings to represent the solar pressure during flight.[16]

Using these models, though, gave rise to various problems with the filter. The solutions that it yielded were inaccurate and for a number of reasons. Experimenting with the filter eventually bore reasonable results. However, during the week before Pathfinder entered the Martian atmosphere, the standard shapes chosen to represent the backshell became a serious problem. Navigators tried several remedies, including deleting certain data. In the end, they chose to model the backshell as a flat plate over the entire data arc for all orbit determinations done after the fourth course correction and leading up to entry into the atmosphere.[17] Small non-gravitational forces altering spacecraft trajectories—such as solar pressure—have always been the bane of space navigation. No matter how eely they are to master, or at least to anticipate, JPL navigators managed to detect and handle them in an expeditious manner. No probe had been lost as a result, at least not yet.

Missing at Mars, again. Exploration of the Red Planet continued with Mars Surveyor '98, which consisted of the Mars Climate Orbiter and the Mars Polar Lander. The Orbiter would study Martian weather and climate change and serve as a communication relay for the Polar Lander. Without the relay, the Lander would have to transmit directly to

---

[14] "Nav Flight Ops for Mars Pathfinder," 649 & 661; "Guidance and Nav for Pathfinder," 550-551; Presentation, Sam W. Thurman, Jeffrey A. Estefan, and Shyam Bhaskaran, "Doppler Navigation of Interplanetary Spacecraft using Different Data Processing Modes," March 1993, 6, 7, 8 & 13 in F247/B30, JPL508.
[15] "Nav Flight Ops for Mars Pathfinder," 643; "Guidance and Nav for Pathfinder," 546 & 551; Thurman, *Pathfinder Project*, 10 & 15; Kallemeyn, "Nav Ops," 7 & 9; Spencer, 1-2.
[16] "Nav Flight Ops for Mars Pathfinder," 643 & 645; "Guidance and Nav for Pathfinder," 546; Thurman, *Pathfinder Project*, 10 & 15; Kallemeyn, "Nav Ops," 7 & 9; Spencer, 1-2.
[17] "Nav Flight Ops for Mars Pathfinder," 646.

Earth either losing half the data or, if relayed via Mars Global Surveyor, forcing that probe to cut short its own observations. Both probes were to aid in seeking out likely sites for future robotic landers. Development of the Orbiter and Lander took place under very tight funding constraints. Their combined development cost (including launch expenses) was about the same as that for Pathfinder.[18] Still, while Pathfinder had a severely limited budget, as Lincoln Wood pointed out, "they were in Fat City compared to the Mars 1998 missions."[19] This was the very essence of Faster, Better, Cheaper.

The Mars Climate Orbiter launched first. On September 23, 1999, it fired its main engine for 16 minutes to enter orbit around Mars. As with the Global Surveyor, the orbit-insertion maneuver was the most critical flight segment. It required that the probe fly above an altitude of at least 85 km to survive. Five minutes into the orbit insertion burn the spacecraft passed behind the planet as seen from Earth. When it was due to emerge from behind Mars, the DSN could not reacquire its signal. The next day, NASA abandoned its search for the Mars Climate Orbiter on the presumption that it had flown too close to the planet. It had crashed into the planet, had broken up in the atmosphere, or had flown off into space.[20]

NASA immediately established an investigative panel headed by NASA Marshall's Arthur G. Stephenson. Its top priority was the Mars Polar Lander, the subject of its November 1999 interim report. Its final report, issued in March 2000, included lessons learned. JPL initiated two separate appraisals: one, an internal peer review consisting of the Mars Climate Orbiter team and other JPL personnel; the other, an independent panel selected by JPL.[21]

The original director of the JPL "Special Review Board" was Tony Spear. Dan Goldin had hailed Spear as "a legendary project manager at JPL" who had "helped [to] make Mars Pathfinder the riveting success that it was. If we could bottle his experience we would do so." Spear was unable to serve, however, so JPL veteran John Casani took over. Other JPL participants included mission designer Charley Kohlhase and navigators Norm Haynes and Frank Jordan. Additional members came from APL, Lockheed Martin, and George Washington University.[22]

The internal peer review group consisted of navigators and mission designers. They included Dave Smith, formerly supervisor of the Trajectory Group and manager of the JPL Mission Design Section; Jim Campbell, a former group supervisor and Voyager navigation chief; Dave Curkendall, past manager of the Navigation Section; Fran Sturms,

---

[18] NASA, Press Kit, "Mars Climate Orbiter Arrival," September 1999, http://www2.jpl.nasa.gov/files/misc/mcoarrivehq.pdf (accessed January 18, 2011), 3-4 & 10; Hecht and Adler; Ceguerra, 13; *Report Loss MPL and DS2*, 1.
[19] Wood, interview, 52-53.
[20] MCO Board, *Phase I Report*, 7; Hecht and Adler.
[21] Mars Climate Orbiter Mishap Investigation Board, *Report on Project Management in NASA* (March 13, 2000), http://science.ksc.nasa.gov/mars/msp98/misc/MCO_MIB_Report.pdf (accessed June 17, 2010); "Edward C. Stone Presentation to Caltech Board of Trustees, Mars Status, December 7, 1999," F217/B17, JPL269.
[22] Dumas to Elachi; NASA, Press Release, "Administrator Praises Work of Review Teams," March 13, 2000, http://www.iki.rssi.ru/mpfmirror/msp98/news/news68.html (accessed January 31, 2011); Memorandum, Dumas to Elachi, "Change of Chairman for the Special Review for the Mars Climate Orbiter Loss on September 23, 1999," September 28, 1999, F217/B17, JPL269; JPL Board, *Loss of MCO*, 1.

retired supervisor of the Trajectory Group; Bill Kirhofer, retired navigation leader for the Ranger, Pioneer, and Galileo missions; John McKinney, from the DSN; Glenn Cunningham, retired manager of the Mars Surveyor Operations Project; and Dave Spencer, who had been a manager for the Mars 2001 Project and a member of the Mars Pathfinder navigation team.[23] The JPL investigations, like those undertaken by NASA, focused on making proposals relevant to the forthcoming Mars Polar Lander flight. In the words of JPL Deputy Director Larry Dumas: "The Laboratory's first priority now must be the successful initiation of planetary operations for the Mars Polar Lander, so this review should be conducted with this objective in mind.[24]

The commonly accepted cause of the Orbiter loss was that Lockheed Martin Astronautics had used "English" units (e.g., inches, feet, and pounds), while JPL used metric units for a key spacecraft operation. Jeff Hecht of the *New Scientist* called it a "schoolkid [sic] blunder." Press criticism was rampant and not without cause.[25] It was the first mission failure for which one could blame navigation in the five decades of deep-space navigation at JPL. "Well, it certainly was a failure. I mean, at the root cause, yes, it was a failure of navigation," Shyam Bhaskaran admitted. "It was a difficult time," Lincoln Wood recalled. "We're not used to having failures in things that we do in the section here. There had never been a failure before."[26]

There certainly were mitigating factors. The units mismatch arose from the splitting of navigation functions between Lockheed Martin and JPL with no one person responsible for the entire navigation process.[27] Assigning navigation to a contractor was not new. Boeing, not JPL, had navigated the Lunar Orbiters. Viking was a success story, too. On that project, JPL had the lead, and most of the Flight Path Analysis Group team leaders were JPL staff,[28] which provided a continuity of communication between JPL and contractor navigators. The Mars Climate Orbiter division of navigation labor was far more bureaucratic—a trend that began with Viking—and failed to appoint any single person responsible for quality control of the entire navigation process.

Another mitigating cause, according to the NASA mishap board, was the "less than adequate" staffing of the navigation team. The Mars Program shared a common set of navigators who supported all missions simultaneously: the Climate Orbiter, the Polar Lander, and the Global Surveyor. The board urged giving the Polar Lander a larger navigation team. The team normally was quite small, but grew during critical times. Only large complex projects, such as Galileo or Cassini, could afford a full-time navigation team. The financially limited Mars probes had to make do with much less. Pathfinder had had a very small navigation team, and it was a success. "Not only was that the bar now," Cangahuala noted, "but now they wanted to push even further."[29]

---

[23] JPL Board, *Loss of MCO*, 2 & Appendix 1, 28-29.
[24] Dumas to Elachi.
[25] MCO Board, *Phase I Report*, 16; Ceguerra, 2; Hecht, 6; Hecht and Adler; Oberg.
[26] Bhaskaran, interview, 50; Wood, interview, 49.
[27] JPL Board, *Loss of MCO*, 22 & 23.
[28] O'Neil and Rudd, 1-3.
[29] MCO Board, *Phase I Report*, 22-23; Efron, interview, 13; Wood, interview, 42; Bhaskaran, interview, 69; Cangahuala, interview, 28.

Lincoln Wood remembered the Mars Program navigators as being pretty busy. "They were off in a separate building. They didn't have a lot of interaction with the rest of the section." They included Eric Graat, who did "all the orbit determination analysis" for the Climate Orbiter. Pat Esposito, the part-time navigation team chief, split his time between the Climate Orbiter, the Polar Lander, and the Global Surveyor. Vijayaraghavan Alwar did maneuver analysis for all three missions. In all, five people constituted the total navigation crew for the three missions: one team chief, one maneuver analyst, and three full-time orbit determination analysts, one for each project.[30]

"The team sizes kept dropping with Pathfinder and then Mars Climate Orbiter," Al Cangahuala recalled. "One of the challenges we have in navigation," he added, "is to justify how many people are needed to safely navigate these missions. We're different from hardware areas where you just say, 'You either buy the widget or you don't, and either you certify it, or you don't.' With people it's harder."[31] "It's easy to cut back," Shyam Bhaskaran reflected. "What I would say is that you can do this with a small number of people, but you rely more on luck. The more resources you throw at it, and the more rigorous you are, the less you have to rely on luck. You can catch problems and errors as they come up, if they happen, whereas if you have a smaller team, you're relying much more on luck that those things will not happen."[32]

The specter of small forces. Explaining away the disappearance of the Mars Climate Orbiter by a mere mismatch of measuring units overlooks the crucial failure to detect and compensate for the small forces acting on the probe. The attitude control system—the same system used on the Mars Observer and Mars Global Surveyor—replaced traditional thrusters with reaction wheels that aimed the craft in different directions. They build up momentum over time, and they release this stored momentum through a process called desaturation, which imparts a certain amount of force on the spacecraft and alters its course.[33]

Dealing with the momentum buildup was different and more challenging for the Climate Orbiter than for earlier missions. For instance, the Mars Observer velocity changes had tended to be much smaller and had occurred less frequently (about once a week). The Orbiter desaturations took place 10 to 14 times more often than the navigation team had foreseen. The cause of these frequent and unexpected changes was the asymmetry of the craft's solar panel, which increased dramatically the amount of torque from solar radiation pressure that the reaction wheels had to counterbalance.[34]

The project had come up with a way to deal with the solar panel asymmetry. "They recognized that with the solar panels sticking way up, there would be an imbalance in the center of force from solar radiation effects which would not be the same as the center of mass," Lincoln Wood explained. "If you were to flip the spacecraft every 24 hours, you would average that out to be roughly zero. They called that the

---

[30] Wood, interview, 44.
[31] Cangahuala, interview, 29.
[32] Bhaskaran, interview, 51.
[33] Hecht, 6; JPL Board, *Loss of MCO*, 35; "Nav and MGS Mission," 372.
[34] JPL Board, *Loss of MCO*, 35; MCO Board, *Phase I Report*, 6.

barbecue mode. The barbecue mode was apparently adopted originally without any particular consideration for navigation issues, although it would have been very beneficial, but adopted for other reasons. Then somewhere along the way it was scrapped, again for non-navigational issues without navigation being thought about too much. This mode was not used. The spacecraft, as a result, had a very unusually large level of non-gravitational accelerations."[35] The NASA review board also noted the lack of navigator involvement in the spacecraft development process, including the key Preliminary Design and Critical Design Reviews.[36]

The Achilles heel of Mars Climate Orbiter navigation was how the project dealt with these small forces. The project specified that the file relating to small forces would be in metric units, not those of Lockheed Martin. As a result, subsequent processing of the small forces data by the navigation software underestimated the effects of the small forces by a factor of 4.45, which is the factor for converting pound-seconds to newton-seconds.[37]

The JPL Special Review Board calculated that the cumulative effect of the small forces acting on the Climate Orbiter would have produced an error at Mars of 10,000 km. The four trajectory corrections removed all but 169 km of that error. That difference, they deemed, led to "a major discrepancy in the mission's navigation" that led the probe far too close to Mars.[38]

En route to Mars, navigators' instincts told them that something was wrong with the data. The NASA investigators found evidence that, during the first four months of the Orbiter's cruise, desaturation files had been omitted from orbit calculations because of multiple format errors and incorrect specifications. It took four months to fix the problems; navigators were not able to use them until April 1999. "It was evident fairly quickly," Wood related, "that the results from that were not consistent with the Doppler solutions." They reported the problem to project management and Lockheed Martin. The problem, unfortunately, never really was solved. Within a week it became apparent that the files contained anomalous data that indicated incorrect estimates of the course perturbations caused by the desaturations. These errors prevented navigators from being able to detect and investigate what would become the root cause of the probe's disappearance.[39]

Communications between JPL navigators and Lockheed Martin reflected the cultural shifts within navigation brought about by Viking. In particular, the NASA panel faulted navigators for the means by which they communicated with project management and Lockheed Martin. The panel acknowledged that they had expressed their concerns, but management had not responded because, in their words, "They did not use the existing formal process for such concerns." JPL had a special form to invoke a so-called

---

[35] Wood, interview, 46.
[36] MCO Board, *Phase I Report*, 18-19; JPL Board, *Loss of MCO*, 21. MCO Board, *Phase I Report*, 21, also pointed out the lack of communication with the navigators.
[37] MCO Board, *Phase I Report*, 16. The Mars Polar Lander Red Team similarly identified small forces as the culprit. Presentation, C. Jones, Mars Polar Lander Red Team Final Report, November 23, 1999, F152/B20, JPL273.
[38] JPL Board, *Loss of MCO*, 1.
[39] Wood, interview, 43 & 45; MCO Board, *Phase I Report*, 17-18; JPL Board, *Loss of MCO*, 22.

incident surprise and anomaly procedure. The navigators had failed to follow the rules about filling out that form to document their concerns.[40] The time-rooted informality of navigators—such as direct communication via e-mail—butted against well-meaning (one presumes) bureaucratic procedures.

Yet another contribution to the Climate Orbiter's demise was the decision not to undertake the fifth course correction scheduled to occur before orbit insertion. The maneuver was a contingency plan to raise the Orbiter to a safe altitude virtually at the last minute. The NASA mishap board identified this decision as a contributing cause.[41] The course-correction decision pitted current and former navigators against each other. The two key people were Sam Thurman, who had left the navigation section several years earlier, and Pat Esposito. "There was also a little bit of a history between Sam and Pat Esposito," Wood recollected. On the Climate Orbiter, "Sam was kind of Pat's boss, and Pat was the navigation team chief, and Pat was very concerned about these orbit solutions."[42]

Esposito's concern arose from the discrepancies among the orbit determination results obtained from just Doppler versus Doppler plus range versus range only. "I think Sam tended to regard Pat as kind of a conservative worrywart individual who wasn't up on the latest techniques of tightly-weighting ranging, which was to some extent Sam's invention," Wood remembered. "There was a belief that Pat was just too conservative. I don't think Pat, in voicing concerns, was listened to properly by the project. There were various meetings held where he was in some cases simply not invited to meetings that pertained to navigation."[43]

The time to decide whether to execute the fifth trajectory correction occurred just a few hours before orbit insertion. Esposito had assumed that the maneuver was a viable option and "strongly encouraged" Thurman to execute it to raise the craft's altitude. "He wanted to do that just to build in some safety to make sure that if the orbit solutions were poorer than he thought, that nothing disastrous would happen," Wood explained. "Well, Pat wasn't invited to the critical meeting where whether to do the maneuver or not was decided on." By the time Esposito found out about the meeting, it was too late. "Sam told him that the case was closed."[44]

In looking over the NASA board report, Wood did not deem the fifth trajectory correction to be as feasible as Esposito had believed, "although on paper it should have been viable." He partly blamed the project's limited budget: "there were issues with whether it could be done without jeopardizing the orbit insertion a limited number of hours later. There were also issues if you went in a little bit high, you wouldn't be in quite the right orbit for doing the aerobraking. You might not be able to achieve the science orbit when you wanted to achieve it. So there were some reasons for not doing the fifth maneuver." The decision was made not to perform the final course adjustment,

---

[40] MCO Board, *Phase I Report*, 21; Oberg.
[41] MCO Board, *Phase I Report*, 19-20.
[42] Wood, interview, 47.
[43] Wood, interview, 47.
[44] Wood, interview, 48. Esposito did not reply to telephone and e-mail requests for an interview.

"but without receiving all of the inputs that really should have been received," Wood concluded.[45]

Supplemental data types. Could navigators have observed and compensated for the unexpected course aberration created by the reaction wheel system? In fact, the NASA mishap board acknowledged that the faulty analysis of the small forces files and the mismatch of units probably would have become noticeable early in the mission—and therefore manageable—had the navigation team had a data type other than Doppler and range. The board explicitly recommended the use of so-called "supplemental tracking data types" to "enhance or increase" navigational accuracy for the upcoming Mars Polar Lander. Their specific recommendation was near simultaneous tracking as well as "a three-way measurement" and a "difference[d] range process" as a check on the two-way Doppler for the Polar Lander.[46]

Noticeably absent from the NASA report was any specific mention of ΔDOR. It would have provided information, lacking from either Doppler or range, on the position of the spacecraft relative to the plane of the sky, and that knowledge would have made the Orbiter's course oddness clear. Doppler indicates the spacecraft velocity along the line of sight, while range measures the distance along the same line of sight. They do not measure *directly* any movement perpendicular to the line of sight direction, but ΔDOR *does* measure it directly.[47] What had happened to ΔDOR?

In part, ΔDOR was too expensive given the tight budget noose around the Climate Orbiter. "During the nineties, after Mars Observer," Al Cangahuala recalled, "they mothballed the ΔDOR" because of its cost. "That's what we were told. Cost. We said: 'Well, this really helps.' And they said: 'Well, do you absolutely need it?' And we said: 'Well, if nothing goes wrong, no; but we won't know that nothing goes wrong without some sort of insurance.'" "At the time," Cangahuala added, "I was just a technical peon waving my fist at the sky, because it really was the wrong thing to do. It was the loss of an insurance policy."[48]

As we saw in Chapter 18, when presented with a substitute for ΔDOR that cost less, Galileo and Magellan project managers dropped it in favor of the alternative. The abandonment of ΔDOR permanently for the Mars effort is not surprising. Cost-cutting was the chief priority, and project managers and others believed that they could get to Mars without the benefit of its high precision and high cost. In short, Mars exploration did not need (and did not want to pay for) any insurance of mission success.

A series of studies by navigators also supported the shift away from ΔDOR. In 1993, with NASA Headquarters money, navigators Jeff Estefan, Vince Pollmeier, Dan Scheeres, and Sam Thurman pitted ΔDOR against other data types aided by the new filter. The first phase looked only at two-way X-band Doppler and range. The second phase compared the navigational accuracy of X-band ΔDOR with two-way X-band

---

[45] MCO Board, *Phase I Report*, 19-20; Wood, interview, 48.
[46] MCO Board, *Phase I Report*, 30.
[47] Bhaskaran, interview, 75.
[48] Cangahuala, interview, 24 & 25.

Doppler and ranging, but only for missions to the inner planets: the cruise phases of the Mars Observer and Pathfinder and Cassini's Venus-Earth cruise segment. The study concluded that ΔDOR "data are not providing any significant information regarding the spacecraft trajectory that is not already present in the Doppler and ranging data." Doppler alone could provide navigational accuracies of 50 to 150 nanoradians, and Doppler and range combined yielded accuracies of 25 to 100 nanoradians. Adding ΔDOR to the two provided little improvement: accuracies of 20 to 100 nanoradians. Therefore, the best accuracies resulted from Doppler alone.[49] Their conclusions meshed with the low-cost navigational philosophy that was part and parcel of the Discovery Program requirement to cut costs and more generally with Goldin's Faster, Better, Cheaper management philosophy.

During the Galileo flight, Bhaskaran distinctly recalled attending a meeting that led to the decision to decommission the ΔDOR from Galileo. "We talked about it, and discussed it, and looked at the results." The group decided—based on the advice of navigators—that they could fly the mission without ΔDOR. Once Galileo managers decommissioned the ΔDOR, no missions used it: neither the Mars Global Surveyor nor Pathfinder, which landed perfectly fine without it; neither the Stardust comet nor the NEAR asteroid missions, both navigated by JPL.[50]

Lessons learned? Although both NASA and JPL Mars Climate Orbiter investigations had concentrated (rightly so) on the impending December 3, 1999, landing of the Mars Polar Lander, it too was lost on arrival at Mars along with the two Deep Space 2 craft stowed onboard. The losses now totaled four probes. The loss of science was incalculable. Communication ended, according to plan, as the Lander prepared to enter the Martian atmosphere. The mission design called for a resumption of communication once the Lander and its cargo were on the surface. Repeated efforts to contact all three probes continued for several weeks but to no avail.[51]

NASA named a mishap board, headed by Lockheed Martin's Thomas Young. JPL appointed a Special Review Board under the direction of Caltech's Arden Albee. The probable cause of the accident was the premature shutdown of the descent engines as a result of an inherent software problem. The intentional lack of telemetry hindered the

---

[49] Specifically, the money came from RTOPs [Research and Technology Operating Plan] 60, 61, and 63. In this case, the RTOP assigned funding from Headquarters to JPL for the purpose of carrying out a specific research project led by a Principal Investigator. Memorandum, Jeff A. Estefan, Vincent M. Pollmeier, Daniel J. Scheeres, and Sam W. Thurman to Carl S. Christensen, "TDA (430) Navigation Study X-Band Doppler and Ranging Error Analysis—Inner Solar System Mission Scenarios," March 16, 1993, F247/B30, JPL508; Memorandum, Jeffrey A. Estefan, Vincent M. Pollmeier, Daniel J. Scheeres, and Sam W. Thurman to Carl S. Christensen, "TDA (430) Navigation Study X-Band Radio Navigation Error Analysis: ΔDOR Analysis and Additional Mission Scenarios," March 16, 1993, F247/B30, JPL508.
[50] Bhaskaran, interview, 75-76.
[51] *Report Loss MPL and DS2*, 2; Suzanne E. Smrekar and Sarah A. Gavit, "Deep Space 2: The Mars Microprobe Project and Beyond," 36-38 in Francisco Anguita, ed., *The First International Conference on Mars Polar Science and Exploration*, LPI 953 (Houston: Lunar and Planetary Institute, 1998). Despite what appeared to be its parachute, no trace of the lander had materialized by 2005. David Tytell, "Mars Polar Lander Still Missing," *Sky & Telescope* 111 (January 2006): 22.

crash investigations. The NASA panel zeroed in on the lack of telemetry during the critical entry, descent, and landing phase. Its absence "was a defensible project decision, but an indefensible programmatic one." The project had dropped telemetry to reduce costs in the belief that it did not contribute directly to landing safely on Mars. The decision had the approval of both NASA Headquarters and JPL. The NASA investigators advised adding telemetry coverage on future missions, specifically on the upcoming flight of the Mars '01 Lander (later renamed Mars Odyssey).[52]

The Young panel focused largely on management issues and laid blame for the loss of both the Orbiter and Lander on a lack of funding: "The [Mars '98] project was significantly underfunded from the start for the established performance requirements." They contrasted the Pathfinder development cost of about $200 million with the $190 million spent on the Mars Orbiter and Lander combined. "This clearly indicates the significant lack of sufficient budget for Mars '98," the Young report declared.[53]

On the other hand, the panel did not fault the overriding Faster, Better, Cheaper management philosophy. Indeed, they supported it as "the right path for NASA's present and future. FBC [Faster, Better, Cheaper] has produced highly successful missions, such as Mars Pathfinder. More importantly, no other implementation philosophy can affordably accomplish NASA's ambitious future goals within a feasible budget and schedule." Failures were due not to any inherent fallibility of Faster, Better, Cheaper, but to its implementation. NASA and JPL had not "completely made the transition to FBC." Moreover, "project managers have their own and sometimes different interpretations [of what constitutes Faster, Better, Cheaper]. This can result in missing important steps and keeping lessons learned from others who could benefit from them." Thus, the flight failures were failures of implementation, not an unworkable management concept.[54]

The Young Report repeatedly disparaged navigators' performance on the Mars Polar Lander. The panel again admonished the handling of small forces. Immediately after launch stray light from the spacecraft interfered with its star camera, which the Polar Lander needed to adjust its attitude. Reminiscent of the Climate Orbiter, the result was a dramatic increase in the number and frequency of small forces acting on the probe throughout the cruise phase and adversely affecting navigational accuracy.[55]

The revised mission design had called for the navigation team, with additional staff in response to the Orbiter disappearance, to model the effects of small forces based on spacecraft telemetry rather than Doppler. The unexpected data caused by the star camera was challenging to model, and navigators did not make meaningful progress in taming it until the final weeks before encounter. This freshly acquired knowledge of the spacecraft's position and velocity, and the small forces acting on it, went into a recalculation of the fourth trajectory correction.[56]

NASA investigators also criticized the final trajectory correction, which similarly had been a Mars Orbiter weakness. Again, the final course change was scheduled to

---

[52] Mars Team, *Summary*, 13; *Report Loss MPL and DS2*, 7, 13 & 15.
[53] Mars Team, *Summary*, 6.
[54] Mars Team, *Summary*, 8.
[55] *Report Loss MPL and DS2*, 14 & 33.
[56] *Report Loss MPL and DS2*, 33.

occur shortly—only six hours—before orbit insertion. The ability to fly the Polar Lander into the relatively small hole required by the mission design depended on the precision of this final maneuver, which in turn relied on the accuracy of navigation up to that point. Planning ahead for the short period between the maneuver and orbit insertion—and to avoid any potential human error in designing a maneuver from scratch at this critical juncture—the project decided to design a series of maneuvers in advance. At the appropriate time, the project would transmit the maneuver code that was closest to the desired solution. Pathfinder, Global Surveyor, and even Magellan already had trusted in this approach. Not surprisingly, the size of the maneuver was substantially larger than originally planned. The NASA review called for future missions to make the final trajectory correction more flexible, while JPL investigators suggested developing a maneuver strategy that was "less sensitive" to spacecraft performance and small forces.[57]

All the same, Polar Lander navigation *had* heeded the advice of the Climate Orbiter boards and had implemented many proposed changes, but these had no impact on the mission's outcome. For example, both had recommended that navigators implement "a series of independent peer reviews." NASA added a team of senior navigators to the Polar Lander to carry out such reviews. Led by Bhaskaran, the so-called Navigation Advisory Group (NAG) reviewed the Polar Lander navigation plan and instituted major changes. The advisory group performed the same orbit determination analyses as the project navigation team, but in parallel and with minimum interference. The two navigation groups coordinated through meetings during the entire cruise phase.[58]

Navigation also had responded by increasing to three the number of navigators assigned to the Polar Lander. But, as the NASA panel discovered, only two actually were on the project. The rationale was that "24 hour/day navigation staffing" was intended for the brief period before the probe began its entry, descent, and landing. "Such coverage may be difficult even for a team of three navigators and certainly was not possible for the single navigator of MCO [Mars Climate Observer]," the board concluded. They urged adding a third navigator "as soon as possible."[59]

Additionally, following the advice of the Mars Orbiter board, navigators had access to a "supplementary" data type, namely "near-simultaneous tracking" data, for both the fourth trajectory correction and planetary approach. The JPL review panel criticized the overly optimistic assumptions made about the quality and availability of the new unproven navigation data type. They further noted that the staff originally planned to handle the Polar Lander prior to the loss of the Orbiter likely would not have succeeded in aiming the probe correctly had they followed the pre-launch navigation

---

[57] *Report Loss MPL and DS2*, 14, 33 & 34.
[58] MCO Board, *Phase I Report*, 22; JPL Board, *Loss of MCO*, 25-26; L. Alberto Cangahuala, James S. Border, Joseph R. Guinn, Timothy P. McElrath, and Michael M. Watkins, "Mars Polar Lander Operations: Navigation Advisory Group Activities," 1, http://trs-new.jpl.nasa.gov/dspace/bitstream/2014/15021/1/00-1027.pdf (accessed June 19, 2010).
[59] MCO Board, *Phase I Report*, 23.

plan, while simultaneously overseeing both Climate Orbiter aerobraking and the Global Surveyor mapping mission.[60]

Neither NASA nor JPL Polar Lander investigators advised adding ΔDOR specifically as a complementary data type. Yet, now its revival seemed indispensable. That meant installing or upgrading DSN equipment. Cangahuala insisted that bringing back ΔDOR be treated not as "another science experiment" to look for quasars, but as an engineering task. "There was kind of a cultural boundary condition," he explained. "You need to tell me your turnaround time. You need to commit to levels of performance." He added, "It was a little rocky at first, because they weren't sure who was going to pay for it. Should the Mars missions pay? Is this a Mars mission? But it could benefit other missions. Eventually they straightened it out." Small DSN upgrades installed since the mothballing of ΔDOR had lowered the amount of human intervention required to operate the equipment, so the cost of the revived ΔDOR was lower.[61]

Navigators also benefited from better performance. For Mars Observer, ΔDOR accuracy had been 20 to 50 nanoradians. With the new equipment, a projected accuracy level of 5 to 10 nanoradians seemed attainable. The first beneficiary was Mars Odyssey in 2001. "They were hitting like five [nanoradians] pretty easily," Cangahuala recalled. The Mars Exploration Rovers realized accuracies of three to four nanoradians.[62] The loss of the Mars Climate Orbiter finally "made a believer out of them," Curkendall declared. "It took, what, a $150 million mission. Can you imagine going to the project manager and saying, 'Gee, you know what? We ran into the planet.'"[63]

---

[60] *Report Loss MPL and DS2*, 34-36.
[61] Cangahuala, interview, 26-27; Personal communication, James Border, October 4, 2012.
[62] Cangahuala, interview, 27.
[63] Curkendall, interview, 48.

Chapter 22

Autonomous Navigation

In contrast to the resounding Mars failures of 1999, which were direct consequences of Dan Goldin's Faster, Better, Cheaper approach, navigation realized a series of dazzling successes with the transformation of optical navigation that shifted even more functions to probes. This milestone achievement, called autonomous navigation, was possible only because of the wave of electronic miniaturization that finally had begun to take hold of spacecraft design and construction. A demonstration of autonomous navigation was made possible by the New Millennium Program, a NASA Headquarters effort to speed up the development and testing of novel space technologies for future low-cost flights. The program paid homage to the development of high-tech industry favored by President Clinton as well as Goldin. Ironically, the cost-cutting that was fundamental to Faster, Better, Cheaper also was a disincentive to the development of advanced technology, so the New Millennium and its ilk had become a necessity for realizing the technological progress required for future space exploration.

Techniques of greater precision were a prerequisite for studying asteroids and comets. While the exploration of planets and their satellites did not abate, close-up analysis of these smaller objects was a growing enterprise, as discussed in Chapter 19. Their diminutive size made flying by or around them a challenge, and autonomous navigation was the best solution.

Autonomous navigation, usually called autonav, was a logical extension of optical navigation, which was labor-intensive and computer-intensive with the labor and computer located on Earth. Autonav transferred labor from Earth-bound navigators to the spacecraft's computer. The main motivation for creating autonomous navigation was not to replace skilled navigators with software, but to facilitate exploration of small solar-system bodies.

Autonomous navigation effectively overcame the time lag between the acquisition of camera pictures and the computation of an orbit solution from them. This interval was not critical, nor even a factor, during interplanetary cruises or planetary approaches. It was not a necessity for navigating to the distant planets of the outer system, as Galileo's adept ballet through the Jovian satellites attested. Instead, autonomous navigation served specifically the study of asteroids and comets. The use of autonomous navigation continued the links between optical navigators and ground-based astronomers armed with CCD-camera-equipped telescopes. The development of autonomous navigation had momentous import for optical navigation and as the software evolved, it ultimately changed what deep-space navigation meant. An outward sign of that transformation was a new name for the division that was home to navigation, the Autonomous Systems Division.

Automating optical navigation. Autonomous navigation originated within the Optical Navigation Group. Steve Synnott, who became the Group Supervisor in 1982 shortly after the Voyager encounters, initiated the drive toward automating navigation, when he returned to the Group in 1991. He realized that one could use optical navigation in an entirely new way. Traditionally, people thought of opnav as a tool or technique for aiming probes precisely as they approached a target. When "you're screaming into Io or you're screaming into Triton out at Neptune, and you cannot rely on 5,000-kilometer accuracy," Synnott explained, "you have to take the local pictures to get to 10 or 20 kilometers." What if, he wondered, "you could use it in cruise from Earth to Mars or Earth to Jupiter looking at asteroids against the star background." "Instead of requiring—and DSN people might not like to hear this—instead of requiring huge antennas tracking to get range and Doppler and VLBI for months and months and months at a time, I thought you could free up a lot of time from the DSN stations and just have a package on the spacecraft that could navigate itself." Of course, autonomous navigation was no threat to the DSN, and the former remained "an absolute necessity" for landing on a comet, for instance.[1]

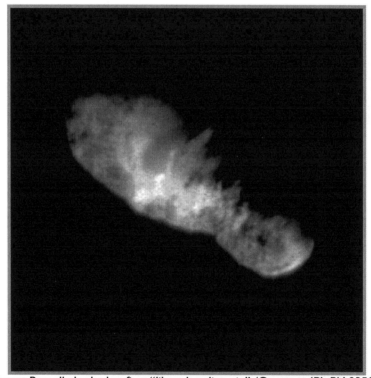

Comet Borrelly looked at first "like a bowling pin" (Courtesy JPL PIA03500)

---

[1] Cangahuala, interview, 31; Synnott, interview, 18, 19 & 21; Sjogren, interview, 2008, 39; Maize, interview, 4.

The development of autonav really took off late in 1993, when Synnott offered a position in the Optical Navigation Group to Shyam Bhaskaran "partly to continue on Galileo optical navigation, but primarily because there were other projects that were coming along." Bhaskaran was relatively new to JPL, but not to navigation. Within six months, he was working on proposals for Discovery class projects to explore comets and asteroids. He also participated in optical navigation on the Deep Space Program Science Experiment (DSPSE), a joint project of the Ballistic Missile Defense Organization (the new name for the Strategic Defense Initiative) and NASA that launched in 1994. It included a flyby and mapping of the asteroid 1620 Geographos in early September 1994 that involved using optical navigation for its asteroid approach. After orbiting the Moon, however, a spacecraft failure precluded the encounter. As Bhaskaran deepened his experience with optical navigation and worked on Discovery proposals, "It became clear," he recalled, "that an autonomous system for orbiting or landing on these bodies would be very advantageous, as opposed to ground-based navigation. That's what started the seed of autonomous navigation for me: working on these small proposals."[2]

The focus, then, was to be orbiting around small bodies, not enhancing cruise navigation. In 1994, with Synnott's support, Bhaskaran submitted a proposal for fiscal year 1995 support and won a grant. "It was a small amount of money, and it was just mostly my own thing, although once again Steve Synnott helped me a great deal." He also received important proposal mentoring from Ed Riedel, another member of the Optical Navigation Group. Bhaskaran then spent a year developing prototype autonav software based on current optical navigation methods.[3]

A millennial opportunity. In 1995, an opportunity to test the prototype software presented itself, when NASA's Office of Space Science and Office of Earth Science jointly established the New Millennium Program. Its goal was to accelerate development and testing of cutting-edge space technologies to be used on future reduced-cost flights. The agency christened the opening project Deep Space 1.

Synnott, Bhaskaran, Riedel, and others proposed autonomous navigation as one of the technologies to undergo testing to the JPL New Millennium project manager, E. Kane Casani. He agreed. Autonomous navigation was "no longer a pie-in-the-sky or amorphous concept. We were now on a mission," Bhaskaran recalled. "We have to actually fly this thing for real. And we have actual money, yes, to work on this." Now, as the project progressed, more people joined—Shailen Desai, Bob Werner, and Brian Kennedy—to form the core autonav team with substantial help coming from such optical navigators as Bill Owen, who later became the Optical Group Supervisor, and Phil Dumont. Ed Riedel was the New Millennium Navigation Team chief.[4]

---

[2] Bhaskaran, interview, 1-7 & 14; Bernard Kaufman, Jay Middour, Robert Dasenbrock, and Richard Campion, "An Overview of the Astrodynamics for the Deep Space Program Science Experiment Mission (DSPSE)," *Acta Astronautica* 35 (May-June 1995): 661-668.
[3] Bhaskaran, interview, 15.
[4] Synnott, interview, 21; Bhaskaran, interview, 22; Shyam Bhaskaran, Joseph E. Riedel, and Stephen P. Synnott, "Autonomous Nucleus Tracking for Comet/Asteroid Encounters: The Stardust Example," 463 in

Deep Space 1 tested twelve new technologies. Of the twelve, the Miniature Integrated Camera and Spectrometer, the Solar Electric Ion Propulsion System, and autonav were most directly relevant to the craft's navigation. The propulsion system, a xenon ion engine, posed peculiar navigational challenges. Typically, probes spent long periods coasting through space without firing their thrusters, the only exceptions being course-correcting maneuvers. In contrast, the ion engine was active constantly over most of the flight path.[5]

The main non-gravitational forces perturbing the spacecraft's trajectory were solar radiation pressure, thrust from the propulsion system, and attitude corrections by the hydrazine-propelled reaction control system. Modeling the solar radiation pressure was reasonably uncomplicated. The solar panels received most of the force, and for practical reasons these remained perpendicular to the Sun at all times. Modeling the propulsion thrust and attitude control jets was far more complicated, because their actions and effects were not constant. In order to keep up with them, the autonomous navigation program constantly updated its models of these forces. The Earth-bound navigation team, though, preferred to work directly with raw telemetry. Also, for various technical reasons, the Deep Space 1 propulsion halted once a week. It had become clear that avoiding constant engine thrusting while the DSN tracked the probe benefited orbit determination accuracy.[6]

Autonomous navigation relied on the chief science instrument, the onboard CCD-based Miniature Integrated Camera and Spectrometer (MICAS). The device had no shutter and could image in the visible, ultraviolet, and infrared light spectra. The autonav team used only one of the two visible light channels. The MICAS was fastened to a telescope that was attached to the spacecraft, so that the entire craft had to swing around to look at a given portion of the sky. The other visible light channel was for the experimental Active Pixel Sensor array, which had a lower resolution than the MICAS (a 256 versus a 1,024 square-pixel array).[7]

Onboard navigation had become possible in no small part because of advances in computer technology. Computers were becoming faster, more powerful, smaller, and lighter. The Deep Space 1 computer was a radiation-hardened RAD6000 system developed by IBM for the Air Force Research Laboratory. NASA used it on a myriad of other probes, including Mars Pathfinder, the Mars Polar Lander, the Mars Climate Orbiter, Mars Odyssey, MESSENGER, Genesis, Stardust, and Dawn as well as the Spirit and Opportunity Mars rovers. The 33-MHz computer had 96 megabytes of Rapid Access Memory (RAM). About 50 megabytes of that memory was available for storing science and navigation information. Consequently, during most of the mission, both science and navigation dominated half of the Central Processing Unit's capacity.[8]

Felix R. Hoots, Bernard Kaufman, Paul J. Cefola, and David B. Spencer, ed., *Astrodynamics 1997*, Part I (San Diego: Univelt, 1997).
[5] Wood, "Nav Evolution, 1989-1999," 887-888.
[6] *DS-1 Nav: Primary Mission*, 4 & 10-12; "Nav for New Millennium," 305.
[7] "DS-1 Nav at Borrelly," 2; "In-Flight Performance," 2.
[8] "Nav for New Millennium," 306-307; "AFRL's RAD6000 Computer," http://www.kirtland.af.mil/shared/

The autonomous navigation system represented a tremendous amount of development, especially "bulletproofing the software." The programs, often referred to as the Navigator or simply AutoNav, consisted of a so-called executive (NavExec) program that interacted with the rest of the flight software, the spacecraft's attitude control system, and the camera. It concerned itself with the a-to-z of navigation: planning picture sequences, analyzing images, estimating the spacecraft trajectory, and calculating course corrections. At predetermined times, it commanded the attitude control system to point the camera (and spacecraft) at a desired target, it told the camera to take a picture, and it analyzed the resulting picture using information on the craft's attitude (telemetry). The software possessed all of the usual elements for performing basic orbit determinations and maneuvers such as models of solar radiation pressure and the small forces exerted by the propulsion and attitude control systems. It also had its own ephemeris based on the JPL Development Ephemeris.[9]

The autonomous navigation system became operational during the cruise phase, as Synnott had imagined its use years earlier. The MICAS imaged so-called beacon asteroids, one at a time, relative to background stars using a variation of the single-frame mosaic technique developed for Galileo. Navigators identified about 80 asteroids to serve as beacons, and the Table Mountain Observatory alone observed 31 of these in the hope of improving their orbits by a factor of 3 or 4. Observations typically took place once a week over a span of 4 to 6 hours. After recording and processing the new images, the autonomous system generated a new orbit determination based on a combination of new and old data collected over the past 30 days. The accuracy of the navigation solutions depended on a number of factors—the brightness of the asteroids, their distance, and their angular separation—not the least of which was the precision of the asteroid ephemerides stored on the spacecraft.[10] Don Yeomans provided the ephemerides for both the beacon and target asteroids. "So we sent out requests for additional observations," Yeomans related, "and updated their orbits, and we made sure that they were well taken care of." Navigation initially intended to use just the Hipparcos star catalog, but the Naval Observatory's Astrographic Catalog and ESA's Tycho catalog provided a denser set of stars, so they used the former in conjunction with the Tycho catalog.[11]

---

media/document/AFD-070404-100.pdf (accessed March 17, 2009); BAE Systems, "RAD6000 Space Computers," http://www.baesystems.com/BAEProd/groups/public/documents/bae_publication/bae_pdf_eis_sfrwre.pdf (accessed March 17, 2009).

[9] "OD Performance AutoNav," 1295-1314; "DS-1 AutoNav," 42-52; Synnott, interview, 21; "Nav for New Millennium," 307; "Brief History of JPL OpNav," 342; Wood, "Nav Evolution, 1989-1999," 888. For a detailed discussion of the Deep Space 1 autonav: Joseph E. Riedel, Shyam Bhaskaran, Shailen Desai, Dongsuk Han, Brian Kennedy, George W. Null, Stephen P. Synnott, Tseng-Chan Wang, Robert A. Werner, Elaine B. Zamani, Timothy McElrath, and Mark Ryne, *Autonomous Optical Navigation (AutoNav) DS1 Technology Validation Report* (Pasadena: JPL, March 2001).

[10] "OD Performance AutoNav," 1300 & 1301; "DS-1 AutoNav," 44-46 & 48-49; Wood, "Nav Evolution, 1989-1999," 888.

[11] Yeomans, interview, 21; "High-Accuracy Astrometry," 89-102; Synnott, interview, 23; Sean E. Urban and Thomas E. Corbin, *The ACT Reference Catalog* (Washington: USNO, 1997); Urban, Corbin, and Gary L. Wyckoff, "The ACT Reference Catalog," *The Astronomical Journal* 115 (May 1998): 2161-2166.

Blind at Braille. Deep Space 1 also had a scientific mission. Its original objective was to study the asteroid McAuliffe from a distance of just 20 km on January 20, 1999. Its flight path continued onward to Mars in April 2000 to receive a gravity boost and possibly to perform a flyby of either Deimos or Phobos. Its final science objective was an encounter with comet West-Kohoutek-Ikemura at a distance of about 500 km in early June 2000.[12] This ambitious plan soon yielded to a simpler one: a study of asteroid 9969 Braille during a flyby encounter. Discovered in 1992, the asteroid retained its rather anodyne provisional designation (1992KD) until The Planetary Society's "Name That Asteroid" contest on July 26, 1999, named it after Louis Braille, the sightless inventor of the eponymous writing system for the blind. On July 29, 1999, just three days later, Deep Space 1 had its ill-fated meeting with the aptly named asteroid.[13]

The craft left Earth on October 24, 1998. Navigation based on Doppler and range guided the probe "out of the harbor," in a manner of speaking, and continued for the remainder of the cruise to Braille, as it was the only independent means for evaluating the autonomous system. Navigators knew that they could not rely immediately on the more advanced aspects of the autonomous programs—planning, taking, and processing images, orbit determination—because they had not been validated. As it turned out, this decision was a good idea. Navigators were unable to test the autonav system for several months because of the experimental camera. The MICAS was letting considerable amounts of stray light into pictures and distorting them. This flaw was not new; it had been known as early as eight months before launch, when testing revealed that the CCD channel could not handle brightness amounts above a certain level, coincidentally equal to the brightness of the target asteroid. However, the belief was that the imager would be unusable during only the last few minutes of approach, not for months during interplanetary cruise.[14]

Because of the ongoing camera problems, no orbit determinations took place with it until February 19, 1999. Navigators processed these images on Earth. The trajectory estimate obtained was about 4,000 km off the Doppler results. During the three months between March 1 and June 1, navigators continued to assess the system's ability to perform orbit determinations while addressing two new modeling problems that had surfaced. At one point, the navigational accuracy had degraded to nearly 10,000 km, and later it fell to 13,000 km, as more systematic modeling errors plagued the autonomous software. Gradually, performance began to improve to 6,000 km, by late April to 4,000 km, in early May to 2,000 km, and finally later in May to 1,700 km.[15]

Uploading new software that included several improved models for camera performance and other parameters—and fixing an error in ground data-processing—

---

[12] "Nav for New Millennium," 303-304; "OD Performance AutoNav," 1296.
[13] CNN, "Asteroid target gets new name," July 26, 1999, http://edition.cnn.com/TECH/space/9907/26/space.briefs/ (accessed December 31, 2010).
[14] *DDS-1 Nav: Primary Mission*, 3-4, 27 & 38-39; "In-Flight Performance," 5; Wood, "Nav Evolution, 1989-1999," 889.
[15] *DDS-1 Nav: Primary Mission*, 40-41.

increased the accuracy of trajectory estimates sharply. Two orbit determinations done on June 29 and July 2 showed differences with Doppler results of 662 km and 904 km, respectively. The error level of onboard orbit determinations eventually stabilized between 700 and 1,000 km. This accuracy was not high enough to support autonomous maneuver planning completely.[16]

Navigators saw Braille for the first time three days ahead of encounter, but only by processing multiple images on the ground. Braille had not been visible on the first two attempts (July 24 and July 25), and the third try (July 26) succeeded in revealing an extremely faint "phantom" image, but again only with intensive ground processing. The asteroid appeared to be about 350 km from its expected position. Controllers ordered the probe to execute a trajectory correction 1.5 days before encounter.[17] The reliance on ground processing was not a good sign.

The mission design had called for Deep Space 1 to take 18 pictures of Braille on July 28. The AutoNav acquired, processed, and stored them in a computer file. Then, a software bug caused the craft's computer to reboot and go into safe mode. About three hours later, Deep Space 1 recovered. Luckily, the ground version of the software worked fine, and navigators used its results fruitfully. It was ironic that the autonomous navigation system was blind during the Braille encounter. "The irony was not lost on many of us at the time," Bhaskaran added with a laugh. "So, it was maybe perhaps a poor choice of a name in retrospect. But we weren't the ones who named it." One can attribute the failure at Braille almost entirely to the fact that the MICAS camera did not meet its design requirements.[18]

On to Wilson-Harrington? On September 18, 1999, the Braille encounter over, Deep Space 1 began its extended mission. Its new science goal was to collect information about comet Wilson-Harrington (also designated asteroid 4015 Wilson-Harrington) in January 2001 and comet Borrelly in September 2001. Its engineering objectives included validation of the autonomous navigation and ion propulsion systems. Both technologies performed nearly flawlessly during the first two months of the extended mission. The software was still a bit questionable, but the craft managed to navigate itself, the true test of the AutoNav. Not all systems were as sound, however.[19]

On November 11, 1999, the Stellar Reference Unit failed. This onboard star tracker was the chief instrument for maintaining spacecraft attitude. Deep Space 1 spent the next seven months in an extended safe configuration, while project managers and engineers mulled over a solution. It was clear that the probe could not undertake both encounters, especially given the extremely low supply of hydrazine attitude control gas.[20] Deep Space 1 consumed a large amount of hydrazine maintaining itself in safe mode and

---

[16] *DDS-1 Nav: Primary Mission*, 41 & 43; "DS-1 AutoNav," 46-49; Wood, "Nav Evolution, 1989-1999," 889.
[17] *DDS-1 Nav: Primary Mission*, 49; "In-Flight Performance," 9; "Brief History of JPL OpNav," 342.
[18] *DDS-1 Nav: Primary Mission*, 49; "In-Flight Performance," 10 & 12; "Brief History of JPL OpNav," 342; "DS-1 AutoNav," 49-52; Wood, "Nav Evolution, 1989-1999," 889; Bhaskaran, interview, 29.
[19] *DS-1 Nav: Extended Missions*, 1; *DDS-1 Nav: Primary Mission*, 53-54.
[20] "Radiometric OD Borrelly," 2218.

aiming its high-gain antenna to communicate with Earth. The small amount of remaining fuel had a big impact on trajectory design and operation of the probe as it made its way to Borrelly. In January 2000, the Deep Space 1 science team decided to drop Wilson-Harrington and to redesign the trajectory to take the probe to comet Borrelly for an encounter in September 2001.[21]

The key to bringing Deep Space 1 back from the grave had been to replace the Stellar Reference Unit with technologies already tested during the primary mission, namely the MICAS camera, the ion propulsion, and the autonomous navigation system. Replacing the star tracker with the camera necessitated changing the techniques for maintaining attitude and the flight software, both of which impacted the functioning of the AutoNav. As a result, scheduling navigation picture-taking increased the risk of losing knowledge of the probe's position. Orbit determination had to rely on two-way Doppler and range, as Deep Space 1 propelled itself toward Borrelly, when optical navigation again could play a role. The DSN collected Doppler and range only once or twice every week to conserve hydrazine, because obtaining the data required pointing the spacecraft and its high-gain antenna toward Earth.[22]

One AutoNav change was to prevent the imaging system from being "spoofed" by a spurious bright spot in the field of view. "We didn't want that to happen again," Bhaskaran explained. The improved navigation system "had many more checks and balances" to avoid having bad data once again "ruin our AutoNav system."[23] Also, Bhaskaran and his colleagues developed an extra program affectionately called "the Blobber." It searched a specified area of a camera image and created a list of any contiguous "blobs." The belief was that these blobs represented the comet's nucleus, which was what the AutoNav used for center-finding. Additional software extracted appropriate targeting information for the nucleus-tracking program.[24]

To prepare for the encounter, the autonomous navigation team performed a rehearsal with the planet Jupiter on May 1, 2001. They chose Jupiter because it was large enough to be observed by the camera and bright enough for short exposure times. Additional rehearsals took place on May 8 and June 28 with mixed results that yielded a couple of lessons learned. Additionally, a campaign of ground observations took place, because outgassing and other forces acting on comets tended to hamper one's ability to predict their orbits for even short periods. Once Borrelly reappeared in the sky in May 2001, the Loomberah Observatory in Tamworth, Australia, the Naval Observatory in Flagstaff, Arizona, the Palomar Observatory, and JPL's Table Mountain Observatory made over 200 observations of the comet, which the ephemeris development group processed and delivered to the Deep Space 1 navigators.[25]

The approach navigation, Bhaskaran recalled, "was all done with ground-based navigation. We had standard ground-based optical navigation combined with radio navigation to get us to the right targeting spot for the flyby." The approach phase

---

[21] "Radiometric OD Borrelly," 2220; *DS-1 Nav: Extended Missions*, 1.
[22] *DS-1 Nav: Extended Missions*, 1-2; "Radiometric OD Borrelly," 6.
[23] Bhaskaran, interview, 32.
[24] *DS-1 Nav: Extended Missions*, 4.
[25] Bhaskaran, interview, 32; *DS-1 Nav: Extended Missions*, 27 & 28-30; "DS-1 Nav at Borrelly," 3.

started 40 days before encounter, as the spacecraft acquired its first navigation images. Synnott recalled that Borrelly appeared initially "like a bowling pin." "We didn't see it resolved until the last few hours." When the spacecraft was days or even hours away, "it [Borrelly] was still just a point."[26]

As Deep Space 1 approached Borrelly, the restored ΔDOR system played a key role. The DSN made two ΔDOR observations on September 14 and September 15, one week before encounter. The resulting orbit estimate was in close agreement (within 20-30 km) with the previous radiometric-based computation and aided in predicting the time and place of the encounter. There was, nonetheless, a discrepancy of 1,500 km between the Doppler-range results and the ephemeris prepared from telescopic observations. The validation of the radiometric orbit determinations with ΔDOR showed that the Doppler and range were not the sources of the error. It turned out that the ground observations used one technique for determining the center of brightness (the brightest pixel), while navigators used another (known as a standard Gaussian fit).[27]

Guiding the spacecraft to its target relied on an ephemeris based on data derived from the combination of Earth observatories, the DSN, and the AutoNav system. In effect, to skirt problems with the ground-based ephemeris, navigators tied the spacecraft to Borrelly's ephemeris rather than to a Sun-centered reference frame. This solution involved calculating the position and velocity of Deep Space 1 relative to Borrelly—the basic necessity of navigation—but not relative to the heliocentric reference frame, which was the source of a number of issues.[28]

Another navigation setback occurred for reasons unrelated to the project. A planned trajectory correction did not take place, but not because Deep Space 1 was on course. The cause was a JPL management decision that responded to a crisis taking place on the Eastern seaboard. They shut down the campus in reaction to the events of September 11, 2001, and made it impossible for the commands to be sent. As a result, the probe lost lock on its star and was unable to maintain its attitude.[29]

A maneuver executed around 18 hours prior to encounter on September 21, 2001, put Deep Space 1 roughly 2,200 km from the comet. Using pictures taken 11 hours before encounter, navigators tweaked their knowledge of the spacecraft's encounter time and place. This information became the basis for AutoNav to guide the spacecraft through the flyby. The software started working just thirty minutes before encounter with no help from ground navigators. Post-flight data analysis indicated that the system maintained a lock on the nucleus up to about 2 minutes prior to closest approach as planned. During those thirty minutes, AutoNav commanded fresh pictures

---

[26] Bhaskaran, interview, 32; "DS-1 Nav at Borrelly," 1-2; Synnott, interview, 22.
[27] *DS-1 Nav: Extended Missions*, 16 & 31-32; "Radiometric OD Borrelly," 2227; "DS-1 Nav at Borrelly," 1-2 & 7. For a discussion of just the radiometric navigation of Deep Space 1, see: Timothy P. McElrath, Dongsuk Han, and Mark S. Ryne, "Radio Navigation of Deep Space 1 During Ion Propulsion Usage," International Symposium on Spaceflight Dynamics, June 2000, Biarritz, France, http://trs-new.jpl.nasa.gov/dspace/bitstream/2014/15046/1/00-1052.pdf (accessed January 17, 2009).
[28] "DS-1 Nav at Borrelly," 2-3, 6, 7 & 9.
[29] "DS-1 Nav at Borrelly," 7.

of the comet, performed onboard navigation, and updated its own trajectory. Closest approach was at a distance of 3,514 km.[30]

"That actually worked flawlessly," Bhaskarn reflected. "It worked beautifully for the flyby. It worked better than the simulations at the time. It was our first encounter of a comet; the first AutoNav flyby of an object worked fine. And that was our first success." Synnott considered it to be "the first example I know about at least for small bodies in which the spacecraft took pictures and corrected the estimate of the trajectory and thereby kept track of where the object was going to be in the next picture and kept doing that all the way through the encounter."[31]

The flyby was an unqualified scientific success, too. The spacecraft returned images of the comet nucleus at a resolution of 46 meters per pixel, the highest-quality pictures of a comet to date. The only other close-up comet photographs were those taken by ESA's Giotto spacecraft as it flew past Halley in 1986. Giotto was the first spacecraft to image a small body. Although it had an autonomous onboard system to track Halley's nucleus, the camera failed, probably from impacts with cometary dust, 50 seconds before closest approach. Still, it acquired approach images that revealed the nucleus to resolutions on the order of several hundreds of meters.[32]

Following the Borrelly encounter, Deep Space 1 continued to live on with NASA's approval of the so-called Hyper-Extended Mission. The idea was to continue testing the ion propulsion and other experimental technologies until the craft's decommissioning on December 19, 2001.[33] Autonomous navigation was no longer a concept or a prototype. It had a track record. It was an operational system capable of not only improving navigational accuracy, but of taking control of spacecraft navigation to assure mission and scientific success.

We are stardust. Although Deep Space 1 had shown that AutoNav could function during an encounter (Borrelly), it had not demonstrated any satisfactory ability to navigate during an interplanetary cruise, Synnott's original goal for optical navigation. That was the accomplishment of Stardust, a Discovery class mission launched on February 7, 1999, to collect and bring back samples of the coma of the comet Wild-2, discovered in 1978. It was the first NASA mission dedicated to a comet.[34]

---

[30] "DS-1 Nav at Borrelly," 8-10; Bhaskaran, interview, 32; *DS-1 Nav: Extended Missions,* 30.

[31] Bhaskaran, interview, 32; Synnott, interview, 26.

[32] "DS-1 Nav at Borrelly," 10; Shyam Bhaskaran, Joseph E. Riedel, and Stephen P. Synnott, "Autonomous Target Tracking of Small Bodies During Flybys," 2080, in Shannon L. Coffey, Andre P. Mazzoleni, K. Kim Luu, Robert A. Glover, ed., *Spaceflight Mechanics 2004,* Pt. II (San Diego: Univelt, 2004); Nigel Calder, *Giotto to the Comets* (London: Presswork, 1992); Rüdeger Reinhard and Bruce Battrick, *The Giotto Mission: Its Scientific Investigations* (Noordwijk, NL: ESA, 1986).

[33] *DS-1 Nav: Extended Missions,* 39.

[34] "Nav for New Millennium," 303-320; Joseph E. Riedel, Shyam Bhaskaran, Shailen D. Desai, Dongsuk Han, Brian M. Kennedy, Timothy McElrath, George Null, Mark Ryne, Stephen Synnott, Tseng-Chan Wang, and Robert Werner, "Using Autonomous Navigation for Interplanetary Missions: The Validation of Deep Space 1 AutoNav," IAA L-0807, Fourth IAA International Conference on Low-Cost Planetary Missions, May 2000, Laurel, MD, http://trs-new.jpl.nasa.gov/dspace/bitstream/2014/14409/1/00-0844.pdf (accessed

Stardust navigators relied on a panoply of data types—X-band two-way Doppler and range, differenced Doppler, and camera images—and ran their software on an HP-C3700 workstation. The Stardust AutoNav was quite different from that tested on Deep Space 1. For one, it lacked the ability to do maneuvers. The encounter distance was large enough, and the accuracy requirements were loose enough, that navigators could compute the final approach maneuver on the ground. They used AutNav only to track the nucleus during the closest approach to the asteroid.[35]

Tracking the nucleus required developing a special algorithm to estimate the center of brightness. With a 10-second time lapse between pictures, the algorithm had to run quickly on the Stardust computer. The craft had the same kind of computer as that which flew on Deep Space 1, but it was somewhat slower. In simulations, though, the computer was able to run the program in 2.7 seconds.[36]

Navigators did a "dry run" of the optical navigation setup, as Stardust was flying past asteroid 5535 Annefrank on November 2, 2002, at a distance of about 3,000 km. The large flyby distance minimized any likelihood of impact with the asteroid, but it reduced the size of Annefrank in the images, even at closest approach, to less than 30 pixels. AutoNav started 20 minutes before encounter with an initial spacecraft ephemeris provided from just Doppler and range. The geometry of the encounter, coupled with the asteroid's small size (3 km), resulted in Annefrank being too dim to image. Navigators did some tweaking using the Galileo mosaic technique.[37]

On the way to comet Wild-2, problems with the onboard camera emerged. Spacecraft out-gassing in late July 2003 coated its lens with contaminants. This coating meant having to increase exposure times, and the longer exposures smeared the images. Turning on heaters and reorienting the craft removed the contamination which, all the same, eventually returned and required repeating the process. Among several other problems was the corruption of images by stray light.[38]

Optical navigation started as Stardust began to approach Wild-2. The onboard camera took the first pictures at a distance of 26 million km. The three pictures taken with three-second exposures showed no discernible sign of the comet. Again, though, the mosaic technique succeeded in bringing out an image. Navigators processed subsequent images in the same way to compute trajectory correcting maneuvers. At this point, the chief role of the optical data was to update the Wild-2 ephemeris, which was crucial for properly targeting the object at a distance of 120 to 150 km while traveling at a speed of a bit more than 6 kilometers per second. The roundtrip light-time of 40 minutes precluded image-processing on Earth. The navigators' strategy was to begin taking pictures about 20 minutes before closest approach and to have the autonomous

---

May 3, 2010). Kenneth E. Williams, "Earth Return Maneuver Strategies for Genesis and Stardust," 1925-1944 in Shannon L. Coffey, Andre P. Mazzoleni, K. Kim Luu, and Robert A. Glover, ed., *Spaceflight Mechanics 2004*, Pt. II (San Diego: Univelt, 2005), discusses navigation back to Earth.

[35] "AutoNav Tracking: Stardust," 452; "Stardust OD: Annefrank to Wild," 1864 & 1865; "Brief History of JPL OpNav," 342; Maize, interview, 21; Synnott, interview, 27.
[36] "AutoNav Tracking: Stardust," 455-461 & 462; Synnott, interview, 26; "OpNav for Stardust," 457-458.
[37] "Auto Tracking Small Bodies during Flybys," 2081, 2090 & 2092; "OpNav for Stardust," 455.
[38] "Stardust OD: Annefrank to Wild," 1863, 1864, 1869 & 1870; "OpNav for Stardust," 456 & 457.

software refine and update knowledge of the craft's position relative to the comet's nucleus throughout the rendezvous.[39]

Optical and autonomous navigation systems performed successfully all the way through the Wild-2 encounter at a distance of only 237 km. Stardust also captured the desired science images. The craft completed its mission on January 15, 2006, by delivering its samples to Earth. NASA reprogrammed the spacecraft for another pioneering undertaking: a follow-up rendezvous with Comet 9P/Tempel 1, recently explored by Deep Impact on February 14, 2011. It flew under the name Stardust-NExT, for Stardust New Exploration of Tempel 1 (NExT), until March 24, 2011, when it completed its last transmission to Earth.[40]

Deep Impact. The greatest and most dramatic success of autonomous navigation to date, not to mention the most demanding use of autonomous navigation, has been on Deep Impact, the eighth in the Discovery series. Deep Impact was a dual spacecraft mission that consisted of the so-called Flyby and Impactor probes. They launched conjoined as a single craft on January 12, 2005, on a trip to comet Tempel 1 and arrived on July 4, 2005.[41]

The dance to be performed at Tempel 1 was rather complex. The Impactor, after separating from the Flyby, targeted the comet's nucleus and autonomously guided itself to crash into the target while capturing high-resolution pictures of the nucleus' surface. Meanwhile, the Flyby observed the impact and the resulting plume, took infrared pictures of the ejecta and high-resolution images of the fully-developed crater using its two cameras. AutoNav ran on both the Impactor and Flyby and guided the Impactor on its collision course with the comet's nucleus, while the Flyby took pictures. The idea was to realize a first: to expose the interior composition of the nucleus to provide a unique insight into their substance and structure. In brief, as Earl Maize put it, Deep Impact was like "hitting a bullet with a bullet, and then photographing it at the same time."[42]

During the cruise and approach phases, navigators relied on X-band Doppler, range, and ΔDOR. AutoNav came into play as the craft started to approach the comet. The CCD camera was supposed to begin acquiring navigation pictures about 120 days before encounter (March 6, 2005), but delays postponed their acquisition until April 25, 2005. These initial pictures aided in estimating the craft's trajectory and in improving the Tempel 1 ephemeris. Nick Mastrodemos, Mike Wang, Brian Rush, and Bill Owen worked around the clock in 12-hour shifts during that last week, making hourly

---

[39] "OpNav for Stardust," 456 & 458-460; "AutoNav Tracking: Stardust," 451-453.
[40] "OpNav for Stardust," 455-460; Synnott, interview, 26; "Brief History of JPL OpNav," 342; JPL, "NASA's Venerable Comet Hunter Wraps Up Mission," March 24, 2011, http://www.jpl.nasa.gov/news/news.cfm?release=2011-095 (accessed March 25, 2011).
[41] "Brief History of JPL OpNav," 329. For mission history: Christopher T. Russell, ed., *Deep Impact Mission: Looking Beneath the Surface of a Cometary Nucleus* (Norwell, MA: Springer, 2005).
[42] Maize, interview, 20; "Brief History of JPL OpNav," 343; Bruce Battrick and Monica Talevi, ed., *Rosetta: Europe's Comet Chaser*, BR-215 (Noordwijk, NL: ESA, 2003).

deliveries to the navigation team that included Steve Synnott, George Null, Brian Kennedy, and Ed Riedel. About 10 days before encounter, the optical navigators began to distinguish the nucleus from the surrounding coma.[43]

Dan Kubitschek was the lead for the autonomous navigation team, which included Nick Mastrodemos and Bob Werner, both of whom had worked on Deep Space 1. The goal was the same as on previous missions: to keep the target body in the camera's field of view. The added challenges were to develop new algorithms to target a specific location on the comet's nucleus, to synchronize the timing of imaging sequences on both probes, to photograph the Impactor collision, to track the impact location, and to provide the timing needed to orient the Flyby for passage through the inner coma.[44]

The delicate *pas de deux* required a new campaign of ground observations with CCD-equipped telescopes to improve the Tempel 1 ephemeris. A few observations dated back to its discovery in 1867, but no continuous sightings occurred until a century later. Those made in 1967 and 1972 were the preliminary basis for the comet's ephemeris. A varied community of professional and citizen astronomers around the world acquired images of Tempel 1 using a wide range of imaging and data reduction techniques. For example, both the Keck Observatory in Hawaii and the Hubble Space Telescope made some of the observations. Steve Chesley and Don Yeomans found the data to be a bit iffy for several reasons including, but not limited to, the outgassing from the comet's surface that was difficult to model and even harder to predict. Earl Maize recalled: "a tremendous amount of very sophisticated ephemeris development going on as well." "I think that if you have to put a gold ribbon on something," he judged, "it's going to be the AutoNav, but the ephemeris development and the ground-based tracking were also significant efforts."[45]

The autonomous navigation system controlled the pointing of the Flyby camera and guided the Impactor to its target. It interacted with the attitude control software on both probes as well as with their cameras. Here was a small precursor of the future AutoGNC, optical navigation software capable of controlling a spacecraft's flight. The Deep Impact version of AutoNav had all three modules for processing images, estimating the spacecraft's orbit, and computing maneuvers. Rather than utilizing a starry background, it determined the probes' trajectories from the attitude control system and its star tracker. The spacecraft initiated the autonomous software two hours before impact. On the Impactor, it commanded photography, performed orbit determinations, and computed and executed three targeting maneuvers during the last 90 minutes.[46]

---

[43] Taylor and Hansen, 8, 9, 42, 43 & 62; "Deep Impact Nav Performance," 46, 47 & 55; "Ground-Based OD for Deep Impact," 1179-1202; Bhaskaran, interview, 38, 44 & 52; "Brief History of JPL OpNav," 343.
[44] Bhaskaran, interview, 44; Maize, interview, 20; "Deep Impact AutoNav," 382.
[45] Karen Jean Meech, Michael F. A'Hearn, Yan R. Fernandez, Carey M. Lisse, Hal A. Weaver, Nicolas Biver, and Laura M. Woodney, "The Deep Impact Earth-Based Campaign," *Space Science Reviews* 117 (March 2005): 297-334; "Deep Impact Nav Performance," 48 & 51; Stephen R. Chesley and Donald K. Yeomans, "Comet 9P/Tempel 1 Ephemeris Development for the Deep Impact Mission," 1271-1282 in Srinivas Rao Vadali, L. Alberto Cangahuala, Paul W. Schumacher, Jr., and Jose J. Guzman, ed., *Space Flight Mechanics 2006*, Pt. II (San Diego: Univelt, 2006); Maize, interview, 21.
[46] "Impactor Targeting," 1795; "Deep Impact AutoNav, 385; Nickolaos Mastrodemos, Daniel G. Kubitschek, Robert A. Werner, Brian M. Kennedy, Stephen P. Synnott, George W. Null, Joseph E. Riedel,

AuoNav again used the blobber as the center-finding program. The mission design required that the Impactor hit a lit portion of the nucleus, but its size and shape were not known very well. Despite its lack of sophistication, the blobber succeeded in finding a desirable spot on the surface.[47]

From the perspective of the AutoNav team, the most difficult feature of the Deep Impact mission was the lack of exact information about the size, mean radius, albedo, topography, and rotational period and orientation of the nucleus. Navigators developed the nucleus model for the mission in collaboration with Bob Gaskell of JPL. They started with the best images of comets available, namely those of Halley acquired in 1986 by the Giotto spacecraft. The nucleus in those images spanned only a few pixels, however. As the Borrelly and Wild-2 images became available from Deep Space 1 and Stardust, Bob Gaskell's team refined their model. Each comet's nucleus seemed to have a different overall shape. Eventually, the Deep Impact Science Team accepted his Borrelly nucleus model as the baseline model for Tempel 1 navigation analysis for both the Flyby and Impactor.[48]

The autonomous phase of the encounter started 120 minutes before impact. The Impactor autonomous software produced navigation images every 15 seconds and updated the orbit determination once every 60 seconds. The AutoNav system also computed four targeting maneuvers and commanded their execution. The Flyby autonomous software processed nucleus images every 15 seconds and updated its trajectory every 60 seconds in order to point its camera continuously at the nucleus. The release of the Impactor went very well. Seven minutes later, the Flyby S-band receiver locked onto the S-band signal from the Impactor. Telemetry began to flow. AutoNav began flying the two craft. Success![49]

The Deep Impact mission did not end with the Tempel 1 encounter. The surviving Flyby returned to Earth in late 2007 to become a new Discovery series mission dubbed EPOXI (Extrasolar Planet Observation and Deep Impact Xtended mission). The name combined two diverse science objectives with a single spacecraft name: Deep Impact Extended Investigation (DIXI) and Extrasolar Planet Observations and Characterization (EPOCh), a search for planets lying outside our solar system.

The mission design called for a decision by November 2007 on whether to target EPOXI to observe comet 85P/Boethin during a close flyby in December 2008 or comet 103P/Hartley 2 in December 2011. NASA decided to send it to Boethin and initiated a campaign of ground observations to reduce uncertainties in the comet's ephemeris. A maneuver brought the craft toward Earth on its way to Boethin, but the rendezvous failed. The comet was too small and faint for navigators to estimate its orbit accurately. Fresh trajectory alterations put the probe on course for three Earth flybys,

---

Shyam Bhaskaran, and Andrew T. Vaughan, "Autonomous Navigation for Deep Impact," 1251-1270 in Srinivas Rao Vadali, L. Alberto Cangahuala, Paul W. Schumacher, Jr., and Jose J. Guzman, ed., *Space Flight Mechanics 2006*, Pt. II (San Diego: Univelt, 2006); Maize, interview, 21.

[47] Maize, interview, 3 & 21; "Brief History of JPL OpNav," 343; "Impactor Targeting," 1796 & 1804.
[48] "Deep Impact AutoNav," 391, 392 & 405; "Impactor Targeting," 1802.
[49] "Impactor Targeting," 1792; Taylor and Hansen, 30-32; "Deep Impact AutoNav," 401 & 404; "Brief History of JPL OpNav," 343; Maize, interview, 3 & 20; "Deep Impact Nav Performance," 53-54.

the first of which occurred on November 1, 2007. The new trajectory brought the former Deep Impact spacecraft to Hartley 2, with closest approach occurring on November 4, 2010.[50] The exploration of asteroids and comets fit autonomous navigation to a tee. The upgraded optical navigation software maintained the link between optical navigators and ground-based astronomers and their CCD-equipped telescopes. The development of autonomous navigation also took the evolution of deep-space navigation one step further in a new direction with implications for what deep-space navigation meant.

---

[50] "Deep Impact Nav Performance," 39; JPL, "EPOXI Mission Overview" http://epoxi.umd.edu/1mission/index.shtml (accessed January 17, 2011); Tudor Vieru, "EPOXI Will Fly By Comet Harley [sic] 2 Tomorrow," November 3, 2010," http://news.softpedia.com/news/EPOXI-Will-Fly-By-Comet-Harley-2-Tomorrow-164425.shtml (accessed March 7, 2011).

Part Six

A
New
Normal

Chapter 23

Navigating in a New Era

In 2001, as a new decade began, NASA and JPL appeared to be on the road to recovery from the multiple spacecraft losses of 1999. In February 2001, NEAR landed on the asteroid Eros, and in September Deep Space 1 successfully encountered Borrelly. The Mars Global Surveyor was still returning observations of that planet in 2002, when Mars Odyssey, named for Arthur C. Clarke's novel, *2001: A Space Odyssey*, symbolized the rebound of the Mars effort and marked the resurgence of ΔDOR for navigation.

Just a few years later, looking back from the vantage of November 2004, JPL manager Mary Bothwell described "An Exceptional Year at the Jet Propulsion Laboratory."[1] In the past 12 months, NASA had launched the Spitzer Space Telescope (August 2003), Stardust had been to Comet Wild 2 (January 2004), the Spirit and Opportunity Mars Exploration Rovers had landed (January 2004), Cassini had started orbiting Saturn (July 2004), Aura was studying Earth's atmosphere, and Genesis had returned samples of the solar wind (September 2004). The next year Deep Impact encountered comet Tempel 1. More missions followed. The Mars Reconnaissance Orbiter, Phoenix, the Mars Science Lander, and MAVEN all were scheduled to carry on NASA's exploration of the fourth planet.

These triumphs of exploration found NASA and JPL neither halcyon nor prosperous in this post-1999 world. The widely held view was that the NASA panels convened to study the Climate Orbiter and Polar Lander mishaps had put the final stake in the heart of Goldin's Faster, Better, Cheaper management philosophy, despite what those reports actually stated. Indeed, the application of SDI management approaches to NASA—the basis of Faster, Better, Cheaper—lingered, as did the budget reductions of the previous decade. The Global War on Terrorism assured the continuation of the national security crisis and the role of the national security state.

In contrast to what appeared to be old wine in a new bottle, navigation was on the verge of what seemed to be a potentially new era. The DSN was upgrading to an even higher frequency range, so that tracking data and the orbit determinations made from it would be inherently more accurate. But, taking full advantage of this boon required more accurate measures of time, media effects, and Earth motions as well as a better quasar catalog. New software became available to tackle many of the problems of landing on planets or even asteroids. Optical navigation was an imperative for characterizing the shape, orientation, and spin rate of asteroids and comets. Increasingly more navigation was taking place aboard spacecraft, as autonomous navigation merged with guidance and control software. The new capability could aid in navigating close to small bodies as well as landing on them.[2]

---

[1] Mary Bothwell, "An Exceptional Year at the Jet Propulsion Laboratory," November 11, 2004, http://trs-new.jpl.nasa.gov/dspace/bitstream/2014/38810/1/04-3643.pdf (accessed April 23, 2011).
[2] "Next 25 Years," 398-400; "JPL Roadmap," 1927-1929.

A new data type—derived from radio astronomy—promised to raise navigational precision, but it also appeared to threaten ΔDOR, while at the same time the European Space Agency was embracing ΔDOR. Navigational science also moved forward among the outer planets with the Cassini mission, as navigators, optical astronomers, radio astronomers, and science were interwoven in the daunting task of drawing up satellite ephemerides. Science and navigation worked side by side, as if astronomers and the ancient mariners using their ephemerides worked alongside each other at the Royal Observatory in Greenwich.

More cuts, more SDI. Budget and personnel reductions persisted after Sean O'Keefe replaced Goldin in December 2001. O'Keefe already had a reputation for cutting NASA's budget that he had earned while serving as a staff member on and later staff director of the Senate Committee on Appropriations. He also brought along military credentials, having been Comptroller and Chief Financial Officer of the Department of Defense.[3]

In February 2003, NASA lost another Shuttle and its crew. The investigative board concluded that: "the NASA organizational culture had as much to do with this accident as the foam." A GAO report published only a month before the accident had reached the same conclusion. It found the agency's problems to be rooted in its culture and long-standing ways of doing business and that NASA needed to undertake a "major transformation."[4] This language echoed that of Dan Quayle and the National Space Council a decade earlier, but this time the rhetoric led to far milder changes. O'Keefe launched One NASA, an attempt to change the agency's culture by focusing employees on the totality or oneness of the agency ("the NASA family"), rather than on their individual centers, but did not address the agency's fundamentally feudal structure. Other efforts to achieve cultural change ensued,[5] but had NASA really changed?

President George W. Bush, not to be outdone by his father, continued the tradition of announcing mammoth space ventures. In January 2004, he announced his Vision for Space Exploration. It would guarantee a substantial portion of the space agency's budget for human spaceflight. The renewed emphasis on astronauts did not bode well for the budget for interplanetary exploration.

SDI was still alive, but in the guise of the Missile Defense Agency. SDI's influence at NASA seemed to surge with Michael Griffin's appointment to run the agency in April 2005. After two years in the JPL Guidance and Control unit, Griffin spent the next seven years at APL working on a number of Strategic Defense Initiative Organization (SDIO)

---

[3] NASA, Agency News, "Administrator O'Keefe Resigns," http://www.nasa.gov/about/highlights/aok_resigns.html (accessed March 31, 2011).
[4] Columbia Accident Investigation Board, *Report*, Vol. I (Washington: NASA, August 2003), 9 & 97; Presidential Commission on the Space Shuttle Challenger Accident, *Report to the President*, Vol. I (Washington: The Commission, 1986), Chapter VI, "An Accident Rooted in History," 120-151; GAO, *Major Management Challenges and Program Risks: National Aeronautics and Space Administration*, GAO-03-114 (Washington: GPO, January 1, 2003), 5 & 20-21.
[5] "Overview of the One NASA Concept," http://www.onenasa.nasa.gov/About_OneNASA/OVERVIEW.htm (accessed November 10, 2006); Behavioral Science Technology, *Assessment Plan for Organizational Cultural Change at NASA* (Ojai, CA: BST, March 15, 2004).

projects, such as the POLAR BEAR spacecraft. He became SDIO's deputy for technology in 1989 and interacted in various ways with Quayle's National Space Council, until he transferred to NASA in 1991. There he worked on the space station and Bush's Space Exploration Initiative. Critics saw him as just one of the "star warriors" who were joining the agency.[6] Griffin also had CIA links. Between SDIO and NASA, he was president and chief operating officer of In-Q-Tel, a private company funded by the CIA, and he was head of the APL Space Department,[7] which won a number of noteworthy NASA contracts in competition with JPL.

Meanwhile at JPL, a seismic shift occurred in May 2001, as Charles Elachi replaced Ed Stone as JPL Director. Elachi established a new position, JPL Chief Technologist, and renamed the DSN (at least as far as JPL was concerned). Henceforth, it sat in the InterPlanetary Network (IPN) and Information Systems Directorate led by Gael F. Squibb. The dramatic shedding of civil service and contractor jobs that had marked the 1990s and "reinventing government" returned to JPL. On September 7, 2005, Elachi informed employees that NASA required JPL to decrease its budget between 5 and 8 percent for fiscal 2006. In turn, John Beckman, the head of the Engineering and Science Directorate, navigation's institutional home, instructed his Division Managers to cut their workforce by about 5 percent, which meant, as a rule, snipping 5 percent from each Section, by October 17, 2005.[8]

Navigation meets guidance. Navigation itself already had undergone yet another seismic tremor in 2003 that left the section bigger and more heterogeneous than ever before. Within the Engineering and Science Directorate, navigation was in the Autonomous Systems Division managed by Roger Gibbs and his Deputy, Sam Thurman.[9] He was the same Sam Thurman who had pushed for alternatives to ΔDOR.

The new structure combined navigation with guidance and control for the first time. There was some irony in the merger. In the past, the two areas had waged a turf battle over optical navigation, but now they were united under the banner of "autonomous systems," the newest advance in optical navigation. This new Guidance, Navigation, and Control Section (Section 343) was huge, comprising 299 JPL and contractor employees, although the number of groups had increased only from 13 to 15. From January 2005, the Section Manager was Earl Maize, who had been serving as

---

[6] Griffin, interview, 4-7, 9-14, 18-19 & 36-38; McCurdy, *Faster, Better, Cheaper*, 44, 46 & 47.
[7] In-Q-Tel, "In-Q-Tel Names Dr. Michael D. Griffin as President and Chief Operating Officer," August 5, 2002, http://www.iqt.org/news-and-press/press-releases/2002/IQT_08-05-02.html (accessed March 17, 2008).
[8] SpaceRef, "New JPL Director Announces Lab Reorganization," May 2, 2001, http://www.spaceref.com/news/viewpr.html?pid=4715 (accessed February 22, 2011); SpaceRef, "NASA JPL Internal Memo: Elachi to Discuss Lab Outlook at All-Hands Meeting," October 6, 2005, http://www.spaceref.com/news/viewsr.html?pid=18289 (accessed February 22, 2011); SpaceRef, "NASA JPL Internal Email: 3X Workforce Reduction," October 5, 2005, http://www.spaceref.com/news/viewsr.html?pid=18283 (accessed February 22, 2011). The latter is a copy of the e-mail that Beckman sent to all employees in his Directorate.
[9] [JPL] "Laboratory Organization Charts [2006]," http://hspd12jpl.org/files/LabOrgCharts.pdf (accessed February 22, 2011).

Deputy Program Manager on Cassini since September 1993. Al Cangahuala became his Deputy.[10]

The merger with guidance and control acknowledged the evolution of deep-space navigation. Navigation started as a strictly ground-based endeavor, while guidance and control took place onboard spacecraft. This distinction caused the two to develop in different technical divisions of JPL. "Even though they have a lot of common mathematics and that sort of thing," Wood explained, "they were never really organizationally close together until quite recently."[11] Optical, then autonomous, navigation gradually were moving certain navigation functions from Earth to spacecraft. The next evolutionary stage was to merge guidance and control capabilities with autonomous navigation.

Managing this greatly expanded navigation area had its challenges. One needs to keep the size of a group, the smallest management unit, at a reasonable level yet functionally consistent. "You want to keep the disciplines more or less cohesive within a group, so that the group supervisor can manage people both technically and administratively," Maize explained. "So, about twenty people is about right, although we have groups as large as forty."[12]

In 1998, to handle the enormous size of the navigation area even then, the work had been divided between "Outer Planet" and "Inner Planet" (Inner Planet Navigation and Gravity until 2003) groups. Mission designers already had made this partition back in the 1970s. The demarcation, Maize explained, "is based on just sizing. One large navigation group is just too much to handle." "Typically," he added, "the outer-planet navigation tends to deal with planets like Jupiter and Saturn, where you're dealing with orbiting missions, such as Galileo and Cassini, for example, and the ephemerides of small bodies orbiting around them, and things of that sort. Whereas the inner planets more recently have been more concentrating on orbiters, Mars in particular, with landing and issues of those sorts. So, there are some particular branches of navigation that are applicable to one and not to the other that also make the delineation somewhat less than artificial."[13]

A KinetX response. A major perturbation in the arc of JPL navigation had JPL navigators competing against other JPL navigators who were now in private practice. A not insignificant number of experienced first-rate navigators steered a course away from NASA to employment with KinetX founded in 1992. The company's founders were former Lockheed engineers with substantial satellite and SDI experience. They included Christopher Bryan, Rick Sarmento, and Kjell Stakkestad, who had worked on SDI projects. Their first contract was with Motorola to help on the Iridium satellite

---

[10] OrgChart, Navigation and Mission Design Section 312, April 8, 2002, JPL556; OrgChart, Section 343, February 2, 2005, JPL556; Wood, interview, 9; Maize vita, nd, "Maize," DSN Files; Cangahuala vita, nd, "Cangahuala," DSN Files.
[11] Wood, interview, 10.
[12] Maize, interview, 1.
[13] Orgchart, Navigation Section, July 1, 1998, JPL556; Personal communication, Lincoln Wood, May 30, 2013; Maize, interview, 1.

telephone project. Over the next few years, KinetX expanded its involvement in Iridium to include various software services, hardware development, and operation of the satellite constellation. KinetX soon supplied these and other services to such corporations as Lockheed-Martin, Boeing, General Dynamics, Aerojet, and TRW. The greatest portion of the company's total experience was on military programs, until KinetX branched out into deep-space navigation.[14]

The first KinetX hire was Bobby Williams, a 24-year veteran of JPL navigation who had participated in or had led Viking, Pioneer Venus Orbiter, and TOPEX/Poseidon navigation teams. He also had been involved in gravity field analysis for Mars and Venus probes, including as a science co-investigator. More recently, he was navigation manager for the Deep Space 1 and NEAR missions. This last achievement earned him numerous accolades, including a NASA Outstanding Leadership Medal and the honor of having an asteroid named after him.[15]

As Director of Space Navigation and Flight Dynamics (SNAFD) for KinetX, Williams quickly began recruiting from among JPL navigators and mission designers. Accordingly, as the company's website boasted, the SNAFD team "has over 130 accumulative years of mission design and execution." KinetX also calls on a Senior Advisory Council with "over 420 additional years of experience." The Council's task "is to independently review and assess the quality and reliability of SNAFD solutions." Its 13 members are some of the best and brightest of JPL navigators: Pete Breckheimer, Joseph Brenkle, Dave Curkendall, Sam Dallas, Len Efron, Jordan Ellis, Hal Gordon, Don Gray, Bill Kirhofer, Jim Miller, Paul Penzo, and Bill Sjogren. Their head, Jim McDanell, was once the manager of the JPL navigation section. The Council convenes weekly for breakfast in La Cañada-Flintridge within eyeshot of the old Crest Building where navigators worked decades earlier.[16]

The exodus of navigators was a boon to NASA, because it meant desirable personnel cuts, but it was a sore point for JPL navigation managers. These were the wrong people for NASA to be losing, and sadly they were only part of the forced migration of talent and expertise from the space agency and, more generally, the federal government. The loss of talent, expertise, and experience continues to be felt across government. Sjogren described his own uneasy situation: "I'm an outcast. I was warned by the section manager." He said, "You know you're still associated here at JPL. Yet you're over here working with Bobby Williams on this KinetX thing. And those guys were trying to get projects. In fact, they stole the Pluto mission from JPL."[17]

The selection of KinetX for the New Horizons flight to Pluto and the Kuiper Belt plus the MESSENGER mission to Mercury led to some jealousy within the space contracting industry. Williams reported that KinetX had been "getting some heat" for beating out JPL for the contracts. "A lot of people don't like what we are doing," he

---

[14] KinetX, "Management Team"; KinetX, "Company History," http://www.kinetx.com/company.aspx?p=history (accessed April 4, 2011); KinetX, "Services," http://www.kinetx.com/services.aspx (accessed April 4, 2011).
[15] KinetX, "Management Team"; Yeomans, interview, 20.
[16] KinetX, "Management Team"; KinetX, Space Navigation and Flight Dynamics, "Mission Design," http://www.kinetx.com/services.aspx?p=nav (accessed April 4, 2011); Sjogren, interview, 2008, 38-39.
[17] Sjogren, interview, 2008, 37-38.

said. "It changes the status quo." The engagement of a private company, KinetX, for navigation services on the MESSENGER and New Horizons probes was perhaps the most dramatic of the changes that affected navigation during this decade. JPL's hegemony over NASA deep-space navigation always faced competition from corporations, but now the competition came from former colleagues.[18]

KinetX won the MESSENGER and New Horizons contracts because they had Bobby Williams. In both cases, KinetX provided navigational services to APL, which had beat JPL for the contracts. APL chose KinetX for New Horizons mainly because of their past experience with Bobby Williams and other KinetX navigators. The selection of KinetX for MESSENGER was less direct. NASA Headquarters had selected Sean Solomon, Director of the Department of Terrestrial Magnetism at the Carnegie Institution in Washington, DC, to lead MESSENGER as its Principal Investigator. Solomon previously had worked with Williams over several years when he was at JPL, for example on the Magellan Project Science Group and the Radar Investigation Group (1982-1994) and the Mars Observer/Mars Global Surveyor Laser Altimeter Team (1986-2005). "We had worked closely with Bobby for several years," Solomon told reporters. "He changed to KinetX, and we stayed with him. We made our decision based on our belief that KinetX had the right staff and the right software to do the job."[19]

MESSENGER. Launched on August 3, 2004, the MESSENGER probe, the seventh in the Discovery series, set out to study the planet closest to the Sun. Its flight path included an Earth gravity-assist followed by two flybys of Venus and three of Mercury. In March 2011, MESSENGER arrived at Mercury and entered orbit to undertake a year-long detailed scientific study of the planet. Navigation relied on the gamut of data types: Doppler, range, ΔDOR, and optical images. JPL's Radio Metric Data Conditioning group delivered Doppler and range to the KinetX navigators every thirty minutes during the first hours of the MESSENGER voyage.[20]

The navigation team, under the leadership of former JPL navigator Tony Taylor, was part of a multi-mission navigation group that allowed the project to adjust its size as events dictated. Mars Program navigation had operated along the same lines, but with far fewer people. During launch, the MESSENGER team consisted of eight navigators working in shifts to provide continuous support from twelve hours before to about fifteen hours after launch. A few days later, the number of navigators decreased to five and, during most of the cruise phase, even fewer. Exceptions were planetary flybys (for

---

[18] Hal Mattern, "Whether Guiding a Satellite or Navigating a Probe to Pluto, KinetX's Success Has Been Largely Defined by…Getting There," May 26, 2003, http://www.kinetx.com/news.aspx?a=20030523 (accessed March 31, 2011); Andrew J. Butrica, "Deep Space Navigation, Planetary Science, and Astronomy: A Synergetic Relationship," 496 in Steven J. Dick, ed., *NASA's First 50 Years: Historical Perspectives*, SP-4704 (Washington: NASA, 2010).
[19] "Biographical Information: Sean Carl Solomon," nd, http://www.dtm.ciw.edu/users/scs/Solomon_CV_Aug2010.pdf (accessed April 4, 2011); "Sean Solomon Biography," October 27, 2004, http://www.dtm.ciw.edu/component/content/131?task=view (accessed April 4, 2011).
[20] "Early Nav Results," 1235-1236 & 1243-1244.

gravity assists) and trajectory correcting maneuvers. On approach to Mercury, the team size grew to five for orbit insertion then settled at about four for the orbital phase.[21]

Optical navigation as usual employed the onboard science camera, in this case the so-called Mercury Dual Imaging System (MDIS). The latter consisted of wide-angle and narrow-angle CCD cameras, both of which pointed to the same place. The system underwent testing during Earth and Venus flybys, when range and Doppler sufficed. Optical navigation at Venus was not a necessity, especially as the opacity of its atmosphere limited the precision of the center-finding software, but the acquisition and processing of Venus navigational images served as an excellent checkout of the system.[22]

Optical navigation helped to improve navigational accuracy on approach to the three Mercury flybys and during orbit insertion. Approaches varied from one mission segment to another. During the interplanetary cruise, for example, navigators relied on images of planets against a background of stars. As the probe entered orbit around Mercury, navigation depended on pictures of landmarks, namely craters on the planet's surface. Also, while MESSENGER was in orbit, navigators refined their model of the planet's gravitational field using Doppler. The model improvements enhanced the accuracy of orbit determinations and contributed to the project's scientific goals, which included modeling the planet's gravitational field.[23]

An important part of cruise navigation was the use of ΔDOR to ensure the accuracy of flight path estimates. Tony Taylor, Bobby Williams, and other KinetX navigators tested the ΔDOR during the Earth flyby (August 2005). They used standard Doppler and range for the first Venus flyby, allowing them to validate the ΔDOR under non-critical circumstances. Subsequently, the ΔDOR enhanced navigational accuracy for the close Venus and Mercury flybys, the lowest of which was at only 200 km above the planet's surface. The navigation team additionally called on ΔDOR for the Mercury flybys in particular as an independent check on the orbit solutions derived from radiometric and optical data.[24]

New frontiers. KinetX also supplied navigation services to the New Horizons flight to Pluto. The flight was part of the New Frontiers program that included the Juno probe launched to Jupiter on August 5, 2011. Juno will orbit the gas giant in an innovative, highly elliptical 11-day polar orbit to avoid the planet's highest radiation regions. The solar-powered craft will circle Jupiter for about one year (32 orbits). Mission designers timed the one-year mission to occur between solar conjunctions to simplify operations and to avoid Doppler degradation.[25]

---

[21] "Early Nav Results," 1235-1236 & 1243-1244.
[22] Sjogren, interview, 2008, 39; James V. McAdams, David W. Dunham, and Robert W. Farquhar, "Trajectory Design and Maneuver Strategy for the MESSENGER Mission to Mercury," *Journal of Spacecraft and Rockets* 43 (2006): 1054-1064; "Early Nav Results," 1246; "First OpNav Images."
[23] "Early Nav Results," 1245; "First OpNav Images"; Deborah Domingue and Christopher T. Russell, *The MESSENGER Mission to Mercury* (New York: Springer, 2007), 22, 28, 228, 237, 251, 261-262, 268, 271, 275, 335, 583, 587, 590.
[24] "Early Nav Results," 1234, 1245 & 1246.
[25] "Juno New Frontiers," 3-4.

The Discovery Working Group, led by APL's Bob Farquhar, conducted the first study of the New Frontiers Pluto mission in 1990. Farquhar would become the New Horizons Mission Director from 2002 to 2006, and from 1999 to 2006 he was also MESSENGER Mission Manager. Then, in 2008, he joined KinetX, after being honored by the National Air and Space Museum with the Charles A. Lindbergh Chair for Aerospace History. Farquhar concerned himself with mission design (along with Yanping Guo at APL) and management. The New Horizons navigators included Jim Miller, Dale Stanbridge, and Bobby Williams. At JPL, Miller had worked with Ken Rourke on developing VLBI techniques for navigation. The former Viking navigation team leader also had planned navigation for Galileo's approach to Jupiter and had conducted navigation studies for possible future missions.[26]

New Horizons launched on January 19, 2006, on a long journey to Pluto, where it was scheduled to make its closest approach on July 14, 2015, at a distance of 12,500 km (7,767 miles). Its first planetary flyby of Jupiter took place on February 28, 2007. Subsequently, the specter of unmodeled small forces soon haunted the probe, and their correction required executing a trajectory maneuver on June 30, 2010. KinetX navigators determined that a small amount of force was being created by thermal photons emitted from the craft's RTG power source, as they reflected off the backside of its high-gain antenna. At the time of the maneuver, New Horizons was more than 2.4 billion km (1.49 billion miles) from Earth, nearly the distance of Uranus's orbit. This was the fourth such course adjusting maneuver of its flight.[27]

While New Horizons made its way to Pluto, its mission grew more complicated, as scientists learned more about that dwarf planet. In 2005, shortly before New Horizons launched, the Hubble Space Telescope Pluto Companion Search Team imaged two new moons, named Nix and Hydra. Then, in 2011 and 2012, two more small moons, Kerberos and Styx, respectively, came to light. Scientists now speculate that Pluto may have as many as 10 satellites and one or more ring systems.[28] Even before the discovery of these moons, though, navigating New Horizons during its approaches to Pluto and Charon was going to be daunting. Some difficulties necessarily will arise from the vast distance from Earth, about 30 astronomical units, and the roundtrip light time will exceed eight hours, making two-way Doppler difficult to acquire.

Because of that vast distance, the probe will be in two-way lock—a requisite for two-way Doppler—for only a few hours during each station pass. The mission design therefore called for collecting three-way Doppler, although subject to degradation from timing differences between stations and other factors. Range will provide a direct measure of the line-of-sight distance from Earth to the probe. Both range and Doppler

---

[26] Farquhar, 426; Solar System Exploration Survey, *New Frontiers in the Solar System: An Integrated Exploration Strategy* (Washington: The National Academies Press, 2003); Barrie W. Jones, *Pluto: Sentinel of the Outer Solar System* (New York: Cambridge University Press, 2010), 186; Yanping Guo and Farquhar, "New Horizons Mission Design," *Space Science Reviews* 140 (2008): 49-74; Yanping Guo and Farquhar, "New Horizons Pluto-Kuiper Belt Mission: Design and Simulation of the Pluto-Charon Encounter," *Acta Astronautica* 56 (2005): 421–429; "New Horizons Pluto Nav," 529-540.

[27] Michael Buckley, "Course Correction Keeps New Horizons on Path to Pluto," July 1, 2010, http://www.kinetx.com/news.aspx?a=20100701 (accessed February 10, 2011).

[28] Ray Villard, "Pluto Could Have Ten Moons," March 18, 2013, http://news.discovery.com/space/astronomy/pluto-could-have-ten-moons.htm (accessed April 4, 2013).

will contribute to estimating the craft's course and refining the ephemerides for Pluto and Charon. But, because range is just marginally useful for determining the spacecraft's trajectory relative to Pluto during approach, the navigation team will use ΔDOR to provide precise position information relative to the line of sight.[29]

Perhaps the stickiest wicket will be the peculiar motions of Pluto and Charon, which move in elliptic orbits about each other. Navigators will need to determine precisely Charon's orbit around Pluto, as well as the flight path of the spacecraft relative to both bodies (and the effects of the other bodies circling Pluto), in order to position the probe for making scientific observations. Knowing the exact location of the center of mass of the two bodies—their barycenter—will be critical to the project's scientific success. In order to make these precise, long-distance measurements, the mission design called on optical navigation using images taken by the two onboard cameras, the Long-Range Reconnaissance Imager (LORRI) and the Multispectral Visible Imaging Camera (MVIC). JPL, with Bill Owen in the lead, will be undertaking the optical navigation and reporting to KinetX.[30] When Pluto first becomes visible in their images, navigators will separate the orbit of Charon from that of Pluto about their common barycenter. The optical navigation method chosen was the acquisition of Pluto and Charon against starry backgrounds. For these pictures to be useful, Charon will have to be more than 100 pixels from Pluto in the images.[31] Navigating the exploration of the Plutonian system is tailor-made for optical navigation because of the sheer astronomical distance between Earth navigators and the spacecraft and the complex gravitational interplay among the bodies circling Pluto.

---

[29] "New Horizons Pluto Nav," 530-531.
[30] William M. Owen, Jr., Philip J. Dumont, and Coralie D. Jackman, "Optical Navigation Preparations for New Horizons Pluto Flyby," 23rd International Symposium on Space Flight Dynamics, October 29-November 2, 2012, Pasadena, CA, http://issfd.org/ISSFD_2012/ISSFD23_IN3_2.pdf (accessed December 10, 2013).
[31] "New Horizons Pluto Nav," 529.

Home of the Mars Science Laboratory project on the JPL campus (Author)

MSL, launched in 2011, landed the Curiosity rover in Gale Crater on August 6, 2012. NASA invited the public on a website to rank nine rover finalist names submitted in an essay contest.

Chapter 24

Cassini, Navigation, and Science

Navigation and science went hand in hand, especially regarding the preparation of ephemerides. Navigators analyzed tracking data to compute orbital solutions for spacecraft as well as to improve the flight's specialized ephemeris. The improved ephemeris in turn increased the accuracy of the navigators' estimates of the craft's trajectory. The separation between the astronomers and the ancient mariners who used their ephemerides was long gone. The national security crisis had brought about the convergence of science and navigation. The Cassini mission illustrated the ways in which navigators worked in tandem with radio and optical astronomers to guide the craft among Saturn's many moons in order for project scientists to make their observations and to avoid collision with those same bodies.

Cassini was the flagship of large-scale planetary voyages of discovery in an era of smaller, less expensive projects with more limited scientific objectives. A joint venture of ESA and NASA, Cassini was the first craft to orbit Saturn.[1] ESA's Huygens probe scored a first when it landed on Titan in January 2005. Navigating the spacecraft was problematical if for no other reason than the inordinate number of gravity assists and close flybys of satellites whose precise positions and motions were known only approximately from observations by ground telescopes. The mission's scientific success hinged on the accuracy of the satellite ephemerides.

Throughout Cassini's flight among Saturn's many small moons, navigators continually updated their special satellite ephemerides. Optical navigators played a crucial role. They called on ground-based optical and radio astronomers for observations as well as the new Hipparcos and Tycho star catalogs and the International Celestial Reference Frame, the latest standard celestial reference frame adopted by the IAU in 1998, to hone their ephemerides.

Cassini navigators eventually benefited from an experiment to collect measurements made by the Very Long Baseline Array (VLBA), a scientific research facility managed by the NRAO. The proposal to use VLBA data came from the new DSN overseer at NASA Headquarters, a trained radio astronomer. The inclusion of VLBA into navigators' quiver of data types was not unproblematic, and to some navigators it seemed to pose a threat to the newly revived ΔDOR system. Time and budget cuts proved otherwise.

One of the technical factors shaping the future of VLBA and ΔDOR as well as other data types was the DSN upgrade in progress. Over the decades, the DSN advanced from one frequency band to the next, realizing greater performance along the way. The migration from the L band to the S band in the 1960s and then to the X band along the way yielded a nearly 20 dB signal improvement. Moving into the Ka band

---

[1] For a project history: David M. Harland, *Mission to Saturn: Cassini and the Huygens Probe* (New York: Springer, 2002).

would add another net 5 to 6 dB. The World Administrative Radio Conference in 1979 reached a rare broad-ranging agreement on frequency usage that in part allocated certain Ka-band frequencies for deep-space communication. Its adoption by the 150 nation members of the International Telecommunication Union in Geneva, Switzerland, provided the legal framework for the DSN's Ka-band upgrade.[2]

Ka-band navigation, communication, and telemetry became the wave of the future. The DSN already has started tracking Juno's flight to Jupiter in this higher frequency range. Navigators will assess the Jovian gravity field at both X-band and Ka-band frequencies—dual-frequency calibration—to enhance their measurements' precision. Other projects that opted for Ka-band equipment were Cassini, Deep Space 1, and the (canceled) Pluto Fast Flyby. In addition, every NASA probe that launches in 2016 and later years will have Ka-band capability.[3] Was a new normal emerging?

A pilot project. In 2001, Barry Geldzahler became the new Headquarters manager of the DSN. He was, unlike his predecessors, a radio astronomer. When Geldzahler was an astrophysics graduate student at the University of Pennsylvania, he acquired some observational experience at the NRAO in Green Bank, West Virginia. There he became a mentee of the NRAO radio astronomer Kenneth Kellermann and met "80 percent of the world's radio astronomers." Geldzahler accompanied Kellermann when he became director of the Max Planck Institute for Radio Astronomy in Bonn, Germany, and there met the other 20 percent.[4]

Subsequently, Geldzahler worked on the SDI-related project known as the LACE (Low-power Atmospheric Compensation Experiment) satellite as well as on Clementine. The latter started in 1990 when NASA Administrator Truly sought to collaborate with the Star Wars agency on a joint endeavor. Geldzahler, while employed by APL, worked in the NEAR initial operations team. When he came to NASA on April 23, 2001, his responsibilities included the DSN, the Planetary Data System, and the Galileo, NEAR, and Stardust missions.[5]

---

[2] Norman F. deGroot, "Ka-Band (32 GHz) Allocations for Deep Space," 104-109 in TDAPR 42-88.
[3] "Juno New Frontiers," 3, 4 & 6; Richard S. Grammier, "The Juno Mission to Jupiter," ISTS 2006-o-06V, 25th International Symposium on Space Technology & Science, June 7, 2006, Kanazawa, Japan, http://trs-new.jpl.nasa.gov/dspace/bitstream/2014/40780/1/06-2171_A.pdf (accessed April 17, 2009), 4; Philip D. Potter, "Use of Ka-Band for Radio Metric Determinations," 59-66 in TDAPR 42-58; Hansen and Kliore, 110; David Morabito, Stanley Butman, and S. Shambayati, "The Mars Global Surveyor Ka-Band Link Experiment (MGS/KaBLE-II)," in TDAPR 42-137; William A. Imbriale, *Large Antennas of the Deep Space Network* (Pasadena: JPL, February 2002), 225-226 & 239; Leif J. Harcke, P. F. Yi, Miles K. Sue, and Harry H. Tan, "Recent Ka-Band Weather Statistics for Goldstone and Madrid," 1-12 in TDAPR 42-125; Geldzahler, interview, 27.
[4] Geldzahler, interview, 1-3 & 6; Personal communication, Geldzahler, May 24, 2012.
[5] Geldzahler, interview, 11-12, 13-14, 17 & 21; NRL, *A Clementine Collection: Moon Glow* (Washington: NRL, June 1994), esp. 2-3. For the Ultraviolet Plume Instrument and LACE, see, among others, Herbert W. Smathers, Donald M. Horan, J. G. Cardon, Erick R. Malaret, John E. Brandenburg, Richard E. Campion, and Robert R. Strunce, Jr., *UVPI Imaging from the LACE Satellite: The Nihka Rocket Plume* (Washington: NRL, July 12, 1993); Donald M. Horan, R. E. Perram, and Robert E. Palma, *The NRL LACE Program Final Report* (Washington: NRL, April 27, 1993).

As a veteran of SDI and APL projects, Geldzahler's career reflected NASA's continuing reliance on expertise developed by the Strategic Defense Initiative Organization. More importantly for navigation, his radio astronomy background aided him in instigating a new engagement between navigation and science. Geldzahler boldly proposed a new type of "direct" measurement, one acquired from the NRAO's VLBA.

The VLBA consists of ten identical, 25-meter dishes on baselines that span the continent—a distance of 8,000 km—from Mauna Kea, Hawaii, in the west to St. Croix, Virgin Islands, in the east. The Science Operations Center in Socorro, New Mexico, controls these antennas remotely and correlates the data received from the ten antennas, so that they function in effect as a single antenna. VLBA controllers can increase the array's sensitivity by adding the Green Bank Telescope and the VLA. The array's scientific focus is precision astrometry, which it can achieve with accuracies of around 10 millionths of an arcsecond: the state of the art.[6]

Navigator Tomás Martin-Mur recalled Geldzahler's proposal to use the VLBA to improve navigation. "He's very supportive on deep-space navigation," Martin-Mur explained. The proposal, however, raised a red flag among navigators: was this *another* attempt to replace ΔDOR? NASA's decommissioning of ΔDOR had not happened too many years earlier. There were pros and cons for using both VLBA and ΔDOR. VLBA did not need spacecraft DOR tones, nor did it disrupt telemetry like ΔDOR, and VLBA could use weaker and therefore a larger number of quasars than ΔDOR. VLBA also would relieve the DSN of some of the navigation burden and allow trackers to collect more data. Geldzahler also argued for reducing mission risk with the VLBA. "There's always a possibility that if I turn on these DOR tones, what if something happens and they stay on? When those DOR tones are on, I can't transmit data. I'm done." With the VLBA, Geldzahler continued, "I don't have to use them. I have an alternate asset. Now, I'm going to tell you it's controversial, and I don't know how this is going to come out." "Now we're not going to get rid of Delta DOR," Geldzahler affirmed, "because if the VLBA goes away or something craps out, ops [operations] guys always have a Plan B."[7] For the moment, ΔDOR seemed to be safe.

In 2003, in order to investigate the feasibility of using the VLBA, Geldzahler organized the VLBA Spacecraft Navigation Pilot Project in collaboration with the NRAO's Jim Ulvestad and Jon Romney. Ulvestad was familiar with both JPL and the VLBA. Before starting at the NRAO in 2001, he was at JPL for 12 years (1984-1996), where he was a key member of the team that used the VLA for receiving Voyager 2 telemetry during its 1989 Neptune encounter. Funded jointly by NASA and the NRAO, the project ran from August 28, 2003, to September 30, 2004, and carried out a total of

---

[6] NRAO, "Very Long Baseline Array," http://science.nrao.edu/vlba/ (accessed March 17, 2009); Wenjing Jin, Imants Platais, and Michael A.C. Perryman, ed., *A Giant Step: From Milli- to Micro- Arcsecond Astrometry: Proceedings of the 248th Symposium of the International Astronomical Union held in Shanghai, China, October 15-19, 2007* (New York: Cambridge University Press, 2008), esp. Edward B. Fomalont, "Session 3: Astrometry with Radio Interferometers."
[7] Martin-Mur, interview, 35-36; Geldzahler, interview, 29, 31 & 44; "VLBA for Nav," 2.

13 test runs using the Mars Global Surveyor, Mars Odyssey, Stardust, Cassini, and both Mars Exploration Rover spacecraft.[8]

The first VLBA test took place in January 2004 with the Mars Exploration Rover Opportunity (MER-B) during its cruise to Mars. The objective was to evaluate the accuracy of VLBA and ΔDOR under comparable conditions by determining the positions of the Opportunity spacecraft and nearby quasars using both VLBA and ΔDOR. Both systems acquired the same quasar, chosen because of the precise knowledge of its position within the International Celestial Reference Frame (ICRF). Jim Border, Tomás Martin-Mur, and others on the navigation team achieved an accuracy of 2.4 nanoradians with ΔDOR based on observations made over a five-day period. The VLBA acquired data while the craft was in the final weeks of its flight. However, navigators processed the VLBA data after Opportunity landed. "It was more of an evaluation of the kind of capabilities that the VLBA could provide, or the kind of accuracy it could provide," Martin-Mur explained. "It was not used during operations."[9]

This initial VLBA demonstration revealed a number of problems. The Pilot Project was held up at the outset by "an unexpected organizational delay" in starting the collaboration with JPL. The correlation of the first few VLBA measurements revealed a fault in the correlator model that dislocated spacecraft positions. Eventually, the culprit turned out to be a model in Goddard's CALC software that the experimenters had used for reducing VLBA data.[10]

Once navigators had surmounted these and a number of other problems, their results suggested that, over the same baseline length (10,000 km), the VLBA had an angular accuracy of 1 nanoradian, slightly more than a factor of two over that of ΔDOR. Also, adding the VLBA results to the orbit-determination process significantly increased the overall accuracy of the solutions. Uncertainties remained, though, arising mainly from errors in the radio source catalog and the atmospheric model. The navigators recommended placing a high priority on finding more Ka-band quasars to have an adequate number for future VLBA navigation.[11] In other words, the VLBA could aid navigators by doing the science that it was intended to do.

ΔDOR v VLBA. As the Pilot Project drew to a close, a meeting of managers and navigators took place on December 10, 2004, to settle the question of the VLBA replacing ΔDOR. VLBA was under attack, at least initially. The same factors—errors

---

[8] VLBA Nav Project, *Final Report*, 1-2, 3 & 4; NSF Press Release, "NSF Announces the Appointment of James Ulvestad as the New Director of its Division of Astronomical Sciences," February 1, 2010, http://www.nsf.gov/news/news_summ.jsp?cntn_id=116331 (accessed April 9, 2011); NRAO Archives, "Finding Aid to the Papers of James S. Ulvestad, 1984-1999," http://www.nrao.edu/archives/Ulvestad/ulvestad.shtml (accessed April 9, 2011).
[9] "Determination Separation with VLBA," 2, 5, 11 & 15; "VLBA for Nav," 3; Martin-Mur, interview, 35.
[10] VLBA Nav Project, *Final Report*, 3 & 9; E-mail, Gabor Lanyi to Jonathan Romney and Timothy McElrath, "Initial Assessment of the VLBA S/C-Quasar Differential Angular Measurement using MER-B Data," June 4, 2004, F347, DSN Files; "Determination of Separation with VLBA," 6.
[11] VLBA Nav Project, *Final Report*, 1; "Determination of Separation with VLBA," 11 & 14; "VLBA for Nav," 4.

arising from the reference frame and media effects—limited the accuracy of both VLBA and ΔDOR, and an accuracy improvement from the present 1 to 2 nanoradians to 0.5 to 1 nanoradian seemed feasible with either VLBA or ΔDOR.[12]

Navigators saw VLBA as a desirable supplement to Doppler, range, and ΔDOR, and there was a consensus, including among NRAO managers, that VLBA should be a complement or backup to ΔDOR. A drawback to VLBA was the turnaround time of several weeks, about a month in 2006. Data tapes and disks had to be shipped, and data correlation could take a week. In contrast, the turnaround for ΔDOR was now 8-12 hours. Navigation required turnarounds of hours, not weeks. Long turnaround times made measurements useless to navigation, as Martin-Mur pointed out, "because the spacecraft is now in a different place."[13]

Jim Ulvestad told the gathering that only a limited number of hours were available each year for NASA use, roughly 240 hours (around one day per month) for catalog work (finding, measuring, and evaluating useful radio sources) and 50-100 hours for spacecraft tracking. "More than this," he explained, "would have a negative impact on the VLBA's service to its prime customers and would be seen as problematic by the VLBA's key stakeholders—the science community and the NSF."[14]

The timing of observations was a critical issue. Martin-Mur pointed out: "Normally radio astronomers are not as particular about the time of the day in which they want to take an observation. But navigators may be." "But when navigation needs data," he added, "we need data. . . . The astronomers can sometimes do the same experiment either today or tomorrow or one week from today. But when navigation needs it, we need it now."[15]

In addition, the VLBA could not provide data during critical mission periods requiring 24-hour-a-day tracking for the simple reason that the VLBA could not observe Mars for 24 continuous hours. "It's on one side of the Earth only," Lincoln Wood explained, "so it's not available for any given spacecraft all the time." Navigators "may be happy just to have half an hour here, half an hour there, and half an hour there," he added.[16] NRAO management, moreover, objected to NASA using the VLBA during critical flight segments because of its complexity and cost. Instead, they preferred to offer the VLBA on a "best efforts basis."[17]

There was general agreement that the VLBA offered two vital ways to aid NASA deep-space missions. One was to develop a catalog of radio sources at Ka-band, a prerequisite for Ka-band navigation and, in particular, Ka-band ΔDOR. This activity was the one closest in line with the astronomy-focused charter of the NRAO and VLBA. In fact, the VLBA already was involved in improving the current catalog of radio sources. The second VLBA contribution was in tracking and navigating spacecraft. Many constraints limited the availability of the VLBA for navigation, but, under certain circumstances, navigators might take advantage of it on occasion. Scheduling

---

[12] Lichten, Meeting Notes.
[13] "VLBA for Nav," 2-3; Martin-Mur, interview, 36.
[14] Lichten, Meeting Notes.
[15] Martin-Mur, interview, 43.
[16] Wood, interview, 72; Martin-Mur, interview, 43.
[17] Lichten, Meeting Notes.

observations had to be flexible, and the data turnaround time of a week or longer had to be acceptable. The obstacles probably could be overcome, Lincoln Wood opined, "but whether you wind up with a cost-effective result is another question."[18]

Navigators saw an opportunity to improve ΔDOR by upgrading it to the Ka band, as had been done for DSN Doppler. In February 2005, Geldzahler asked them for feedback on *not* adding Ka-band capability to the ΔDOR system. Instead of Ka-band ΔDOR, for example, missions could use optical navigation. Steve Lichten discussed the question with his DSN and navigation colleagues and received comments from navigators Al Cangahuala, Tim McElrath, Tomás Martin-Mur, Jim Border, and Chuck Naudet. They gave four reasons for instituting a Ka-band ΔDOR system. The first was the necessity of making measurements in two frequency ranges to support precession, nutation, UT1, source catalog, station location, and the radio-planetary frame tie efforts. Navigation had to switch from S/X to X/Ka observations, because radio interference from commercial applications was mushrooming and making the S band "unusable for astrometry." Also, there was the probability that the DSN would drop the S band once spacecraft no longer used it.[19]

Navigators also pointed out the improved performance inherent in Ka-band signals. These suffered less from media effects and low Sun-to-spacecraft observing angles, a crucial need for missions to Mercury and Venus. A Ka-band ΔDOR system also could be a backup for optical navigation, which performed weakly in or near bodies with thick atmospheres, such as Venus or Titan, or at Mars during dust storms. Optical navigation also was in danger from reduced-budget missions that might find it too expensive in terms of cost, mass, and power requirements, and spin-stabilized craft simply could not do optical navigation. For that reason, for instance, approach navigation for the spin-stabilized Mars Science Laboratory relied on ΔDOR, Doppler, and range, but not optical navigation.[20]

The fourth argument navigators gave in favor of Ka-band ΔDOR was the ability to develop it incrementally. If the investment in Ka-band ΔDOR were relatively small, then it was worth doing because of the large performance benefits it offered. They suggested maintaining the X-band ΔDOR, along with Doppler and range, and preparing for an eventual shift to the Ka-band. An incremental low-cost, low-risk approach spanning three to five years seemed reasonable, especially as no current missions had a requirement for Ka-band ΔDOR. The implementation of Ka-band ΔDOR was going to happen at a much slower pace, over ten years or more, but at least the DSN had started building a quasar catalog to support it.[21]

Petit Grand Tour. Meanwhile, Cassini navigators were benefiting from VLBA observations. Launched on October 15, 1997, Cassini undertook a long journey before reaching Saturn because of its Venus-Venus-Earth-Jupiter Gravity Assist (VVEJGA) trajectory.

---

[18] Lichten, Meeting Notes; Lichten, "White Paper," 1 & 8-11; Wood, interview, 72.
[19] Lichten, "Ka-Band."
[20] Lichten, "Ka-Band."
[21] Lichten, "Ka-Band"; Personal communication, Barry Geldzahler, February 13, 2014.

Once it reached Saturn in July 2004, the craft began a four-year tour of that planet and its satellites. Gravity assists, which played a decisive role in guiding Cassini to Saturn, also were an integral part of the satellite tour, making it the most complex gravity-assist tour flown to date. Even so, most of Saturn's icy satellites—Mimas, Enceladus, Tethys, Dione, Rhea, Hyperion, Iapetus, and Phoebe—were too small to support a gravity assist, thereby adding to navigational concerns. For the most part the altitude of these satellite encounters was about 500 km. Titan, to be explored further by the Huygens probe, was the satellite scheduled for the greatest number of flybys.[22] To make an understatement, the satellite tour was a great series of navigational challenges.

Overall, navigators performed two key functions for Cassini. One was the customary determination of the position and motion of the spacecraft and later of the Huygens probe. Collecting Doppler also aided in studying satellite gravitational fields, like that of Rhea, and the determination of Titan's ephemeris—to within a few kilometers—rested significantly on the collection of Doppler.[23]

The other critical function that navigators performed was the creation and refinement of a specialized ephemeris of Saturn and its satellites. The mission's scientific goals dominated the satellite tour's design. Some science required trajectories that were time sensitive; other science necessitated multiple flybys, all within the mission's four-year time limit. The foundation for the success of the scientific mission as well as the navigation of the satellite tour as a whole was the specialized satellite ephemeris, which underwent constant improvement.

This tailor-made ephemeris was known as SAT136.[24] For centuries, astronomers had observed and measured the positions and motions of Saturn's satellites. Navigating with these observations would have resulted in uncertainties of hundreds of kilometers in their locations. The accuracy of the existing mathematical models for their motions also was insufficient for the precise navigation required by the satellite tour. Therefore, prior to the craft's final approach to Saturn, navigator Bob Jacobson developed a satellite ephemeris for Cassini.

Jacobson replaced the star locations with the International Celestial Reference Frame positions given in the second edition of the Tycho Catalog and replaced astronomers' mathematical models with the integration of large amounts of various kinds of data. Some data came from tracking spacecraft, namely Doppler from Pioneer 11 and Doppler and range from the Voyagers. The sources of other data were the Naval Observatory, the Hubble Space Telescope, and JPL's Table Mountain Observatory. Bill Owen, Steve Synnott, and George Null had their own CCD camera built and installed on Table Mountain's 0.6-meter telescope, so that they could obtain highly accurate astrometric observations of asteroids and other targets of interest.

---

[22] "Cassini Tour Navigation Strategy," 2219; "OD Results," 185; "Cassini OD Saturn Satellite Tour," 21.
[23] "Cassini OD Saturn Satellite Tour," 22-23 & 32; "Cassini OD Satellite Tour First Eight Orbits," 936; "Cassini Tour Navigation Strategy," 2218-2219; "Brief History of JPL OpNav," 340-341.
[24] "Cassini Tour Navigation Strategy," 2218, 2219 & 2225; "OD Results," 185; "Cassini OD Satellite Tour First Eight Orbits," 935; "Cassini OD Saturn Satellite Tour," 21

Table Mountain played a central role in developing the JPL ephemerides, and optical navigators, such as Bill Owen, were constant users.[25]

Jacobson and the Satellite Working Group frequently interacted with the navigation team during weekly meetings and provided navigators with periodic ephemeris updates. Optical navigation was a primary source for ephemeris upgrades during the satellite tour, and Cassini was better equipped than Galileo for optical navigation. Cassini had two CCD cameras, each with a significantly larger number of pixels, so that they acquired sharper, more detailed images. The camera's computer had a compression algorithm that reduced data volume by 75 to 80 percent. On top of that, Cassini had even greater bandwidth than Galileo, and unlike the latter its high-gain antenna did not fail. Nonetheless, Cassini did lack a scan platform, so the entire craft had to turn in order to point its camera or antenna in the correct direction, and that maneuver left the spacecraft out of contact with Earth while making observations.[26]

Steve Gillam, Bill Owen, Mike Wang, and other optical navigators were highly successful in honing the satellite ephemeris during the four-year tour of Saturn and its moons. The mission design called for the satellite ephemerides of the major moons to become increasingly more precise as Cassini approached Saturn then its moons. Before Saturn approach, those ephemerides were to be known to an accuracy of somewhere between 180 to 1,700 km, depending on the particular satellite. During the approach, the optical navigators had to reduce this uncertainty to less than 100 km. Once Cassini was touring the satellites, their charge was to reduce that uncertainty even more, to less than 10 kilometers.[27]

The incorrigibles. Uncooperative moons tested the limits of optical navigation and navigators. The factors that limit one's ability to find the center of a satellite accurately are its shape, albedo (brightness), and the unevenness of its surface topography. Titan's thick and changeable atmosphere hindered navigators' ability to find its center in images, while drastic differences in icy Iapetus' albedo hampered finding image centers. The latter is nearly white in one hemisphere and virtually black in the other (the enigmatic region known as Cassini Regio). Finding the center of Hyperion was complicated by its irregular nonspherical shape and unpredictable (truly chaotic) axis of rotation. Hyperion and Iapetus were so frustrating that optical navigators imaged them only as a last resort.[28]

Titan perhaps was the most incorrigible satellite. A significant error of 40 km in Cassini's location in relation to Titan surfaced following the first encounter with that moon on October 26, 2004. The mission design called for the craft to fly past Titan at

---

[25] Robert A. Jacobson, "The Orbits of the Major Saturnian Satellites and the Gravity Field of Saturn from Spacecraft and Earth-Based Observations," *The Astronomical Journal* 128 (2004): 492-501; "Cassini Tour Navigation Strategy," 2222; "OD Results," 188; "High-Accuracy Astrometry," 89-102; Synnott, interview, 23.
[26] "Cassini OD Satellite Tour First Eight Orbits," 948; "Brief History of JPL OpNav," 340.
[27] "Cassini Tour Navigation Strategy," 2218-2219; "Brief History of JPL OpNav," 340-341.
[28] "Cassini OD Satellite Tour First Eight Orbits," 938; "Cassini OD Saturn Satellite Tour," 35; "Brief History of JPL OpNav," 341.

an altitude of 1,200 km, but the spacecraft missed by 26 km and flew lower into the satellite's atmosphere. Why had this ephemeris difference not come to light earlier?[29]

It's complicated. Titan is in a resonant orbit with Hyperion. Even though Hyperion's mass is relatively small, it has an effect on Titan. The density of Titan's atmosphere stymied optical navigators' ability to find the satellite's center. Furthermore, during this first Titan encounter, Cassini was under the control of its Reaction Control System, and its jets might have been firing at the time. There also was some atmospheric drag on the probe, and the DSN had acquired no Doppler or range during the flyby. In the end, it seemed that the optical navigation data was simply not sensitive enough to detect a 20 km ephemeris change.[30]

The VLBA Pilot Project recently had ended in success. To help with its ephemeris issues, the Cassini project requested VLBA data to help measure the mass of Iapetus. Navigators were uncertain of the satellite's mass, and the mass ambiguity would have a significant effect on the Huygens probe's trajectory into Titan's atmosphere. The mission design called for Huygens to fly relatively slowly past Iapetus at a distance of 64,000 km. Shortly after Cassini started its orbit around Saturn, navigators' estimate for the mass of Iapetus unexpectedly and dramatically shifted and called into question the values in use. As a safety measure, Cassini project managers elected to increase the craft's Iapetus flyby distance from 64,000 to 126,000 km to lessen the effect of the uncertainty of the satellite's mass.[31]

Cassini managers requested the VLBA data to help in sorting out this ephemeris problem. The NRAO responded to the special proposal, although they still considered the technique and measurements to be experimental, and made the solicited observations over a seven-day period in October 2004.[32] The Pilot Project also had made similar observations in June and September 2004. The VLBA data helped Cassini navigators and ephemeris developers to improve their orbit determinations of Cassini and the Saturn ephemeris.[33]

Although the Pilot Project had hoped to develop VLBA as a navigational tool, this Cassini episode demonstrated its value for improving ephemerides, which, after all, were at the heart of the VLBA's scientific *raison d'être*. Subsequently, based on a proposal submitted by Myles Standish in JPL's ephemerides group, the VLBA carried out a campaign of Cassini observations between October 2006 and April 2009. Those responsible for collecting and analyzing the data included JPL navigators Bill Folkner, Gabor Lanyi, Dayton Jones, Jim Border, and Bob Jacobson and the NRAO's Ed Fomalont and Jon Romney.[34]

The goal of the observations was to improve navigators' knowledge of the position of Saturn within the International Celestial Reference Frame by observing Cassini and nearby quasars with the VLBA. The JPL ephemerides for the outer planets lacked the precision of the inner planets for a number of reasons discussed in Chapter

---

[29] "OD Results," 199-200; "Cassini OD Satellite Tour First Eight Orbits," 946.
[30] "OD Results," 194-195 & 200-201; "Cassini OD Satellite Tour First Eight Orbits," 946.
[31] "VLBA for Nav," 4 & 5; "Cassini OD Satellite Tour First Eight Orbits," 953.
[32] VLBA Nav Project, *Final Report*, 7.
[33] "VLBA for Nav," 4 & 5; "Cassini OD Satellite Tour First Eight Orbits," 955.
[34] Standish, NRAO Proposal; "VLBA Astrometric Observations of Cassini," 31-35.

14. Although the outer-planet ephemerides had been improving by incorporating increasingly more accurate data from optical navigation, radar, and the VLA, very little high-precision spacecraft tracking data was available for the outer planets. The Pioneers and Voyagers furnished only a small amount of such data, and Galileo had suffered from the loss of its high-gain antenna. Cassini was the first opportunity to collect highly precise data on an outer planet for an extended period. The VLBA data would improve the ephemeris for Saturn as well as the link between the inner and outer solar system within the JPL planetary ephemerides. The data also would establish a link between the star-based ephemeris reference frame and the radio reference frame. Navigators hoped to use the same approach later during the Juno mission to refine Jupiter's ephemeris and to tie it to the radio reference frame.[35]

The VLBA observations made between 2006 and 2009 yielded highly precise results, but just as importantly, they indicated how to improve accuracy by reducing certain error sources. The ephemeris constructed from the VLBA observations, DE422, was significantly better than the widely used DE405 supplied in 1997, which itself was a dramatic upgrade of DE200. DE415, issued in April 2006, was the first to include Cassini data. DE422 benefited not only from VLBA data, but also from 18 months of tracking Venus Express, Mars Reconnaissance Orbiter, Mars Express, and Mars Odyssey as well as telescopic observations of the outer planets made with CCD cameras. The prolongation of the Cassini mission until 2017 assured ongoing improvement of Saturn's ephemeris, because future VLBA data would cover a greater portion of Saturn's orbital period.[36]

---

[35] "VLBA Astrometric Observations of Cassini," 30, 31, 32 & 37; Standish, NRAO proposal; "Juno New Frontiers," 3-4.

[36] "VLBA Astrometric Observations of Cassini," 37; Memorandum, E. Myles Standish to Distribution, "The Effect of Future VLBA Measurements upon the Covariance of Saturn: A Preliminary Assessment," May 1, 2006, F347, DSN Files; Geldzahler, interview, 30.

Chapter 25

Mars Redux

Mars exploration in the era after the losses of 1999 was different, if for no other reason than for the revival of ΔDOR. The renaming of the 2001 Mars Surveyor as Mars Odyssey in honor of Arthur C. Clarke and the science fiction film based on his work denoted the optimism that surrounded Mars missions. In 2000, NASA grounded its Mars Exploration Program in the community of planetary scientists. The phrase often used to describe its goals was "Follow the Water."[1] NASA put a new queue of missions in place—Odyssey, two Mars rovers (Spirit and Opportunity), and the Reconnaissance Orbiter—and initiated the Mars Scout program to send small, low-cost spacecraft to Mars. The only Scout flights approved before NASA canceled the program in 2010 were Phoenix and MAVEN, which launched in August 2007 and November 2013, respectively.

The new Mars enterprise called on a range of technological and navigational innovations. For starters, ΔDOR was back. Mars Odyssey was the first to benefit from its restoration which occurred while the craft was flying to the fourth planet. The success and high accuracy of ΔDOR instilled its inclusion on all subsequent flights to Mars as well as to comets and asteroids, and now the European Space Agency (ESA) embraced it. The VLBA also played a role in navigating a craft to Mars, the first operational demonstration of VLBA navigation. Yet another flight would try out noncoherent Doppler.

The demands of assuring the risky entry, descent, and landing of probes on the planet's surface and placing Mars rovers within a specified limited landing ellipse after a rough passage through the atmosphere necessitated new navigational solutions. These came in the form of new software solutions. One, called ARDVARC, allowed navigators to monitor spacecraft progress in nearly real time by looking at Doppler changes. Yet another software change added new capabilities to optical navigation. Optical navigators, who finally succeeded in designing and using their own camera, instead of adapting to a science camera, merged their autonomous navigation software with guidance and control programs. This merger created a new data type that not only gave a "direct" measurement, but also was suited ideally for asteroid and cometary exploration. At the same time, the navigation section came to include guidance and control people alongside navigators, as we saw in Chapter 23.

2001: A Mars odyssey. Mars Odyssey navigators achieved a high degree of precision in inserting the probe into planetary orbit. Cruise navigation resulted in its arrival on target over the Martian North Pole at an altitude less than 1 km off from the 300 km

---

[1] G. Scott Hubbard, Firouz M. Naderi, James B. Garvin, "Following the Water: The New Program for Mars Exploration," *Acta Astronautica* 51 (July-November 2002): 337-350.

target altitude.² This exactitude hid the fact that Odyssey might have become another Mars Climate Orbiter, had it not been for ΔDOR.

Errors arising from several sources plagued orbit solutions during the cruise to Mars. The greatest errors came from the probe's own attitude control system that used reaction wheels like the Climate Orbiter. Every 18 to 25 hours, the Odyssey discharged reaction wheel momentum—mainly a consequence of solar pressure—by burning its reaction control system thrusters. The thrusters moved the craft in a direction perpendicular to the Earth-spacecraft line of direction, so it was not observable directly by either Doppler or range.³ These were the very same small forces acting in the same direction that had damned the Climate Orbiter.

In addition, before the availability of ΔDOR, anomalies became apparent shortly after launch. To be sure that they were modeling media effects, solar pressure, and small forces (the desaturations) adequately, navigators calibrated thruster activity twice during the cruise phase. The first calibration occurred a month after launch and indicated anomalies in the Doppler and range, including one that made its appearance immediately after launch. The Multi-Mission Navigation Team had calculated Odyssey's flight path and sent the information to Odyssey's navigation team. They discovered that the Delta-II had injected the spacecraft into a trajectory that would have taken it considerably distant from its intended target.⁴

NASA Climate Orbiter investigators had recommended using "supplemental tracking data types" to "enhance or increase" navigational accuracy. Odyssey navigators called on the new ΔDOR system. NASA did not initiate its implementation until 2000, however, and it still was not ready for operation until the fall of 2001, as Mars Odyssey was heading into orbit insertion. While JPL's small ΔDOR team was building and testing the new system, they simultaneously were operating a prototype version on Mars Global Surveyor. Odyssey navigators eventually scheduled 48 ΔDOR measurements, more than double the 20 that they originally had requested. As a result, they achieved an accuracy of better than 5 nanoradians.⁵

On October 29, 2001, awash in Mars Odyssey's success, Charles Elachi declared that ΔDOR would be an integral part of Mars Exploration Rover (MER) navigation. The MER navigation team chief judged the ΔDOR measurements to be "the most valuable" in achieving the requisite levels of precision. Subsequently, the Mars Reconnaissance Orbiter, Deep Impact, Phoenix, and the Mars Science Laboratory—among several other missions—embraced ΔDOR as well.⁶

Elachi and the Mars Exploration Rover project were not the only ones impressed by ΔDOR's performance. ESA was equally impressed, enough to invest in the requisite technology and software. A number of ESA missions had been taking advantage of the DSN's ΔDOR system since 1980. For example, ESA navigated Ulysses during its 1992

---

² "2001 Mars Odyssey OD," 394-397.
³ "2001 Mars Odyssey OD," 394-397.
⁴ "2001 Mars Odyssey OD," 397-400, 402 & 404.
⁵ MCO Board, *Phase I Report*, 30; Lichten, "White Paper," 3.
⁶ Lichten, "White Paper," 3.

approach to Jupiter and the Mars Express in 2003 with ΔDOR.[7] The space agency decided to acquire its own capability, probably in response to NASA's mothballing of its ΔDOR system.

To that end, the Europeans built the Cebreros (DSA-2) 35-meter X-band and Ka-band antenna west of Madrid. It became operational in November 2005, in time for an initial ΔDOR test on Rosetta (sent to study comet 67P/Churyumov–Gerasimenko) and Venus Express (ESA's first flight to that planet) plus appropriate quasars. More demanding experiments took place in January and March 2006 using the Mars Express probe. ESA's first operational use of ΔDOR was on Venus Express in March and early April 2006. Analysis showed that the quality of the measurements was only slightly inferior to those obtained using JPL's 34-meter antennas. Not surprisingly, the most accurate were made by the DSN's 70-meter dishes. Nonetheless, the single most important navigational parameter, the minimum altitude above Venus at arrival, showed a dramatic degree of accuracy: only 3 km higher than the predicted 386 km.[8]

Watch the ARDVARC. Success at Mars continued with the rovers dubbed Spirit and Opportunity (originally MER-A and MER-B, respectively). They had to enter the atmosphere at the correct time with enough accuracy to enable each lander to alight within a specified landing ellipse, in this case one measuring about 70 km by 5 km, that project managers had deemed to be both safe for landing and scientifically interesting. This landing ellipse, according to navigator Louis D'Amario, was "the single most important requirement levied on navigation."[9]

Achieving this precision called for adding ΔDOR to the mix of X-band two-way Doppler and range. "It was not a difficult decision to decide that it was pretty important to have [ΔDOR] for the Mars Exploration Rover," Lincoln Wood recalled. The accuracy of the ΔDOR typically was around 1.5 nanoradians, about a factor of three better than pre-launch assumptions. Moreover, the ΔDOR processing time had been cut to 8 hours or less.[10]

Once the rovers arrived at Mars, navigators faced the ordeal of the entry, descent, and landing phases. A heatshield and a parachute slowed down the craft, which next fired rockets to reduce its speed. Airbags smoothed the last landing leg. Several factors complicated navigating these flight segments. One was the necessary transition from two-way to one-way Doppler. The stability of the spacecraft radio transmitter

---

[7] "Delta-DOR," 70; ESA, "Delta DOR," http://www.esa.int/SPECIALS/Operations/SEMZ2TO2UXE_0.html (accessed April 8, 2011).
[8] ESA, "Cebreros - DSA 2," http://www.esa.int/SPECIALS/Operations/SEMVSDSMTWE_0.html (accessed April 8, 2011); ESA, "Delta DOR"; Nick James, Ricard Abello, Marco Lanucara, Mattia Mercolino, and Roberto Maddè, "Implementation of an ESA Delta-DOR Capability," *Acta Astronautica* 64 (June-July 2009): 1041-1049; "Delta-DOR," 73-74.
[9] *MER Navigation*, 1-2 & 6.
[10] *MER Navigation*, 2, 7 & 8; Wood, interview, 60-61; Brian M. Portock, Eric J. Graat, Timothy P. McElrath, Mike M. Watkins, and Geoff G. Wawrzyniak, "Mars Exploration Rovers Cruise Orbit Determination," 529-530 in *A Collection of Technical Papers: AIAA/AAS Astrodynamics Specialist Conference, August 16-19, 2004, Providence, RI*, vol. 2 (Reston, VA: AIAA, 2004).

oscillator, changes in the probe's velocity and direction, and variations in temperature and pressure complicated navigation further. Perhaps more worrisome was the fact that the indicator for parachute deployment was the DSN's loss of lock on the one-way Doppler. The dynamics of the craft's movement after the chute opened guaranteed that the DSN would lose its lock on the craft. Nonetheless, Tim McElrath and his navigation team successfully detected each entry, descent, and landing event for both Spirit and Opportunity.[11]

Navigators took advantage of the fact that various changes in the spacecraft's motion from one flight segment to the next played havoc with the X-band two-way Doppler. For example, the opening of the chute caused a sudden drop in deceleration, while the lander's swinging motion left its own wild signature in the Doppler. Each distinct Doppler signature caused by each entry event was visible by observing real-time Doppler residuals displayed by a computer program known as ARDVARC (Automated Radiometric Data Visualization and Real-time Correction).[12]

ARDVARC was born from a program called ARGOS (Attitude Reckoning from Ground Observable Signals) developed specifically for the joint NASA-ESA Ulysses solar mission launched in 1990. The probe was undergoing some troubling movements after having flown past Jupiter for a gravity assist. As Al Cangahula described it, the spacecraft was "like a pizza box with a high-gain antenna pointing at the Earth, and with long, wiry booms sticking out. The solar radiation pressure on those booms was nutating it a little bit. And that wasn't good." If the craft was nutating too much, one could take steps to correct it. Project managers charged Tim McElrath, one of the Ulysses navigators, to devise a real-time monitoring tool to do two things. One was to monitor the Doppler; the second was to sound an alarm, if the nutation exceeded a certain level. ARGOS worked.[13]

Cangahuala now felt "a little frustrated." ARGOS enabled real-time Doppler monitoring, so now navigators were wondering when they would be able to do "real-time onboard navigation" with the DSN. "Could you listen to a signal from the DSN and figure out your own orbit? Well, we still don't quite do that today." ARGOS simply was not going to be the basis of this future software. "It's just wired wrong. It's not flexible enough to allow for that," Cangahuala explained. Therefore, with management approval, he and Ted Drain, Dan Burkhart, and Vince Pollmeier set out on a project named Automated Real-Time Spacecraft Navigation. The ARTSN (pronounced like "artisan") project's goal was to demonstrate the feasibility of accurate orbit estimates in close to real time. They succeeded in 1997.[14]

Now they had to decide whether to build it or to start anew from scratch. "We chose the latter. We made our point." They started afresh, salvaging the residual monitor from ARTSN. "Out of that, we built a real-time residual monitor that doesn't do orbit determination. It just subtracts the computed from the actual observable that

---

[11] *MER Navigation*, 28 & 31; "MER EDL Nav," 783-791.
[12] *MER Navigation*, 1-3 & 28; Edgar Satorius, Polly Estabrook, J. Wilson, and D. Fort, "Direct-to-Earth Communications and Signal Processing for Mars Exploration Rover Entry, Descent, and Landing," 1-35 in IPNPR 42-153; "MER EDL Nav," 789-795.
[13] Cangahuala, interview, 17.
[14] Cangahuala, interview, 22.

generates the residuals." ARDVARC, he explained, is "actually what you usually see in the darkroom during like Mars EDL [entry, descent and landing] and Earth returns. There's a cloud of black background and lime-green dots. That's ARDVARC. It's very professionally satisfying."[15]

It provides a visual display of Doppler changes that reflect shifts in a spacecraft's trajectory, and it has simplified certain aspects of navigation greatly. It found its first application on Mars Global Surveyor by helping navigators to manage delicate aerobrake maneuvers. The program quickly came into regular use for aerobraking, and its ability to show Doppler changes greatly aided in precisely landing the Mars Exploration Rovers. Because ARDVARC had a nearly real-time monitor that displayed Doppler residuals, Cangahuala explained, "you just watch the ARDVARC."[16]

The navcam orbiter. All of the tools that navigators had at their disposal—high-precision Doppler, ARDVARC, the new dynamic filter, ΔDOR, and optical navigation—came together for the latest search for water on the Red Planet, the Mars Reconnaissance Orbiter. Launched in August 2005, the Orbiter arrived in March 2006 and joined five other probes—Mars Global Surveyor, ESA's Mars Express, Mars Odyssey, and the two Mars Exploration Rovers—in studying Mars. The Orbiter flew low above the planet to carry out its science mission and, according to one report, "On more than one occasion, the MRO spacecraft performed a collision avoidance maneuver increasing its relative range to the Odyssey spacecraft."[17]

The high-precision navigation was typified by the fact that the craft used only two of four planned trajectory corrections to arrive at Mars. The navigators reported: "This is a significant interplanetary navigation achievement and is owed to the attention to detail in spacecraft and dynamical modeling. This also saved about two-thirds of the planned propellant expenditure for interplanetary cruise." The mapping mission required shaving the probe's initial elliptical orbit into a circular orbit using aerobraking. "It all went just absolutely flawlessly," Earl Maize recalled.[18]

It was Shyam Bhaskaran's "first real navigation lead job." The navigation team was small, no more than three people, the others being Dolan Highsmith and Moriba Jah. ARDVARC again handled the delicate aerobraking by allowing navigators to see Doppler changes in almost real time. The real navigation milestone, however, was the CCD camera for optical navigation. It was the first of its kind, designed and built expressly for optical navigation. Steve Synnott, the Principal Investigator for the so-

---

[15] Cangahuala, interview, 23.
[16] Cangahuala, interview, 24 & 29; "MGS Mapping OD," 8.
[17] C. Allen Halsell, Angela L. Bowes, M. Daniel Johnston, Daniel T. Lyons, Robert E. Lock, Peter Xaypraseuth, Shyam Bhaskaran, Dolan E. Highsmith, and Moriba K. Jah, "Trajectory Design for the Mars Reconnaissance Orbiter Mission," 1591-1606 in Daniel J. Scheeres, Mark E. Pittelkau, Ronald J. Proulx, and L. Alberto Cangahuala, ed., *Spaceflight Mechanics*, Pt. II (San Diego: Univelt, 2003); "MRO: Launch to Science Orbit," 345.
[18] "MRO Nav," 167; "MRO: Launch to Science Orbit," 340; Maize, interview, 23.

called navcam experiment, led its creation and design. The camera was sensitive enough to show stars as faint as magnitude 11.0.[19]

Synnott had the idea of building the navcam as a low-cost, lightweight camera system. What he built and flew on the Mars Reconnaissance Orbiter was a prototype to show that such a camera would be useful for future Mars missions, especially for improving approach navigation. He had been working on the idea for 15 years. Synnott wanted a small camera that could do as good a job of optical navigation as "these things that were a meter long and 40 kilograms," like Galileo's camera. "I just started developing with some people in [the Observational Systems Division] 38, with the optical engineers, some possible designs," Synnott recalled.[20]

NASA then lost the two Mars probes. "In part, people worried, because we hadn't done the navigation accurately enough," Synnott explained. People associated with the Mars Technology Program Office and the Mars effort in general asked him, "Steve, you're in optical navigation. Why didn't we have a camera on board?" "It wasn't my decision," Synnott replied. "I was asked many times, 'Do you have a camera that could fly on these missions that would not impact the mass of the spacecraft much, not take much in the way of power resources or volume or anything else?' I said, 'Yes. It just so happens I have that. I've been worrying about this for 15 years.'"[21]

Finding money to build the prototype camera was another matter all together. "I kept proposing these ideas of either a small camera to do navigation in cruise or to do autonomous navigation in particular." He obtained "little piles of money," say $15K a year. Over the years, Synnott and an optical engineer designed the camera. "I talked on the phone so long with that guy—I'd go over to his office, he'd come over here, and we'd talk on the phone." The design period spanned "probably more like a decade." "We'd try an option or two, and then the money would run out. Then the next year I'd get some more."[22]

The money came from Research and Technology Operating Plan (RTOP) funds (money provided to JPL by NASA Headquarters) and various small program offices. "Whatever the source, over a period of a decade or so, we had formulated this camera design." As a result, when NASA lost the two Mars probes, and people asked Synnott if he had a camera to propose, he could answer, "Yes, here it is." "The timing couldn't have been better," he explained. "We had the design. On the little bit of money we had, all we had was a paper design. But within a period of six months, it went from a preliminary paper design to being a full-fledged program that started to formulate every detail of the hardware." The design went from paper to metal from 2000 to 2002[23] and ultimately flew on the Mars Reconnaissance Orbiter.

AutoGNC lands on Mars. Phoenix, launched in August 2007, marked several navigation firsts. Cruise navigation relied on Doppler, range, and ΔDOR and demonstrated

---

[19] Bhaskaran, interview, 46; "MRO Nav," 171; "Brief History of JPL OpNav," 343.
[20] Bhaskaran, interview, 37 & 38; Synnott, interview, 23.
[21] Synnott, interview, 24.
[22] Synnott, interview, 24.
[23] Synnott, interview, 24 & 25.

operational VLBA navigation for the first time. When Phoenix arrived at Mars on May 25, 2008, it also validated the latest incarnation of autonomous navigation, AutoGNC (for autonomous guidance, navigation, and control), in landing the craft. AutoGNC came into play during the most daunting flight segment, the landing phase, which extended from four days before entry into the atmosphere to planetary touchdown. The intervening series of flight events required the execution of 601 spacecraft commands, and the command sequence had to operate in sync with physical events whose timing was known in advance.[24]

AutoGNC was written in a new computer language—Virtual Machine Language (VML)—developed by NASA specifically to simplify spacecraft operations. VML had special properties that made it ideal for writing the AutoGNC program. NASA initiated development of VML in 1997, and by 2010, it had implemented five versions. The first, VML 0, flew on Stardust and the ill-fated Mars Climate Orbiter and Polar Lander. The second version, VML 1, was in use on Mars Odyssey, the Spitzer Space Telescope, and Genesis. Phoenix utilized VML 2.0, the same iteration as the Mars Reconnaissance Orbiter, Dawn, Juno, GRAIL, and MAVEN. VML was especially useful for the Phoenix landing segments, because it allowed one to use complex logic in command sequences. It also provided programmable time tags to delay actions by a calculated amount or until a calculated time.[25]

The AutoGNC program had its technological roots in a number of older missions. Initially called the AutoNav Mark3, it originated in a 2006 decision to make two fundamental changes to the autonomous navigation software. The first alteration rewrote the so-called Executive in VML to adapt it to the multiplicity of missions committed to the new code. The Executive was a key portion of the software that planned and directed autonomous navigation activities and interfaced with the rest of the spacecraft software. Previously, navigators had tailored the Executive code to suit the needs of a given flight project. Each new project required discarding and rewriting old code at substantial expense, even though the new mission might share certain characteristics with the preceding one.[26]

The second basic change to the AutoNav Executive was the addition of attitude control and guidance, which derived from such flights as Cassini, Deep Space 1, and the

---

[24] "Reactive Sequencing," 3-4; Geldzahler, interview, 15; Peter H. Smith, "The Phoenix Scout Mission," 34th Annual Lunar and Planetary Science Conference, March 17-21, 2003, League City, TX, http://www.lpi.usra.edu/meetings/lpsc2003/pdf/1855.pdf (accessed July 27, 2010); JPL, "Missions to Mars: Phoenix," http://mars.jpl.nasa.gov/missions/present/phoenix.html (accessed July 27, 2010); Martin-Mur and Highsmith, 1.
[25] Christopher A. Grasso, "The Fully Programmable Spacecraft: Procedural Sequencing for JPL Deep Space Missions Using VML (Virtual Machine Language)," IEEE Aerospace Applications Conference Proceedings, March 2002, http://trs-new.jpl.nasa.gov/dspace/bitstream/2014/36888/1/01-2408.pdf (accessed January 4, 2010); Grasso, "Techniques for Simplifying Operations Using VML (Virtual Machine Language) Sequencing on Mars Odyssey and the Space Infrared Telescope Facility," IEEE Aerospace Applications Conference Proceedings, March 2003, http://trs-new.jpl.nasa.gov/dspace/bitstream/2014/10556/1/02-2600.pdf (accessed January 4, 2010); "Reactive Sequencing," 3; Marc D. Rayman, Thomas C. Fraschetti, Carol A. Raymond, and Christopher T. Russell, "Dawn: A Mission in Development for Exploration of Main Belt Asteroids Vesta and Ceres," Acta Astronautica 58 (2006): 610; "AutoNav Mark3," 4844.
[26] "AutoNav Mark3," 4843; "Reactive Sequencing," 7.

Mars Exploration Rover.[27] This capability was essential in guiding Phoenix to the Martian surface. AutoNav already had shone during Deep Impact's encounter with Comet Tempel-1. The software had taken over the guidance and control of the craft through its interaction with the flight software and attitude control systems. What had begun as an entirely Earth-based enterprise—deep-space navigation—was becoming an independent spacecraft-based tool for the exploration of the solar system.

The initial idea of AutoGNC, however, was to support a trip to a small body—an asteroid, a comet, or a Martian moon—that would bring back a sample from the surface. Alternatively, as Ed Riedel, Shyam Bhaskaran, Brian Kennedy, Dan Kubitschek, Nick Mastrodemos, Steve Synnott, and others envisioned the software, it would be the combined onboard navigation, tracking, and rendezvous system for undertaking a Mars sample return mission, including rendezvous with an orbiter to return the sample canister to Earth. The delay in sending and receiving signals between Earth and a probe precluded intervention by a real-time operator, and the gravitational and other uncertainties associated with small bodies necessitated a guidance and control system capable of responding rapidly.[28] AutoGNC was on its way to joining ΔDOR as one of many new tools and techniques for navigating future interplanetary flights.

Navigating with the VLBA. Phoenix also showcased the first validation of operational VLBA navigation. Cruise navigation relied on ΔDOR, which provided measurements that tied the craft's trajectory to the highly precise quasar-based celestial reference frame, the same reference frame used by the VLBA. In this way, ΔDOR aided navigators in using VLBA observations. Martin-Mur, though, decided to replace the quasar in the VLBA measurements with a second spacecraft, a form of dual-spacecraft tracking known as same-beam interferometry.[29]

Having the two probes within the same beamwidth obviated timing and other errors arising from slewing the antenna between a spacecraft and a quasar. Of course, inaccuracies in the quasar's position translated into errors in the spacecraft's position. But, as Martin-Mur explained, "If you connect the spacecraft to something that is orbiting Mars, then you remove the errors of the position of Mars, and you remove the errors of the position of the source, and you utterly tie your spacecraft to Mars. I think that could be very accurate." Using the VLBA instead of the DSN allowed navigators to see the two craft in the same beam earlier in the flight. In addition, because beamwidth is inversely proportional to antenna diameter, the 25-meter VLBA antennas had a wider beamwidth than the DSN dishes, which were 34 and 70 meters across.[30]

---

[27] "Reactive Sequencing," 7 & 10; "AutoNav Mark3," 4841.

[28] "Reactive Sequencing," 8 & 16; "AutoNav Mark3," 4835, 4838 & 4839.

[29] Geldzahler, interview, 15; Peter H. Smith, "The Phoenix Scout Mission," 34th Annual Lunar and Planetary Science Conference, March 17-21, 2003, League City, TX, no.1855, http://www.lpi.usra.edu/meetings/lpsc2003/pdf/1855.pdf (accessed July 27, 2010); JPL, "Missions to Mars: Phoenix," http://mars.jpl.nasa.gov/missions/present/phoenix.html (accessed July 27, 2010); Martin-Mur and Highsmith, 1.

[30] Martin-Mur and Highsmith, 2 & 3; Martin-Mur, interview, 42-43.

Martin-Mur pointed out another advantage of dual-spacecraft navigation. "Because the spacecraft signal is much stronger, much stronger than the signal from a quasar, you need less bandwidth to record the spacecraft signal and the data that you have to transfer between the stations and the processing center. You can do spacecraft to spacecraft faster than you can do quasar to spacecraft. For the Phoenix experiment, they wanted an experiment in which they processed the data. They acquired the data from a station that processed it within something like 12 hours. That will be something that will be useful for them."[31]

As the second spacecraft for his test of VLBA same-beam interferometry, Martin-Mur chose Odyssey and the Mars Reconnaissance Orbiter. Eight VLBA tracking sessions took place between March and May of 2008. The NRAO, however, did not deliver the VLBA data until after Phoenix landed. Martin-Mur and his colleague Dolan Highsmith then proceeded to process the data. Initial processing at JPL showed large inaccuracies that traced back to NRAO data reduction procedures. The VLBA corrected the problems and delivered the rectified data. "When we got Phoenix in exactly the same antenna beam as the Mars Reconnaissance Orbiter and Mars Odyssey," Geldzahler recalled, "we were able to measure the relative offsets between Phoenix and these other two orbiters to 150 feet." With Mars over 241 kilometers (150 million miles) away, that meant an accuracy level of a tenth of a nanoradian. "The year 2020 is when we predicted that we would be able to go to a tenth of a nanoradian," specifically for a Mars sample return mission, but Phoenix achieved that level over a decade earlier, Geldzahler noted, albeit not in time to serve for time-critical navigation activities.[32]

Martin-Mur concluded that the experiment had confirmed the utility of VLBA observations for spacecraft-to-spacecraft tracking of spacecraft approaching Mars. He subsequently carried out simulations to analyze their practicality for the Mars Science Laboratory and MAVEN. He found that, using the VLBA and DSN in combination for Mars approach, he could improve the accuracy of navigation solutions compared to those computed from just Doppler, range, and ΔDOR. The greatest benefit, he believed, was in improving trajectory predictions for initializing the entry, descent, and landing sequence.[33]

The exotic and noncoherent world of MSL. Martin-Mur additionally wanted to get VLBA observations for the Mars Science Laboratory (MSL). Also known as Curiosity, MSL launched in November 2011 and arrived in August 2012.[34] In contrast to Phoenix and the Mars Exploration Rovers—where VLBA data processing occurred after landing on Mars—Martin-Mur wanted "the data to be available to us while we are still approaching Mars." That way, "we can see whether we're in the right path." "It's not that we need the data to run the mission," he explained, "but having this data will allow us to achieve

---

[31] Martin-Mur, interview, 42.
[32] Martin-Mur and Highsmith, 4 & 5; Geldzahler, interview, 26-27; "Level 1 Requirements: Navigation," February 2, 2006, F348, DSN Files.
[33] Martin-Mur and Highsmith, 6-8 & 9.
[34] Martin-Mur, interview, 42; "Application of Noncoherent Doppler," 1-6.

the requirements for the mission earlier. It will allow us to sleep better when we're close to Mars, knowing that we are in the right trajectory." Two-way Doppler, range, and ΔDOR will suffice to fulfill mission navigation requirements, but "having the VLBA will also help us to confirm that we are in the right orbit."[35]

Martin-Mur also was considering a more exotic data type, one-way Doppler transmitted from the spacecraft. Its use would simplify ground operations and provide better reception of telemetry. The reconsideration of one-way Doppler arose as a result of the recent development of a high-precision ion-clock. Simulations indicated that replacing two-way Doppler and range with one-way Doppler aided by such a clock would meet the navigation requirements for a hypothetical Mars mission similar to that of the Mars Science Laboratory.[36]

One-way Doppler recently had been a feature of the COmet Nucleus TOUR (CONTOUR), a Discovery class probe designed to flyby, study, and image the nuclei of comets Encke, Schwassmann-Wachmann-3, and d'Arrest, but which failed shortly after its July 2002 launch. APL designed, built, operated, and managed CONTOUR. In the months before total loss of contact on December 20, 2002, CONTOUR demonstrated APL's noncoherent navigation technique. About six weeks of one-way data were acquired while the spacecraft was in the initial orbits about Earth.[37]

APL's noncoherent Doppler was the greatest technical challenge for CONTOUR navigation. It required new software, the incorporation of new procedures into the orbit determination process, and navigators devoted solely to its operation. The small JPL navigation team consisted of five fulltime people from six months before launch until three months after launch. Two alone were dedicated to the noncoherent transceiver technique, leaving three to deal with all other aspects of navigation. This staffing shortage delayed and even prevented some pre-launch analysis, not to mention lateness in validating the noncoherent technique on a previously launched spacecraft (TIMED3).[38]

Noncoherent Doppler was the primary data type for CONTOUR, and the DSN received it almost continuously. APL developed a process for JPL navigators to correct the noncoherent Doppler for their purposes that involved using the spacecraft telemetry to obtain a somewhat coherent two-way Doppler. This processed Doppler went into the JPL orbit determination software being run in Linux on a workstation.[39]

In the end, navigators concluded that CONTOUR had been "one of the most uncertain and challenging launches supported by JPL navigation." Certainly the noncoherent data was problematic, but they also faulted the "limited time and

---

[35] Martin-Mur, interview, 42.

[36] "Application of Noncoherent Doppler," 1-3; J. Prestage and G. Weaver, "Atomic Clocks and Oscillators for Deep-Space Navigation and Radio Science," *Proceedings of the IEEE* 95 (November 2007): 2235–2247; Shyam Bhaskaran, "The Application of Noncoherent Doppler Data Types for Deep Space Navigation," 54-65 in TDAPR 42-121.

[37] "Navigating CONTOUR," 1474 & 1475. On CONTOUR: APL, "CONTOUR (Comet Nucleus Tour)," http://www.jhuapl.edu/newscenter/pressreleases/spacemission/smCONTOUR.asp (accessed February 8, 2011).

[38] "Navigating CONTOUR," 1475.

[39] "Navigating CONTOUR," 1475, 1476 & 1478; J. R. Jensen and R. S. Bokulic, "Highly Accurate Noncoherent Technique for Spacecraft Doppler Tracking," *IEEE Transactions on Aerospace and Electronic Systems* 35 (July 1999): 963-973.

resources" given to navigation because of "the extraordinarily low cost of the mission." They warned: "future projects like this should be aware of, understand, and accept the increase of risk that inevitably accompanies low-cost, bare-bones operations."[40] The failures of the Mars Climate Orbiter and Polar Lander were still recent. On the other hand, dealing with the noncoherent Doppler gave JPL navigators valuable and practice in its use for potential future missions.

---

[40] "Navigating CONTOUR," 1482.

Lights! Cameras! Action!

Optical tracking for a "Bumper"—a WAC Corporal atop a German V-2—in 1951 (NASA GRIN P-0631)

Conclusion

Navigation was born out of the national security crisis that began with World War II. That war brought together scientists and engineers, firing ranges and radars, in the development and testing of missiles. JPL, even before the lab took on that name, became yet another civilian institution in the service of the war effort, when it began developing and testing missiles for the Army. This relationship, forged during World War II, lasted into the Cold War. President Eisenhower was right about the "immense military establishment" whose "total influence—economic, political, even spiritual—is felt in every city, every statehouse, every office of the federal government." This narrative has argued that the pervasive influence of national security on the evolution of deep-space navigation has spanned the seven decades from World War II through the Cold War and beyond the collapse of the Soviet Union.

Army contracts introduced JPL to satellite tracking; the ARPA lunar program laid the foundation for deep-space navigation. Similarly, at other research, academic, and governmental institutions, the armed forces fostered the development of tracking and navigational expertise. Then came the Sputniks and the projection of the Cold War into space. The attention paid to the greatly-publicized attempt to put an American on the Moon by 1970—a race that in retrospect appears to have been one-sided—has tended to overlook the real battle in space waged by both sides: the launching of probes to the Moon and beyond. The need to maintain a certain image of the country overseas, a key aspect of the Cold War, led to repeated attempts to put a civilian face on the nation's space program—the Vanguard satellites, the establishment of NASA—despite its deep and widespread roots in the national security crisis. The fact that civilian scientists had become so thoroughly embedded in defense research facilitated the erection of this façade. Fighting the Cold War in space also put navigators under immense pressure to ensure the often elusive goal of mission success.

During the 1980s, the national security crisis inspired the establishment of the Strategic Defense Initiative, a grandiose space project that continually underwent reinvention. Support for SDI faded once the Soviet Union began to implode, but a new kind of national security crisis was emerging. After North Korea tested a ballistic missile that landed near Alaska, President Clinton revived the SDI idea by signing the National Missile Defense Act of 1999. Danger from Iran and elsewhere kept the missile defense idea going, while the escalating impact and scale of bomb and insurgent attacks assured the ongoing vitality of the national security crisis into the next century.

The SDI organization quickly came to replace the Manhattan Project as a management model. The three decades after 1980 saw no end to enormous, expensive space initiatives, whether the Space Station, the National Aero-Space Plane, Space Exploration Initiative, or the Vision for Space Exploration. The *Challenger* accident and the Hubble Space Telescope mess revealed basic flaws in how NASA operated. The *Columbia* mishap again indicated systemic defects in the agency's operations. Dan Quayle and the National Space Council removed the sitting NASA Administrator and put in their own man in his stead, Dan Goldin, who remained at NASA longer than any other Administrator. At this point, the agency brought in people who had worked on the

Strategic Defense Initiative or who had worked on SDI projects, and Goldin began the process of infusing NASA with SDI management notions under his personal mantra of Faster, Better, Cheaper. The influx of SDI-experienced talent into NASA continued into the next century, even as the threat environment and the nature of the national security crisis evolved, and even after the loss of Mars Climate Orbiter and the Mars Polar Lander had discredited Faster, Better, Cheaper.

In addition to the central role of the national security crisis in shaping the evolution of deep-space navigation, one must consider the impact of budgetary reductions and their corollary, the loss of personnel, expertise, and institutional memory. President Nixon, following the termination of the Apollo flights, approved development of the Space Shuttle. Steadily, Space Shuttle costs eroded away at planetary exploration into the 1980s, a decade that began with a severely pared-down NASA program of solar system exploration. Although President Reagan initiated an impressive number of new, mammoth space undertakings, none favored exploration. The trend toward smaller, cheaper space flights began with the creation of the Planetary Observer and Mariner Mark II mission classes and the restructuring of the Explorer series (Goldin's Discovery probes). Goldin reinvented the nation's space program by applying his version of SDI management ideas. Flight projects had to be smaller and less expensive. Space exploration focused narrowly on Mars and the search for life (but not intelligent life). These cut-rate projects starved navigation expenditures and the size of navigation teams with terrible results.

While SDI notions were at work inside NASA, budget and personnel cuts became more general with the adoption of New Public Management by the Reagan and Clinton administrations. This Faster, Better, Cheaper on a national scale translated into the shedding of jobs and the expertise and institutional memory desperately needed to manage public programs whether by NASA or the Department of Defense. A basic tenet of New Public Management was the idea that government ought to be run "like a business." The Reagan administration kicked off the commercialization of space. The DSN hired itself out to foreign space agencies, and navigators, too, were "for hire" to overseas space agencies and Earth-orbiting satellites. The trends toward budget and personnel cuts and navigators being "for hire" came together with the movement of navigational expertise from JPL to private industry (KinetX).

As Forman, Hoddeson, and Cravens argue, the convergence of engineering and science—both the physical and social sciences—characterizes our modern world. Navigation is no exception. Deep-space navigation often is presented as an applied or engineering version of celestial mechanics (science). In this sense, navigation is an example of engineering science, a branch of learning and practice that deals with the application of scientific knowledge to practical problems. But, navigation did not just apply science; it advanced science. Additionally, in agreement with Hoddeson, navigation is multidisciplinary. By its very nature, it entailed the application of more than one scientific discipline.

First and foremost, navigation involves astronomy: celestial mechanics, observational astronomy, radio astronomy, and radar astronomy. JPL navigators began by borrowing and adapting values, practices, and techniques from astronomers. This is the stuff of engineering science. They ended up determining the ephemerides, physical constants, measuring units, and mathematical models for almanac astronomers. This is different. For decades, astronomers applied their science to the making of ephemerides in the form of almanacs. The observations made by mariners when determining their position at sea with the aid of instruments and those ephemerides did not serve as a basis for improving the ephemerides' accuracy. The relationship between science, application, and maritime navigation was unidirectional. Deep-space navigation, in contrast, was bidirectional. Navigators' orbit determinations of spacecraft *did* serve to improve their ephemerides, not to mention their values for physical constants and mathematical models of celestial bodies. The result was a kind of informational feedback loop between observation, science, and application. This relationship between science and navigation was as if astronomers and the ancient mariners using their ephemerides worked alongside each other at the Royal Observatory in Greenwich. So, navigation was not simply engineering science.

Navigation had a similar relationship with geodesy and geophysics that grew out of the need to know the exact locations of tracking antennas. Acquiring that knowledge meant situating those antennas within a standard three-dimensional geodetic framework. Just as JPL and other NASA navigators had created their own ad hoc system of physical constants, NASA adopted its own geometric model for all spaceflight missions. Again, as JPL's ephemerides efforts had improved the official ephemerides, navigators' enhanced station locations enabled them to show the need to correct Army and Air Force topographic maps of the Moon. In the areas of geodesy and geophysics, as they had in the field of celestial mechanics, JPL navigators succeeded in seizing on the science and improving its application. One also finds navigators engaging with ionospheric physics and radio science, cartography and meteorology, to understand media effects. In this respect, navigation was multidisciplinary. The application of these multiple sciences involved navigators with a range of international scientific institutions: the International Astronomical Union, the Bureau International de l'Heure, the International Polar Motion Service, and the International Union of Geodesy and Geophysics, to name a few. JPL navigators stood at the convergence of multiple sciences and their application. They also stood at center stage in the multinational arena that navigation had become. Halley Pathfinder may have been not just an exemplary model of multinational cooperation in navigation, but emblematic of deep-space navigation's growing trend over the decades toward becoming an international endeavor.

As this book has shown, navigators were mission project scientists who carried out research on celestial mechanics, gravity, and relativity, and performed radio occultations. In addition, they made scientific discoveries in the course of analyzing Doppler (mascons, for example) and doing optical navigation (volcanism on Io, new satellites of the outer planets). Navigation and science went hand in hand: navigators applied science to determine spacecraft positions and velocities, and they carried out research to garner data that improved the results of their computations. Missions, though, offered the possibility of making opportunistic discoveries in areas of science not related to navigation.

Consistent with Hoddeson's model, JPL is nothing if not a government-supported research laboratory organized hierarchically into divisions, sections, and groups with a matrix organizational structure overlay. At the same time, reflective of Caltech policy, an "egalitarian policy" applied to technical staff regardless of their academic degrees. Work and knowledge were the basis of employee evaluations. A cornerstone of this culture was the "campus" atmosphere that encouraged engineers to work on technical problems of their choosing and to publish their results. The rationale for these practices was to develop employee competence in preparation for future missions. Informal training courses given by colleagues contributed to both this preparation for future missions and the "campus" atmosphere. The budgetary largesse of the space race enabled the ready availability of funding as well as a degree of relaxed accounting. The establishment of the DESCANSO center of excellence somewhat prolonged the "campus" atmosphere of the 1960s.

There are similarities and differences between navigation and radar astronomy. For starters, they both emerged thanks to the Cold War and NASA funding. Most radar astronomers were trained as electrical engineers, not scientists. Most navigators were trained in engineering or mathematics, not science (astrodynamics rather than astronomy). Radar astronomy was a science driven by the availability of radars capable of planetary exploration (technology). It was through these instruments and their associated techniques of analysis, not through direct sensory observation, that radar astronomers conducted their experiments, much like navigators and the DSN. Radar astronomers and navigators alike analyzed not sensory experience, but wave patterns of electromagnetic signals which analysis by computer software made "visible." Thus, not only were the instrumentation and techniques of radar astronomy and navigation dependent on technology, so was the very content of the science.[1]

Like navigation, planetary radar astronomy historically has remained at the intersection of science and engineering. Radar astronomers' attendance at meetings of both the IAU and the International Union of Radio Science (URSI) during the 1960s reflected the dichotomous nature of radar astronomy perched between radio engineering (URSI) and astronomical science (IAU). Navigation, though, is at the intersection of multiple sciences and engineering. Radar astronomy is a set of techniques (engineering) used to generate data whose interpretation yields answers to scientific (planetary astronomy and geology) questions. Navigation also is a set of techniques. Some are mathematical, such as numerical integration, the method of least squares, and the root mean square, and are common to other engineering and scientific fields. Navigators do not use these techniques directly to generate data to resolve scientific questions, but rather to determine spacecraft positions and velocities. Whereas radar astronomy investigations end in a scientific result, the result of navigation is an application of science. This is a key difference between science and engineering science.

Navigation and radar astronomy, like subatomic physics for that matter, depend on observations and instrumentalities. The chief instrument for navigation is the Deep Space Network. Like deep-space navigation, its origins are rooted in the military needs

---

[1] Butrica, See the Unseen, vii, ix & 262-263.

of the Cold War. Unlike the instruments constructed for Big Science, such as physicists' particle accelerators, the DSN was not intended to undertake scientific research. Its primary tasks were and remain utilitarian: spacecraft communication, telemetry, and navigation. The DSN became an instrument of scientific research, as we saw in Chapter Seven, thanks to navigation. Navigation similarly made an engineering use of a scientific instrument (used onboard science cameras for optical navigation). The crisscrossing of scientific and engineering purposes is not unfamiliar in the world of engineering science since at least World War II.

Deep-space navigation has become increasingly more precise year after year. What has been behind this apparently inexorable march of progress?[2] Certainly, technological change has been a key factor. The DSN's upgrades from one frequency range to the next brought about inherent improvements in accuracy. NASA's missions of exploration became increasingly more complex and navigationally challenging over the decades. From lunar impacts to planetary flybys, the agency graduated to orbiting around planets, weaving among the numerous and wacky satellites of Jupiter and Saturn, landing on another planet, and landing on an asteroid. The increasing navigational difficulty of missions was another factor pulling and pushing navigators toward greater accuracy. Other than the DSN upgrades, though, there was nothing deterministic about the steadily improving precision of deep-space navigation. The improvements mostly were a direct result of navigators' efforts to match their "fitter's universe" with the "real universe."

Navigators' models were mathematical representations of the "real universe" that benefited from measuring known error sources—media effects, station locations, and non-gravitational forces acting on the spacecraft—and developing mathematical formulas that described their impact. The more that the "fitter's universe" described by these models and their corrections resembled the "real universe," the greater was the navigational accuracy. This was the formula behind the growing precision of JPL navigation. It took patience, knowledge, creativity, loads of mathematics, and nerves of steel, because a navigation mistake could scuttle an entire mission and all of its science.

To the degree possible, navigators made observations of spacecraft with the DSN and, to the extent possible, attempted to measure the contributions from all parameters—media effects, station locations, and non-gravitational forces—including error sources. Various kinds of measurements (Faraday rotation, timing and polar motion information) went into the calibration of the spacecraft observations. A different approach that promised to deliver far more accurate estimates of spacecraft positions involved arranging the spacecraft observation in a new way, one that canceled or at least mitigated error sources from media effects and/or station locations. This "direct" technique replaced making numerical "corrections" with experimental design, as argued in the Introduction and Chapter 13.

---

[2] Donald MacKenzie, *Inventing Accuracy: A Historical Sociology of Nuclear Missile Guidance* (Cambridge: MIT Press, 1990) provides an informative discussion of the political and other factors that shaped the accuracy of submarine missiles.

The search for new "direct" or self-calibrating methods lead in a number of directions, such as DRVID, dual-frequency calibration, ΔVLBI (with Quasi VLBI along the way), and Delta DOR, the latter being the most successful in delivering higher precision. Because it altered the "experimental design" of spacecraft observations so radically, one must include optical navigation among the successful "direct" methods. It was inherently immune to error sources from media effects and station locations, but not resistant to the bugs native to spacecraft television and CCD cameras nor impervious to poor spacecraft design. Autonomous navigation took this "direct" method a step further by performing basic navigational functions on spacecraft. Adding the ability to interact with the guidance and control system was perhaps the ultimate in redesigning observations to reduce errors.

As a result of optical navigation and its evolution, the meaning of what is deep-space navigation at JPL has changed. It started as an activity carried out entirely on the ground. Radio signals constituted the only information processed. Optical navigation began to move some of the process—and equipment—to spacecraft. At first, image processing was a labor-intensive and computer-intensive effort. Computerizing some of the labor facilitated the processing of pictures and the locating of centers (the gist of one technique). Further software improvements and advances in spacecraft computers paved the way for the transfer of many basic navigation steps to the onboard computer. An executive program, conceptually not unlike the executive of the early mainframe orbit determination program, oversaw the planning and execution of picture-taking, image analysis, and orbit computations. Onboard navigation was becoming self-directed and capable of guiding and controlling spacecraft.

Finally, this survey of deep-space navigation seems to have something to say about the role of technology in this application-oriented multidisciplinary scientific endeavor. Clearly, it is dependent on the DSN as its chief instrument for making observations, but the VLA and VLBA, as well as the Hubble Space Telescope and several ground-based astronomical facilities, also contributed their share of observations. The other equally obvious technology is the computer. Computers both limited and enabled what navigators could accomplish.

Prior to around 1980, the mainframe ruled. The computational limitations of those mammoth machines also imposed themselves on navigators. Computer time was expensive, and the actuarial question of "paying" for computer time was entangled in policy decisions made by NASA Headquarters. The early mainframes could handle only a small number of parameters. It was not possible to update station locations or other parameters during flights, for example. Precisely because of this lack of computational power, though, navigators turned their software into a form of ersatz laboratory for studying geodesy (station locations) and celestial mechanics by solving for station locations and physical constants after the mission had flown. They performed this post-flight analysis by "replaying" the flight with their software, but this time taking the craft's trajectory as the "given" and solving for these parameters. The advent of double-precision computing was not a solution, although it did help significantly to increase navigational accuracy. It was only with the revolution in computing—minicomputers, work stations—that navigators began to enjoy the capability of updating station locations and other parameters during flights and realized greater accuracies as a result. The revolution in computing also enabled the development of new generations of programs to accomplish critical navigational tasks, such as ARDVARC (Mars Global Surveyor) and the "blobber" (Deep Space 1), as well as more general-purpose orbit-determination programs, such as MIRAGE and MONTE.

Shown here is the original Goldstone antenna built for the ARPA Pioneer launches as it appeared in early JPL technical reports (JPL Archives, 333-317)

Oral Histories

Oral histories from multiple sources went into the writing of this book. Interviews appear in the notes in an abridged form: interviewee's surname, interview, page number(s). In cases of multiple interviews of the same individual, the year or date of the interview distinguishes them.

The first groups of oral histories are those that I carried out for this work, the Deep Space Navigation History Project, and the Planetary Radar Astronomy History Project. They are housed in the Historical Reference Collection at NASA Headquarters and include notes of telephone interviews. The second grouping, consisting of oral histories conducted by others, are organized by provenance.

Distinguishing among the multiple interviews of Thomas W. Hamilton is a bit problematic. In two instances, the interview date is not known; these are cited in the form Hamilton, interview, Waff or Hamilton, interview, Alonso. The latter refers to the interview by José Alonso at JPL. The former denotes a transcription of a Hamilton interview taped by Craig Waff and conducted by an unknown interviewee on an unknown date. The transcript is in F242, DSN Files, NHRC. Hamilton, interview, 1992, indicates the interview by Alonso at JPL on February 19, 1992. The interview of Hamilton conducted for this work is cited as Hamilton, interview, 2010, without page numbers, because it was not transcribed for a lack of funds.

Deep Space Navigation History Project

Interviewee	Interview Location	Interview Date
Shyam Bhaskaran	JPL	November 14, 2008
James S. Border	JPL	May 14, 2010
Laureano A. Cangahuala	JPL	April 20, 2007
David W. Curkendall	Altadena, CA	March 20, 2009
Saterios S. Dallas	Atheneum, Caltech	March 18, 2009
Leonard Efron	JPL	June 1, 2009
Bernard J. Geldzahler	NASA Headquarters	July 15, 2009
Thomas W. Hamilton	JPL	May 11, 2010
J. Frank Jordan	JPL	April 19, 2007
Earl Maize	JPL	April 23, 2007
Tomás Martin-Mur	JPL	November 7, 2008
Neil A. Mottinger	JPL	March 9, 2009
Neil A. Mottinger	JPL	March 10, 2009
William L. Sjogren	JPL	November 4, 2008
E. Myles Standish	JPL	April 16, 2007
Stephen P. Synnott	JPL	March 12, 2009
Catherine L. Thornton	Eagle Rock, CA	April 18, 2007
Lincoln J. Wood	JPL	November 13, 2008
Donald K. Yeomans	JPL	March 17, 2009

Planetary Radar Astronomy History Project.

Interviewee	Interview Location	Interview Date
Von R. Eshleman	Stanford University	May 9, 1994
Alan E. E. Rogers	Haystack Observatory	May 5, 1994
Irwin I. Shapiro	Harvard-Smithsonian Center for Astrophysics	September 30, 1993
Irwin I. Shapiro	Harvard-Smithsonian Center for Astrophysics	October 1, 1993

Additional Oral Histories Conducted for NASA History Projects.

Interviewee	Interview Location	Interview Date
Michael D. Griffin	Orbital Sciences Corporation	August 18, 1997
Edward C. Stone	JPL	November 23, 1994

Oral History Collection, JPL Archives.
Interviews by José Alonso

Interviewee	Interview Location	Interview Date
Raymond P. Amorose	JPL	July 31, 1992
Thomas W. Hamilton	JPL	February 19, 1992
Thomas W. Hamilton	JPL	unknown date
Arvydas Kliore	JPL	July 13, 1992
William Melbourne	JPL	February 18, 1992

Interviews by Russell Castonguay

Interviewee	Interview Location	Interview Date
William L. Sjogren	JPL	September 21, 2000

Center for History of Physics, College Park, MD.

Paul Herget, interview by David DeVorkin, Cincinnati Observatory, April 19, 1977, http://www.aip.org/history/ohilist/4664_1.html, and April 20, 1977, http://www.aip.org/history/ohilist/4664_2.html (accessed January 11, 2008)

Kaj Strand, by David DeVorkin and Steven Dick, Washington, DC, December 8, 1983, http://www.aip.org/history/ohilist/23026_1.html, and January 3, 1984, http://www.aip.org/history/ohilist/23026_2.html (accessed January 11, 2008)

Computer Oral History Collection, Washington, DC.

Howard Aiken, by Henry Tropp and I. B. Cohen, unknown location, February 26-27, 1973, Computer Oral History Collection, Archives Center, National Museum of American History, Washington, DC, http://invention.smithsonian.org/downloads/fa_cohc_tr_aike73027.pdf (accessed January 18, 2008)

James Melville Gilliss Library, U.S. Naval Observatory.
    Steven J. Dick, *Oral History Interview with Douglas A. O'Handley, E. Myles Standish, and Henry F. Fliegel: December 4, 1999* (Washington: U. S. Naval Observatory, 1999)
    Steven J. Dick, *Oral History Interview with P. Kenneth Seidelmann: July 20, 2000* (Washington: U.S. Naval Observatory, 2000)
    LeRoy E. Doggett and Steven J. Dick, *Oral History Interview with Raynor L. Duncombe and Julena S. Duncombe: on June 18, 1983 and on Jan. 11, 1988* (Washington: U.S. Naval Observatory, 1988)

Oral History Project, Caltech Archives, Pasadena, CA.
    William H. Pickering, by Mary Terrall, Pasadena, CA, November 7–December 19, 1978, transcript, Oral History Project, California Institute of Technology Archives, Pasadena, CA, http://oralhistories.library.caltech.edu/87/1/OH_Pickering_1.pdf (accessed April 4, 2006)

List of Abbreviations

Abbreviated forms are used throughout this work. Acronyms usually are spelled out the first time they appear in the work.

The following sections indicate shortened forms for archival collections. In general, cited documents use such typical abbreviations as RG for Record Group, F for File or Folder, and B for Box, when possible. An exception is the Herget Papers. Full folder names are those assigned by Herget. The letter "D" indicates the filing cabinet drawer where it can be found. The "Vanguard Box" refers to a cardboard box of documents in Drawer 1.

Researching this work required extensive use of JPL records and serials collections. Shortened forms for JPL archival collections are followed by a section indicating the various truncated names for JPL serial publications.

The largest group of abbreviated forms is the final section. Shortened forms of the works cited in the footnotes are here in alphabetical order followed by the full citation form. Any source mentioned twice or more is represented in this list.

Archival collections.

Archives II	National Archives and Records Administration, College Park, MD
Bellcomm	Bellcomm, Inc. Technical Library Collection, Acc. No. XXXX-0093, SINASMAD
Bonney Files	Bonney Files, Record Group 255, Archives II
DSN Files	Deep Space Navigation History Project Files, NHRC
Eckert Papers	Wallace J. Eckert Papers (CBI 9), Charles Babbage Institute, University of Minnesota
Gibson Reaves	Gibson Reaves, "Astronomy Notebooks, 1943-1949," AR 2002-535, Niels Bohr Library and Archives, American Institute of Physics Center for History of Physics, College Park, MD
Goddard	Dr. Robert H. Goddard Collection, Clark University Archives, Worcester, MA
Herget Papers	Paul Herget Papers, USNOGL
Herrick Papers	Samuel Herrick Papers, Archives of American Aerospace Exploration, Ms78-002, Special Collections Department, University Libraries, Virginia Polytechnic Institute and State University, Blacksburg, VA
Hoffleit Papers	Dorritt Hoffleit Papers, 1906-2005, Arthur and Elizabeth Schlesinger Library on the History of Women in America, Radcliffe Institute for Advanced Study, Harvard University Library, Cambridge, MA
JPLPubs	Jet Propulsion Laboratory Publications Collection, Acc. No. XXXX-0612, SINASMAD
LaRCA	Archives, NASA Langley Research Center, Hampton, VA
NHRC	NASA Historical Reference Collection, NASA Headquarters, Washington, DC
SINASMAD	Smithsonian Institution, National Air and Space Museum, Archives Division
Siry Papers	Notes and Other Records of the Head of the Theory and Analysis Branch, Record Group 255, Archives II
USNOGL	James Melville Gilliss Library, U. S. Naval Observatory, Washington, DC
Vanguard	Records Relating to Project Vanguard, Record Group 255.6, Archives II
Willy Ley	Willy Ley Collection, Accession No. XXXX-0098, SINASMAD
X-33 Files	X-33 History Project Files, NHRC

JPL Archives.

Reference	Collection name
JPL4	Magellan Project Records, 1978-1989
JPL105	Charles E. Kohlhase Collection, 1966-1990
JPL110	Arroyo Center Report Collection
JPL127	Daniel J. Alderson Collection
JPL129	Mariner 7 Pre-Encounter Anomaly Collection
JPL142	Office of the Director Collection
JPL155	Radio Science Collection, 1961-1980
JPL169	Galileo Orbiter Functional Requirements Book, 1979-1983
JPL173	Harris M. Schurmeier Collection, 1970-1986
JPL235	Clarence R. Gates Collection, 1980-1995
JPL267	Mars Pathfinder Flight Systems Management Documents Collection
JPL269	Larry N. Dumas Chronological and Travel Files Collection
JPL273	Mars Climate Orbiter Mars Surveyor 1998 Project Collection
JPL420	Telecommunications and Data Acquisition Science Office
JPL421	History Office Collection, Ranger/Mariner/Surveyor
JPL496	Mars Observer Project Manager Records, 1982-1994
JPL508	Mars Pathfinder Collection
JPL516	Tracking and Data Acquisition Office Collection
JPL520	Mars Observer Project Records, 1982-1994
JPL556	Organization and Personnel Charts, 1941-2008
JPLHC	History Collection
JPLDSN	Deep Space Network Collection, 1956-1996

JPL Serial Publications.

In general, when refering to JPL and other technical publications, citations use such standard abbreviations as TR for Technical Report, TM for Technical Memorandum, TN for Technical Note, EPD for Engineering Planning Document, and P or Pub for Publication.

JPLCBS
   Combined Bimonthly Summaries (1947-1954) are numbered consecutively from 1 to 44.

JPLSPS
   Space Programs Summaries (1959-1970) appear in two series. The first is labeled consecutively from 1 to X. The second (1961-1970) precedes the identifying numbers with "37." Each "37" issue consists of two or more volumes, so the convention JPLSPS 37-10:II indicates *Space Prorams Summary No. 37-10*, Volume II.

DSNPR
   DSN Progress Reports have appeared in two series. The first (1971-1974) carry the prefix TR 32-1526 and are distinguished by volume numbers in roman numerals (I through XIX). Thus, DSNPR:X indicates *DSN Progress Report TR 32-1526*, Volume X. *DSN Progress Reports* in the second series (1974-present) carry a unique issue number preceded by "42." DSNPR 42-20, therefore, is *DSN Progress Report 42-20*. However, only those numbered to 42-56 (April 1980) are titled *DSN Progress Reports*.

TDAPR
   The DSN progress reports continue as *The Telecommunications and Data Acquisition Progress Report* (June 1980 to February 1998) numbered consecutively from 42-57 to 42-132.

TMOPR
   The series resumes with *The Telecommunications and Mission Operations Progress Reports* (May 1998 to May 2001) numbered from 42-133 to 42-145.

IPNPR
The *Interplanetary Network Progress Reports* (August 2001 to present) numbered from 42-146 forward is the latest iteration of the DSN's progress reports. Unlike preceding runs, articles are posted electronically in PDF, and each begins on page 1.

Short Forms for Cited Works.

"1989-1995: Galileo Navigation"
Robert J. Haw, Peter G. Antreasian, Eric J. Graat, Timothy P. McElrath, and Francis T. Nicholson, "1989-1995: A Galileo Navigation Space Odyssey Orbit Determination," http://trs-new.jpl.nasa.gov/dspace/bitstream/2014/21902/1/97-0316.pdf (accessed July 14, 2010)

"2001 Mars Odyssey OD"
Peter G. Antreasian, Darren T. Baird, James S. Border, P. Daniel Burkhart, Eric J. Graat, Moriba K. Jah, Robert A. Mase, Timothy P. McElrath, and Brian M. Portock, "2001 Mars Odyssey Orbit Determination during Interplanetary Cruise," *Journal of Spacecraft and Rockets* 42 (May 2005): 394-405

Acton, "Processing Onboard Optical Data"
Charles H. Acton, Jr., "Processing Onboard Optical Data for Planetary Approach Navigation," *Journal of Spacecraft and Rockets*, 9 (1972): 746-750

Acton and Ohtakay
Charles H. Acton, Jr., and Hiroshi Ohtakay, "Mariner-Venus-Mercury Optical Navigation Demonstration: Results and Implications for Future Missions," AAS Paper 75-088, AAS/AIAA Astrodynamics Specialist Conference, July 28-30, 1975, Nassau, Bahamas

Ad Hoc NASA Standard Constants
Victor C. Clarke, Jr., *Constants and Related Data for Use in Trajectory Calculations as Adopted by the Ad Hoc NASA Standard Constants Committee*, TR 32-604 (Pasadena: JPL, March 6, 1964)

Aller, Barnes, and Abell
Lawrence H. Aller, John L. Barnes, and George O. Abell, "Samuel Herrick, Engineering; Astronomy: Los Angeles, 1911-1974," University of California, *In Memoriam* (March 1976): 58-59

"Analysis for VLBI"
Oldwig H. Von Roos, "Analysis of Dual-Frequency Calibration for Spacecraft VLBI," 46-56 in DSNPR:VI

Anderson
John D. Anderson, *Determination of the Masses of the Moon and Venus and the Astronomical Unit from Radio Tracking Data of the Mariner II Spacecraft*, TR 32-816 (Pasadena: JPL, July 1, 1967)

Anderson and Lau
John D. Anderson and Eunice K. Lau, "Determination of the Relativistic Time Delay for Mariner 9: A Status Report on the JPL Analysis of Normal Points," 431-451 in *MM71 Final Report*, Vol. IV

"Application of Noncoherent Doppler"
Sumita Nandi, "Application of Noncoherent Doppler and Range Data for Mars Approach Navigation," in IPNPR 42-177

Asmar and Renzetti
Sami W. Asmar and Nicholas A. Renzetti, *The Deep Space Network as an Instrument for Radio Science Research*, Pub 80-93, Revision 1 (Pasadena: JPL, April 15, 1993)

Astronomy at Yale
E. Dorritt Hoffleit, *Astronomy at Yale, 1701-1968* (New Haven: Yale University, 1992)

"AutoNav Mark3"
Joseph E. Riedel, Shyam Bhaskaran, Dan B. Eldred, Robert A. Gaskell, Christopher A. Grasso, Brian Kennedy, Daniel Kubitschek, Nickolaos Mastrodemos, Stephen P. Synnott, Andrew Vaughan, and Robert A. Werner, "AutoNav Mark3: Engineering the Next Generation of Autonomous Onboard Navigation and Guidance," 4835-4852 in *A Collection of Technical Papers AIAA Guidance, Navigation and Control Conference, Keystone, CO, August 22-25, 2006* (Reston, VA: AIAA, 2006)

"AutoNav Tracking: Stardust"
Shyam Bhaskaran, Joseph E. Riedel and Stephen P. Synnott, "Autonomous Nucleus Tracking for Comet/Asteroid Encounters: The Stardust Example," 451-468 in Felix R. Hoots, Bernard Kaufman, Paul J. Cefola, and David B. Spencer, ed., *Astrodynamics 1997*, Pt. I (San Diego: Univelt, 1997)

"Auto Tracking Small Bodies during Flybys"
Shyam Bhaskaran, Joseph E. Riedel and Stephen P. Synnott, "Autonomous Target Tracking of Small Bodies During Flybys," 2079-2096, in Shannon L. Coffey, Andre P. Mazzoleni, K. Kim Luu, Robert A. Glover, ed., *Spaceflight Mechanics 2004*, Pt. II (San Diego: Univelt, 2004)

Baugh
Harold W. Baugh, "DSN Time-Synchronization Subsystem Performance," 97-99 in JPLSPS 37-53:II

Benjauthrit
Boonsieng Benjauthrit, "A Brief Historical Introduction to Very Long Baseline Interferometry," 146-153 in DSNPR 42-46

Brancheau, "Chronology"
Burk Brancheau, "Chronology of Systems Division," [date unknown, attributed to late 1980s], copy furnished by Lincoln J. Wood

Brandt and Chapman
John C. Brandt and Robert DeWitt Chapman, *Introduction to Comets*, 2d ed. (New York: Cambridge University Press, 2004)

Breckenridge and Acton, "Detailed Analysis"
William G. Breckenridge and Charles H. Acton, Jr., "A Detailed Analysis of Mariner Nine TV Navigation Data," Appendix F, *MM71 OpNav Demonstration*

Brennan
Jean Ford Brennan, *The IBM Watson Laboratory at Columbia University: A History* (New York: IBM, 1970)

"Brief History of JPL OpNav"
William M. Owen, Jr., Thomas C. Duxbury, Charles H. Acton, Jr., Stephen P. Synnott, Joseph E. Riedel, and Shyam Bhaskaran, "A Brief History of Optical Navigation at JPL," 329-348 in Michael E. Drews and Robert D. Culp, ed., *Guidance and Control 2008* (San Diego: Univelt, 2008)

Burnell, Phillips, and Zanteson
   Henry E. Burnell, Horace P. Phillips, and Richard Zanteson, "Meteorological Monitoring Assembly," 152-159 in DSNPR 42-29

Burrows
   William E. Burrows, *This New Ocean: The Story of the First Space Age* (New York: Random House, 1998)

Butrica, *Beyond Ionosphere*
   Andrew J. Butrica, ed., *Beyond The Ionosphere: Fifty Years of Satellite Communication*, SP-4217 (Washington: NASA, 1997)

Butrica, *See the Unseen*
   Andrew J. Butrica, *To See the Unseen: A History of Planetary Radar Astronomy*, SP-4218 (Washington: NASA, 1996)

Butrica, *SSTO*
   Andrew J. Butrica, *Single Stage to Orbit: Politics, Space Technology, and the Quest for Reusable Rocketry* (Baltimore: Johns Hopkins University Press, 2003)

Butrica, "Overview of Acquisition"
   Andrew J. Butrica, "An Overview of Acquisition, 1981-1990," 199-223 in Shannon A. Brown, ed., *Providing the Means of War: Historical Perspectives on Defense Acquisition, 1945 to 2000* (Washington: U. S. Army Center of Military History and Industrial College of the Armed Forces, 2005)

Butrica, "Voyager"
   Andrew J. Butrica, "Voyager: The Grand Tour of Big Science," 251-276 in Pamela E. Mack, ed., *From Engineering Science to Big Science: The NACA and NASA Collier Trophy Research Project Winners* (Washington: NASA, 1998)

Byers
   Bruce K. Byers, *Destination Moon: A History of the Lunar Orbiter Program*, NASA TMX-3487 (Washington: NASA, 1977)

Carr and Hudson
   Russell E. Carr and R. Henry Hudson, *Tracking and Orbit Determination Program of the Jet Propulsion Laboratory*, TR 32-7 (Pasadena: JPL, February 22, 1960)

"Cassini OD Satellite Tour First Eight Orbits"
   Peter G. Antreasian, John J. Bordi, Kevin E. Criddle, Rodica Ionasescu, Robert A. Jacobson, Jeremy B. Jones, Richard A. MacKenzie, M. Cameron Meek, Frederic J. Pelletier, Duane C. Roth, Ian M. Roundhill, and Jason R. Stauch, "Cassini Orbit Determination Performance during the First Eight Orbits of the Saturn Satellite Tour," 933-962 in Bobby G. Williams, Louis A. D'Amario, Kathleen C. Howell, and Felix R. Hoots, ed., *Astrodynamics 2005*, Pt. I (San Diego: Univelt, 2006)

"Cassini OD Saturn Satellite Tour"
   Peter G. Antreasian, John J. Bordi, Kevin E. Criddle, Rodica Ionasescu, Robert A. Jacobson, Jeremy B. Jones, Richard A. Mackenzie, Daniel W. Parcher, Frederic J. Pelletier, Duane C. Roth, and Jason Stauch, "Cassini Orbit Determination Performance during Saturn Satellite Tour from August 2005 through January 2006," 21-40 in Ronald J. Proulx, Thomas F. Starchville, Jr., R. D. Burns, and Daniel J. Scheeres, ed., *Astrodynamics 2007*, Pt. I (San Diego: Univelt, 2007)

"Cassini Tour Navigation Strategy"
    Duane Roth, Vijay Alwar, John Bordi, Troy Goodson, Yungsun Hahn, Rodica Ionasescu, Jeremy Jones, William Owen, Joan Pojman, Ian Roundhill, Shawna Santos, Nathan Strange, Sean Wagner, and Mau Wong, "Cassini Tour Navigation Strategy," in Institute of Navigation, *Proceedings of the Satellite Division's 16th International Technical Meeting, Portland, OR, September 2003* (Washington: Institute of Navigation, 2003)

Ceguerra
    Anna Ceguerra, "Software Bug Report: Mars Climate Orbiter, Assignment 1 for Verification," March 13, 2001, http://www.cs.usyd.edu.au/~anna/Mars.htm (accessed January 18, 2011)

"Celestial Mechanics"
    John D. Anderson, Leonard Efron, and Sun Kuen Wong, "Celestial Mechanics," 127-134 in *Mariner-Mars 1969: A Preliminary Report*, SP-225 (Washington: NASA, 1969)

"Celestial Mechanics Experiment"
    Jack Lorell, John D. Anderson, J. Frank Jordan, Robert D. Reasenberg, and Irwin I. Shapiro, "Celestial Mechanics Experiment," 13-31 in *MM71 Final Report*, Vol. V

"Celestial VLBI Sources"
    Robert A. Preston, David D. Morabito, James G. Williams, Martin A. Slade, Alan W. Harris, Susan G. Finley, Lyle J. Skjerve, Leroy Tanida, Donovan J. Spitzmesser, Ben Johnson, David L. Jauncey, A. Bailey, R. Denise, J. Dickenson, R. Livermore, Alex Papij, Alan Robinson, C. Taylor, Francisco Alcazar, Benito Luaces, and David Munoz, "Establishing a Celestial VLBI Reference Frame: I. Searching for VLBI Sources," 46-56 in DSNPR 42-46

Chao and Muller
    Chia-Chun Chao and Paul Muller, "A Study of Polar Motion and Earthquakes for the Period 1904-1966," 69-74 in JPLSPS 37-56:II

Chao and Ondrasik
    Chia-Chun Chao and V. John Ondrasik, "The QVLBI Doppler Demonstration Conducted With Mariner 10," 52-63 in DSNPR 42-27

Chapront-Touzé
    Michelle Chapront-Touzé, "Progress in the Analytical Theories for the Orbital Motion of the Moon," *Celestial Mechanics* 26 (January 1982): 53-62

Clarke, *Interplanetary Flight*
    Arthur C. Clarke, *Interplanetary Flight: An Introduction to Astronautics* (New York: Harper & Brothers, 1951)

Clemence, "On the System"
    Gerald M. Clemence, "On the System of Astronomical Constants," *The Astronomical Journal* 53 (May 1948): 169-179

Clemence, "System of Constants"
    Gerald M. Clemence, "The System of Astronomical Constants," *Annual Review of Astronomy and Astrophysics* 3 (1965): 93-112

Constants and Related Information
    William G. Melbourne, J. Derral Mulholland, Sjogren, and Francis M. Sturms, Jr., *Constants and Related Information for Astrodynamic Calculations, 1968*, TR-32-1306 (Pasadena: JPL, July 15, 1968)

Constants from Tracking Lunar Probes
William L. Sjogren, Donald W. Trask, Chalres J. Vegos, and William R. Wollenhaupt, *Physical Constants as Determined From Radio Tracking of the Ranger Lunar Probes*, TR 32-1057 (Pasadena: JPL, December 30, 1966)

Corliss, *Histories of Space Tracking*
William R. Corliss, *Histories of the Space Tracking and Data Acquisition Network (STADAN), The Manned Space Flight Network (MSFN), and the NASA Communications Network (NASCOM)* (Washington: NASA, June 1974)

Corliss, *History of the DSN*
William R. Corliss, *A History of the Deep Space Network* (Washington: NASA, 1976)

Corliss, *Space Probes*
William R. Corliss, *Space Probes and Planetary Exploration* (New York: D. Van Nostrand Company, 1965)

Curkendall, "Precision Navigation Project: Introduction"
David W. Curkendall, "Precision Navigation Project: Introduction," 2-1 in *The Mariner VI and VII Flight Paths and Their Determination from Tracking Data*, TR 33-469 (Pasadena: JPL, December 1, 1970)

D'Amario
Louis A. D'Amario, "Navigation of the Galileo Spacecraft," 115-143 in Cesare Barbieri, Jurgen H. Rahe, Torrence V. Johnson, and Anita M. Sohus, ed., *The Three Galileos: The Man, The Spacecraft, The Telescope* (New York: Springer-Verlag, 1997)

Davies
Richard W. Davies, ed., *Proceedings of the Conference on Experimental Tests of Gravitation Theories*, TM 33-499 (Pasadena: JPL, November 1, 1971)

DE19
Charles J. Devine, *JPL Development Ephemeris Number 19*, TR 32-1181 (Pasadena: JPL, November 15, 1967)

DE69
Douglas A. O'Handley, Douglas B. Holdridge, William G. Melbourne, and J. Derral Mulholland, *JPL Development Ephemeris Number 69*, TR 32-1465 (Pasadena: JPL, December 15, 1969)

DE96
E. Myles Standish, Michael S. W. Keesey, and X X Newhall, *JPL Development Ephemeris Number 96*, TR 32-1603 (Pasadena: JPL, February 29, 1976)

"DE102"
X X Newhall, E. Myles Standish, Jr., and James G. Williams, "DE 102: A Numerically Integrated Ephemeris of the Moon and Planets Spanning Forty-Four Centuries," *Astronomy and Astrophysics* 125 (1983): 150-167

"Deep Impact AutoNav"
Daniel G. Kubitschek, Nickolaos Mastrodemos, Robert Werner, Brian Kennedy, Stephen P. Synnott, George Null, Shyam Bhaskaran, Joseph E. Riedel, and Andrew T. Vaughan, "Deep Impact Autonomous Navigation: The Trials of Targeting the Unknown," 381-406 in Steven D. Jolly and Robert D. Culp, ed., *Guidance and Control 2006* (San Diego: Univelt, 2006)

"Deep Impact Nav Performance"
Raymond B. Frauenholz, Ramachandra S. Bhat, Steven R. Chesley, Nickolaos Mastrodemos, William M. Owen, Jr., and Mark S. Ryne, "Deep Impact Navigation System Performance," *Journal of Spacecraft and Rockets* 45 (January-February 2008): 39-56

"Delta-DOR"
Roberto Maddè, Trevor Morley, Ricard Abelló, Marco Lanucara, Mattia Mercolino, Gunther Sessler, and Javier de Vicente, "Delta-DOR: A New Technique for ESA's Deep Space Navigation," *ESA Bulletin* 128 (November 2006): 69-74

"ΔVLBI Demo Part I"
D. Lee Brunn, Robert A. Preston, Sien C. Wu, Herbert L. Siegel, and David S. Brown, "ΔVLBI Spacecraft Tracking System Demonstration Part I: Design and Planning," 111-132 in DSNPR 42-45

"ΔVLBI Demo Part II"
Carl S. Christensen, Benjamin Moultrie, Philip S. Callahan, Frank F. Donivan, and Sien C. Wu, "ΔVLBI Spacecraft Tracking System Demonstration Part II: Data Acquisition and Processing," 60-67 in TDAPR 42-60

"Delta VLBI Observations"
Frank F. Donivan and X X Newhall, "Delta VLBI Observations of Mars Viking Lander I," *Bulletin of the American Astronomical Society* 13 (March 1981): 555

"Determination Separation with VLBA"
Gabor Lanyi, James Border, John Benson, Vivek Dhawan, Edward Fomalont, Tomás Martin-Mur, Timothy McElrath, Jonathan Romney, and Craig Walker, "Determination of Angular Separation Between Spacecraft and Quasars with the Very Long Baseline Array," in IPNPR 42-162

DeVorkin
David H. DeVorkin, *Science with a Vengeance: How the Military Created the U.S. Space Sciences after World War II* (New York: Springer-Verlag, 1992).

Dick, *Sky and Ocean*
Steven J. Dick, *Sky and Ocean Joined: The U.S. Naval Observatory 1830-2000* (New York: Cambridge University Press, 2003)

Director's Letter
Director's Letter, "Reorganization and Consolidation," no. 12, July 30, 1976, F736/B44, JPL142

Doel
Ronald E. Doel, *Solar System Astronomy in America: Communities, Patronage, and Interdisciplinary Science, 1920-1960* (New York: Cambridge University Press, 1996)

Domingue and Russell
Deborah L. Domingue and Christopher T. Russell, ed., *The MESSENGER Mission to Mercury* (New York: Springer, 2007)

"Doppler and Interferometric"
Eric J. Graat, Mark S. Ryne, James S. Border, and Douglas B. Engelhardt, "Contribution of Doppler and Interferometric Tracking during the Magellan Approach to Venus," 919-939 in Bernard Kaufman, Kyle T. Alfriend, Ronald L. Roehrich, and Robert R. Dasenbrock, ed., *Astrodynamics 1991*, Pt. II (San Diego: Univelt, 1992)

DPODP
Michael R. Warner, *Double Precision Orbit Determination Program*, Vol. 1, EPD 426 (Pasadena: JPL, November 1, 1966)

Drechsel
Horst Drechsel, "The International Halley Watch: An Example for International Cooperation in Astronomy," 25-35 in Carlos Jaschek and Christiaan Sterken, ed., *Coordination of Observational Projects in Astronomy* (New York: Cambridge University Press, 1988)

"DS-1 AutoNav"
Shyam Bhaskaran, Joseph E. Riedel, Stephen P. Synnott, and Tseng-Chang Wang, "The Deep Space 1 Autonomous Navigation System: A Post-Flight Analysis," AIAA 2000-3935, AIAA/AAS Astrodynamics Specialist Conference, August 14-17, 2000, Denver, CO

"DS-1 Nav at Borrelly"
Shyam Bhaskaran, Joseph E. Riedel, Brian M. Kennedy, and Tseng-Chang Wang, "Navigation of the Deep Space 1 Spacecraft at Borrelly," AIAA 2002-4815, AAS/AIAA Astrodynamics Specialist Conference, August 5-8, 2002, Monterey, CA

DS-1 Nav: Extended Missions
Brian Kennedy, Shyam Bhaskaran, J. Edmund Riedel, and Tseng-Chang Wang, *Deep Space 1 Navigation: Extended Missions* (Pasadena: JPL, September 2003)

DS-1 Nav: Primary Mission
Brian Kennedy, J. Edmund Riedel, Shyam Bhaskaran, Shailen Desai, Don Han, Tim McElrath, George Null, Mark Ryne, Steve Synnott, Tseng-Chang Wang, and Robert Werner, *Deep Space 1 Navigation: Primary Mission* (Pasadena: JPL, April 2004)

DSN: Instrument for Research
Nicholas A. Renzetti, Gerald S. Levy, Thomas B. H. Kuiper, Pamela R. Wolken, and R. C. Chandlee, *The Deep Space Network: An Instrument for Radio Astronomy Research*, Publication 82-68, Revision 1 (Pasadena: JPL, September 1, 1988)

"DSN Mark IVA Description"
Robert J. Wallace and Roger W. Burt, "Deep Space Network Mark IVA Description," 255-260 in TDAPR 42-86

Dumas to Elachi
Memorandum, Larry N. Dumas to Charles Elachi, "Formation of a Special Review Board for the Mars Climate Orbiter Loss on September 23, 1999," September 24, 1999, F217/B17, JPL269

Dunne and Burgess
James A. Dunne and Eric Burgess, *The Voyage of Mariner 10: Mission to Venus and Mercury* (Washington: NASA, 1978)

Durham and Purrington
Frank Durham and Robert D. Purrington, *Frame of the Universe: A History of Physical Cosmology* (New York: Columbia University Press, 1983)

Duxbury, "Data from MM69"
Thomas C. Duxbury, "Navigation Data from Mariner Mars 1969 TV Pictures," *Navigation* 17 (1970): 219-225

Duxbury and Acton
Thomas C. Duxbury and Charles H. Acton, Jr., "On-Board Optical Navigation Data from Mariner '71," *Journal of the Institute of Navigation* 19 (Winter 1972-1973): 295-307

"Dynamical Astronomy"
Raynor L. Duncombe, P. Kenneth Seidelmann, and W. J. Klepczynski, "Dynamical Astronomy of the Solar System," *Annual Review of Astronomy and Astrophysics* 11 (1973): 135-154

"Early Nav Results"
Bobby Williams, Anthony Taylor, Eric Carranza, James Miller, Dale Stanbridge, Brian Page, D. Cotter, Len Efron, Robert Farquhar, James McAdams, and David Dunham, "Early Navigation Results for NASA's MESSENGER Mission to Mercury," in David A. Vallado, Michael J. Gabor, and Prasun N. Desai, ed., *Space Flight Mechanics 2005*, Pt. II (San Diego: Univelt, 2005)

Eckert, "Improvement"
Wallace J. Eckert, "Improvement by Numerical Methods of Brown's Expressions for the Coordinates of the Moon," *The Astronomical Journal* 63 (November 1958): 415-418

Eckert, Jones, and Clark
Wallace J. Eckert, Rebecca Jones, and H. Kenneth Clark, *Improved Lunar Ephemeris, 1952-59: A Joint Supplement to the American Ephemeris and the [British] Nautical Almanac* (Washington: GPO, 1954)

Eckert, Walker, and D. Eckert
Wallace J. Eckert, M. J. Walker, and Dorothy Eckert, "Transformations of the Lunar Coordinates and Orbital Parameters," *The Astronomical Journal* 71 (June 1966): 314-332

Ekelund, "History of ODP at JPL"
Manuscript, John E. Ekelund, "History of the ODP at JPL," date unknown, electronic copy provided to author, F234, DSN Files

Ekelund, "JPL OD Software"
John E. Ekelund, "The JPL Orbit Determination Software System," 79-88 in Paul A. Penzo, Bernard Kaufman, Luis Friedman, and Richard Battin, ed., *Astrodynamics 1979*, Pt. I (San Diego: Univelt, 1979)

Ellis
Jordan Ellis, "HEO Multimission Navigation Concept," 261-267 in TDAPR 42-86

Ellis to Stelzried
Memorandum, Jordan Ellis to Charles T. Stelzried, "Navigation of Planetary Orbiters Using Differenced 2Way-3Way Doppler," March 9, 1988, F508/B30, JPL508

Emerson and Wilkins
Brian Emerson and George A. Wilkins, ed., "The IAU System of Astronomical Constants," *Celestial Mechanics* 4 (1971): 128-149

Euler, "Viking Mission Overview"
Edward A. Euler, "Viking Mission Overview: Lessons Learned and Challenges for the Future," in E. Brian Pritchard, *Mars: Past, Present, and Future* (Washington: AIAA, 1992)

Ezell and Ezell, *On Mars*
Edward C. Ezell and Linda N. Ezell, *On Mars: Exploration of the Red Planet 1958-1978*, SP-4212 (Washington: NASA, 1984)

Ezell and Ezell, *The Partnership*
　　Edward C. Ezell and Linda N. Ezell, *The Partnership: A History of the Apollo-Soyuz Test Project*, SP-4209 (Washington: NASA, 1978)

Farquhar
　　Robert W. Farquhar, *Fifty Years on the Space Frontier: Halo Orbits, Comets, Asteroids, and More* (Denver: Outskirts Press, 2011)

Fifth PTTI Meeting
　　Proceedings of the Fifth Annual NASA and Department of Defense Precise Time and Time Interval (PTTI) Planning Meeting (Greenbelt: GSFC, 1974)

"First OpNav Images"
　　Spaceref, "MESSENGER Team Receives First Optical Navigation Images of Mercury," January 9, 2008, http://www.spaceref.com/news/viewpr.html?pid=24457 (accessed December 29, 2010)

Fliegel
　　Henry F. Fliegel, "A Worldwide Organization to Secure Earth-Related Parameters for Deep Space Missions," 66-73 in DSNPR:V

Fliegel and Wimberly
　　Henry F. Fliegel and Ravenel N. Wimberly, "Time and Polar Motion," 77-81 in *TSAC for MM71*

Fowler
　　Wallace T. Fowler, "Department of Aerospace Engineering and Engineering Mechanics, The University of Texas at Austin: A Brief History," 87-98 in Barnes McCormick, Conrad Newberry, and Eric Jumper, ed., *Aerospace Engineering Education during the First Century of Flight* (Reston, VA: AIAA, 2004)

"Galileo OD Earth-2"
　　Shyam Bhaskaran, Francis T. Nicholson, Pieter H. Kallemeyn, Robert J. Haw, Peter G. Antreasian, and Gregory J. Garner, "Galileo Orbit Determination for the Earth-2 Encounter," 343-354 in Arun K. Misra, Vinod J. Modi, Richard Holdaway, and Peter M. Bainum, ed., *Astrodynamics 1993*, Pt. I (Reston, VA: AIAA, 1994)

"Galileo OD for Gaspra"
　　Pieter H. Kallemeyn, Robert J. Haw, Vincent M. Pollmeier, and Frank T. Nicholson, "Galileo Orbit Determination for the Gaspra Asteroid Encounter," 370-380 in *A Collection of Technical Papers: AIAA/AAS Astrodynamics Conference, August 10-11, 1992, Hilton Head Island, SC* (Reston, VA: AIAA, 1993)

"Galileo OD for Ida"
　　Peter G. Antreasian, Frank T. Nicholson, Pieter H. Kallemeyn, Shyam Bhaskaran, Robert J. Haw, and Peter Halamek, "Galileo Orbit Determination for the Ida Encounter," 1027-1048 in John E. Cochran, Jr., Charles D. Edwards, Jr., Stephen J. Hoffman, and Richard Holdaway, ed., *AAS/AIAA Spaceflight Mechanics 1994*, Pt. II (San Diego: Univelt, 1994)

"Galileo OD Satellite Tour"
　　Peter G. Antreasian, Timothy P. McElrath, Robert J. Haw, George D. Lewis, and Timothy Krisher, "Galileo Orbit Determination Results During the Satellite Tour," 1491-1536 in Felix R. Hoots, Bernard Kaufman, Paul J. Cefola, and David B. Spencer, ed., *Astrodynamics 1997*, Pt. II (San Diego: Univelt, 1998)

Garthwaite, Holdridge, and Mulholland
　　Kersti Garthwaite, Douglas B. Holdridge, and J. Derral Mulholland, "A Preliminary Special Perturbation Theory for the Lunar Motion," *The Astronomical Journal* 75 (December 1970): 1133

Gordon and Michel
Harold J. Gordon and John R. Michel, "Midcourse Guidance for Mariner Mars 1964," AIAA 65-402, AIAA Second Annual Meeting, July 26-29, 1965, San Francisco, CA

"GPS Flight Experiment"
William G. Melbourne, Byron D. Tapley, and Thomas P. Yunck, "The GPS Flight Experiment on TOPEX/Poseidon," http://trs-new.jpl.nasa.gov/dspace/bitstream/2014/35607/1/93-1328.pdf (accessed February 24, 2011)

Gray
Donald L. Gray, "VLBI Data Performance in the Galileo Spacecraft Earth Flyby of December 1990," 335-352 in TDAPR 42-106

Green and Lomask
Constance McLaughlin Green and Milton Lomask, *Vanguard: A History* (Washington: Smithsonian Institution Press, 1971)

"Ground-Based OD for Deep Impact"
Mark Ryne, David Jefferson, Diane Craig, Earl Higa, George Lewis, and Prem Menon, "Ground-Based Orbit Determination for Deep Impact," 1179-1202 in Srinivas Rao Vadali, L. Alberto Cangahuala, Paul W. Schumacher, Jr., and Jose J. Guzman, ed., *Space Flight Mechanics 2006*, Pt. II (San Diego: Univelt, 2006)

GSFC, *Significant Accomplishments*
Goddard Space Flight Center, *Significant Accomplishments in Science, 1970*, SP-286 (Washington: NASA, 1972)

"Guidance and Nav for Pathfinder"
Sam W. Thurman and Vincent M. Pollmeier, "Guidance and Navigation for the Mars Pathfinder Mission," *Acta Astronautica* 35, Sup 1 (1995): 545-554

Guinn and Wolff
Joseph R. Guinn and Peter J. Wolff, "TOPEX/Poseidon Operational Orbit Determination Results Using Global Positioning Satellites," 143-158 in Arun K. Misra, Vinod J. Modi, Richard Holdaway, and Peter M. Bainum, ed., *Astrodynamics 1993*, Pt. I (San Diego: Univelt, 1993)

Hagen, "Radio Tracking"
John P. Hagen, "Radio Tracking, Orbit and Communication for the Earth Satellite," *Aeronautical Engineering Review* 16 (May 1957): 62-66

Hagen, "Viking and Vanguard"
John P. Hagen, "The Viking and the Vanguard," 122-141 in Eugene M. Emme, ed., *The History of Rocket Technology: Essays on Research, Development, and Utility* (Detroit: Wayne State University Press, 1964)

Hall, *Essays*
R. Cargill Hall, ed., *Essays on the History of Rocketry and Astronautics: Proceedings of the Third Through the Sixth History Symposia of the International Academy of Astronautics*, Vol. II (Washington: NASA, 1977)

Hall, *Lunar Impact*
R. Cargill Hall, *Lunar Impact: A History of Project Ranger* (Washington: NASA, 1977)

Hall, "Origins"
R. Cargill Hall, "Origins and Development of the Vanguard and Explorer Satellite Programs," *The Airpower Historian* 11 (October 1964): 104-111

Hall, "Project Ranger"
R. Cargill Hall, "Project Ranger: Forging a New Era in Space Science," *Journal of the British Interplanetary Society* 32 (December 1979): 459-462

Hall, *Ranger Chronology*
R. Cargill Hall, *Project Ranger: A Chronology* (Pasadena: JPL, April 1971)

Hall, "Satellite Proposals"
R. Cargill Hall, "Early U. S. Satellite Proposals," *Technology and Culture* 4 (Fall 1963): 410-434

Hamilton and Melbourne
Thomas W. Hamilton and William G. Melbourne, "Information Content of a Single Pass of Doppler Data from a Distant Spacecraft," 18-23 in JPLSPS 37-39:III

Hamilton and Trask, "Introduction," *SPS 37-55*
Thomas W. Hamilton and Donald W. Trask, "Introduction," 12-13 in JPLSPS 37-55:II

Hamilton and Trask, "Introduction," *SPS 37-58*
Thomas W. Hamilton and Donald W. Trask, "Introduction," 65-66 in JPLSPS 37-58:II

Hansen and Kliore
David M. Hansen and Arvidas J. Kliore, "Ka-Band (32-GHz) Benefits to Planned Missions," 110-119 in TDAPR 42-88

Hartmann and Raper
William K. Hartmann and Odell Raper, *The New Mars: The Discoveries of Mariner 9*, SP-337 (Washington: NASA, 1974)

Heacock
Raymond L. Heacock, "Ranger: Its Mission and Its Results," *Space Log* (Summer 1965): 5, F5370, NHRC

Hecht
Jeff Hecht, "Schoolkid Blunder Brought down Mars Probe," *New Scientist*, October 9, 1999, 6

Hecht and Adler
Jeff Hecht and Robert Adler, "Mars Up Close," September 29, 1999, *SpaceDaily*, http://www.spacedaily.com/news/marsmco-99c.html (accessed January 18, 2011)

Helvey
T. C. Helvey, ed., *Space Trajectories* (New York: Academic Press, 1960)

Herget, "Eckert Memoir"
Paul Herget, memoir of Eckert, n.d., attached to letter, Herget to R. F. Powell, December 21, 1971, "Eckert Memorial, 1971-73," D8, Herget Papers

Herget, "Leuschner"
Paul Herget, "Armin Otto Leuschner," National Academy of Sciences, *Biographical Memoirs* 49 (1978): 129-147

Herget and Musen, "Erratum"
   Paul Herget and Peter Musen, "Erratum: A Modified Hansen Lunar Theory for Artificial Satellites," *The Astronomical Journal* 64 (1959): 73

Herget and Musen, "Modified Hansen"
   Paul Herget and Peter Musen, "A Modified Hansen Lunar Theory for Artificial Satellites," *The Astronomical Journal* 63 (1958): 430-433

Herrick, *Tables*
   Samuel Herrick, *Tables for Rockets and Comet Orbits* (Washington: National Bureau of Standards, 1953)

"High-Accuracy Astrometry"
   William M. Owen, Jr., Stephen P. Synnott, and George W. Null, "High-Accuracy Asteroid Astrometry from Table Mountain Observatory," 89-102 in Rudolf Dvorak, Herman F. Haupt, and Karl Wodnar, ed., *Modern Astrometry and Astrodynamics* (Vienna: Verlag der Österreichischen Akademie der Wissenschaften, 1999)

"Higher Density Catalog"
   James S. Ulvestad, Ojars J. Sovers, and Christopher S. Jacobs, "A Higher Density VLBI Catalog for Navigating Magellan and Galileo," 274-300 in TDAPR 42-100

Hoffleit, "DOVAP"
   Dorrit Hoffleit, "DOVAP: A Method for Surveying High-Altitude Trajectories," *Scientific Monthly* 68 (March 1949): 172-178

Hofmann
   Albert H. Hofmann, "Pioneer F and G Mission Support Area," 136-140 in DSNPR:V

Hogan, *Mars Wars*
   Thor Hogan, *Mars Wars: The Rise and Fall of the Space Exploration Initiative*, SP-4410 (Washington: NASA, 2007)

Holdridge
   Douglas B. Holdridge, *Space Trajectories Program for the IBM 7090 Computer*, TR 32-223 (Pasadena: JPL, September 1, 1962)

Hudson
   R. Henry Hudson, *Subtabulated Lunar and Planetary Ephemerides*, TR 34-239 (Pasadena: JPL, November 2, 1960)

Hunter to Distribution
   Memorandum, John A. Hunter to Distribution, "Notes from August 1 Meeting of DSN Advisory Group," September 6, 1979, F191/B12, JPL142

"Impactor Targeting"
   Daniel G. Kubitschek, "Impactor Spacecraft Targeting for the Deep Impact Mission to Comet Tempel 1," 1791-1812, in Jean de Lafontaine, Alfred J. Treder, Mark T. Soyka and Jon A. Sims, ed., *Astrodynamics 2003*, Pt. III (San Diego: Univelt, 2003)

"In-Flight Performance"
   Shyam Bhaskaran, Joseph E. Riedel, Stephen P. Synnott, Tseng-Chang Wang, Robert A. Werner, and Brian M. Kennedy, "In-Flight Performance Evaluation of the Deep Space 1 Autonomous Navigation System," International Symposium on Spaceflight Dynamics, June 2000, Biarritz, France, http://trs-new.jpl.nasa.gov/dspace/bitstream/2014/14457/1/00-0894.pdf (accessed April 11, 2010)

"Innovations in DDOR"
James S. Border, "Innovations in Delta Differential One-Way Range: From Viking to Mars Science Laboratory," http://www.mediatec-dif.com/issfd/Orbitdl/Border.pdf (accessed May 27, 2010)

"Interplanetary OD"
J. William Zielenbach, Charles H. Acton, Jr., George H. Born, William G. Breckenridge, Chia-Chun Chao, Thomas C. Duxbury, Donald W. Green, Navin Jerath, J. Frank Jordan, Neil A. Mottinger, Stephen J. Reinbold, Kenneth H. Rourke, Gary L. Sievers, and Sun Kuen Wong, "Interplanetary Orbit Determination," 20-59 in *Mariner 9 Navigation*, TR 32-1586 (Pasadena: JPL, November 13, 1973)

Jordan
Frank Jordan, "Halley Pathfinder," September 14, 2012, copy provided by author

"Joseph William Siry"
Joseph M. Siry, "Joseph William Siry, 1920-2001," *Bulletin of the American Astronomical Society* 33 (2001): 1582-1583

JPL Board, *Loss of MCO*
JPL Special Review Board, *Report on the Loss of the Mars Climate Orbiter Mission*, D-18441 (Pasadena: JPL, November 11, 1999)

"JPL Ephemeris Development"
Charles L. Lawson, "JPL Ephemeris Development, 1960-1967," February 23, 1981, F159/B13, JPL127

JPL Ephemeris Tapes
Paul R. Peabody, James F. Scott, and Everett G. Orozco, *Users' Description of JPL Ephemeris Tapes*, TR 32-580 (Pasadena: JPL, March 2, 1964)

"JPL Roadmap"
Tomas J. Martin-Mur, Douglas S. Abraham, David Berry, Shyam Bhaskaran, Robert J. Cesarone, and Lincoln J. Wood, "The JPL Roadmap for Deep Space Navigation," 1925-1932 in Srinivas Rao Vadali, L. Alberto Cangahuala, Paul W. Schumacher, Jr., and Jose J. Guzman, ed., *Spaceflight Mechanics 2006*, Pt. II (San Diego: Univelt, 2006)

"Juno New Frontiers"
Steve Matousek, "The Juno New Frontiers Mission," IAC-05-A3.2.A.04, International Astronautical Congress, October 17-21, 2005, Fukuoka, Japan http://trs-new.jpl.nasa.gov/dspace/bitstream/2014/37654/1/05-2760.pdf (accessed April 17, 2009)

Kallemeyn, "Nav Ops"
Pieter Kallemeyn, "Navigation Operations," 4-28 in *Mars Pathfinder Navigation Peer Review, June 8, 1994*, D-13060 (Pasadena: JPL, 1994), F116/B13, JPL267

KinetX "Management Team"
KinetX, "Management Team," http://www.kinetx.com/company.aspx?p=mgmt (accessed March 31, 2011)

Koppes
Clayton R. Koppes, *JPL and the American Space Program: A History of the Jet Propulsion Laboratory* (New Haven: Yale University Press, 1982)

Kraemer
    Robert S. Kraemer, *Beyond the Moon: A Golden Age of Planetary Exploration, 1971-1978* (Washington: Smithsonian Institution Press, 2000)

Lambright
    W. Henry Lambright, *Transforming Government: Dan Goldin and the Remaking of NASA* (Arlington, VA: PricewaterhouseCoopers Endowment for the Business of Government, March 2001)

LE4
    J. Derral Mulholland and Neil L. Block, *JPL Lunar Ephemeris Number 4*, TM 33-346 (Pasadena: JPL, April 15, 1967)

LE6
    J. Derral Mulholland, *JPL Lunar Ephemeris Number 6*, TM 33-408 (Pasadena: JPL, October 15, 1968)

Lichten, "Ka-Band"
    Memorandum, Stephen M. Lichten, "Ka-Band VLBI Navigational Tracking," March 14, 2005, F347, DSN Files

Lichten, Meeting Notes
    Stephen M. Lichten, "Notes from Meeting Held on Dec 10, 2004," F347, DSN Files

Lichten, "White Paper"
    Stephen M. Lichten, "White Paper on Use of Very Long Baseline Interferometry (VLBI) for Deep Space Tracking and Navigation of NASA Missions," January 21, 2005, F347, DSN Files

Lieske
    Jay Lieske, *Newtonian Planetary Ephemerides 1800-2000: Development Ephemeris Number 28*, TR 32-1206 (Pasadena: JPL, November 15, 1967)

Liewer
    Kurt M. Liewer, "DSN Very Long Baseline Interferometry System Mark IV-88," 239-246 in TDAPR 42-93

"Linda Morabito Kelly"
    The Planetary Society, "The Stories Behind the Mission: Linda Morabito Kelly As Told to A. J. S. Rayl in 2002 on the Occasion of Voyager's 25th Anniversary" http://www.planetary.org/explore/topics/space_missions/voyager/stories_kelly.html (accessed August 22, 2008)

Liu, "Results, Phase I"
    Anthony Liu, "Results of the Doppler-Ranging Calibration Experiment, Phase I," 23-28 in JPLSPS 37-46:III

Liu, "Results, Phase II"
    Anthony Liu, "Results of the Doppler-Ranging Calibration Experiment, Phase II," 30-40 in JPLSPS 37-48:II

"Long-Range Plan"
    Presentation, "Long-Range Plan for Flight Mission Operations," attached to memorandum, Bruce Murray to Walter Downhower, Kurt Heftman, and Robert Parks, January 8, 1979, F192/B12, JPL142

Lorell, "LO Gravity Analysis"
    Jack Lorell, "Lunar Orbiter Gravity Analysis," *Earth, Moon & Planets* 1 (February 1970): 190-231

Lorell and Sjogren, *LO Data Analysis*
　　Jack Lorell and William L. Sjogren, *Lunar Orbiter Data Analysis*, TR 32-1220 (Pasadena: JPL, November 15, 1967)

Lorell, Carr, and Hudson
　　Jack Lorell, Russell E. Carr, and R. Henry Hudson, *The Jet Propulsion Laboratory Lunar-Probe Tracking and Orbit-Determination Program*, TR 34-16 (Pasadena: JPL, March 10, 1960)

"Lunar Laser Ranging"
　　Jean O. Dickey, Peter L. Bender, James E. Faller, X X Newhall, Randall L. Ricklefs, Judit G. Ries, Peter J. Shelus, Christian Veillet, Arthur L. Whipple, Jerry R. Wiant, James G. Williams, and Charles F. Yoder, "Lunar Laser Ranging: A Continuing Legacy of the Apollo Program," *Science* 265 (July 22, 1994): 482-490

"M4 Occultation Experiment"
　　Arvydas J. Kliore, Dan L. Cain, Gerald S. Levy, Von R. Eshleman, Frank D. Drake, and Gunnar Fjeldbo, "The Mariner 4 Occultation Experiment," *Astronautics & Aeronautics* 3 (July 1965): 72-80

M4 Occultation Instrumentation
　　Gerald S. Levy, Tom Y. Otoshi, and Boris L. Seidel, *Ground Instrumentation for Mariner IV Occultation Experiment*, TR 32-9984 (Pasadena: JPL, September 15, 1966)

"M10 OpNav Demo"
　　Richard H. Stanton, Hiroshi Ohtakay, James A. Miller, Charles C. Voge, "Demonstration of Optical Navigation Measurements on Mariner 10," AIAA 75-86, AIAA Aerospace Sciences Meeting, January 20-22, 1975, Pasadena, CA

MacDoran, "VLBI"
　　Peter F. MacDoran, "Very Long Baseline Interferometry (VLBI) Earth Physics," 62-73 in *Proceedings of the Fourth Precise Time and Time Interval Planning Meeting* (Greenbelt: GSFC, 1972)

MacDoran, "VLBI Earth Physics"
　　Peter F. MacDoran, "VLBI Earth Physics," 9-16 in JPL Science Office, *Annual Progress Review, July 1, 1972-June 30, 1973* (Pasadena: JPL, August 30, 1973)

MacDoran and Martin
　　Peter F. MacDoran and Warren L. Martin, "DRVID Charged-Particle Measurement With a Binary-Coded Sequential Acquisition Ranging System," 34-41 in JPLSPS 37-62:II

MacDoran and Wimberly
　　Peter F. MacDoran and Ravenel N. Wimberly, "Charged-Particle Calibrations from Differenced Range Versus Integrated Doppler: Preliminary Results from Mariner Mars 1969," 73-77 in JPLSPS 37-58:II

"Magellan Lessons Learned"
　　"Magellan Lessons Learned Presentation to Mars Observer," April 14, 1992, "Lessons Learned / MESUR / MGN / Pathfinder," B18, JPL496

Makemson
　　Maud Worcester Makemson, "Russell Tracy Crawford, 1876-1958," *Publications of the Astronomical Society of the Pacific* 71 (December 1959): 503-505

Malina, "Memoir"
　　Frank J. Malina, "The U. S. Army Air Corps Jet Propulsion Research Project, GALCIT Project No. 1, 1939-1946: A Memoir," 153-202 in Hall, *Essays*

"Maneuver History"
James V. McAdams, David W. Dunham, Larry E. Mosher, J. Courtney Ray, Peter G. Antreasian, Clifford E. Helfrich, James K. Miller, "Maneuver History for the NEAR Mission: Launch through Eros Orbit Insertion," 244-261 in *A Collection of Technical Papers: AIAA/AAS Astrodynamics Specialist Conference, August 14-17, 2000, Denver, CO* (Reston, VA: AIAA, 2000)

"Mariner 4 Flight Path"
Norman R. Haynes, John R. Michel, George W. Null, and Richard K. Sloan, "Mariner 4 Flight Path to Mars," *Astronautics & Aeronautics* 3 (June 1965): 28-33

Mariner IV OD
George W. Null, Harold J. Gordon, and Dennis A. Tito, *The Mariner IV Flight Path and Its Determination From Tracking Data*, TR 32-1108 (Pasadena: JPL, August 1, 1967)

Mariner R Project
The Mariner R Project: Progress Report September 1, 1961-August 31, 1962, TR 32-353 (Pasadena: JPL, January 1, 1963)

Mariner-Venus 1962
JPL, *Mariner-Venus 1962, Final Project Report*, SP-59 (Washington: NASA, 1965)

Mars Team, *Summary*
Mars Program Independent Assessment Team, *Summary Report*, March 14, 2000, http://sunnyday.mit.edu/accidents/mpiat_summary.pdf (accessed June 17, 2010)

Martin-Mur and Highsmith
Tomás J. Martin-Mur and Dolan E. Highsmith, "Mars Approach Navigation Using the VLBA," http://www.nrao.edu/nio/vlba-partners/issfd_2009_tjmm.pdf (accessed March 29, 2011)

Mayo
Alton B. Mayo, "Orbit Determination for Lunar Orbiter," *Journal of Spacecraft* 5 (April 1968): 395-399

McClure to Bayley
Memorandum, Donald H. McClure to William H. Bayley, "NAVNET Project Completion Summary," May 19, 1980, F70/B2, JPL516

McClure to Distribution
Memorandum, McClure to Distribution, "The Navigation Network Project," March 30, 1978, F70/B2, JPL516

McCurdy, *Low-Cost Innovation*
Howard E. McCurdy, *Low-Cost Innovation in Spaceflight: The Near Earth Asteroid Rendezvous (NEAR) Shoemaker Mission* (Washington: NASA, 2005)

McCurdy, *Faster, Better, Cheaper*
Howard E. McCurdy, *Faster, Better, Cheaper: Low-Cost Innovation in the U.S. Space Program* (Baltimore: Johns Hopkins University Press, 2001)

McNamee to Distribution
Memorandum, John McNamee to Distribution, "The Use of Differenced Doppler Data to Support Magellan Navigation During the Mapping Mission," May 25, 1990, F508/B30, JPL508

MCO Board, *Phase I Report*
    Mars Climate Orbiter Mishap Investigation Board, *Phase I Report, November 10, 1999*, ftp://ftp.hq.nasa.gov/pub/pao/reports/1999/MCO_report.pdf (accessed June 17, 2010)

Melbourne and Curkendall
    William G. Melbourne and David W. Curkendall, "Radio Metric Direction Finding: A New Approach to Deep Space Navigation," paper read at the AAS/AIAA Astrodynamics Specialist Conference, September 7-9, 1977, Jackson Hole, WY

Mengel and Herget
    John T. Mengel and Paul Herget, "Tracking Satellites by Radio," *Scientific American* 188 (January 1958): 23-29

"MER EDL Nav"
    Darren T. Baird, Timothy M. McElrath, Brian M. Portock, Eric J. Graat, Geoff G. Wawrzyniak, Philip C. Knocke, Louis A. D'Amario, Mike M. Watkins, Lynn E. Craig, and Josesph R. Guinn, "Mars Exploration Rovers Entry, Descent, and Landing Navigation," 783-796 in *A Collection of Technical Papers: AIAA/AAS Astrodynamics Specialist Conference, August 16-19, 2004, Providence, RI* , vol. 2 (Reston, VA: AIAA, 2004)

MER Navigation
    Louis A. D'Amario, *Mars Exploration Rover Navigation* (Pasadena: JPL, December 2005)

"MGS Aerobraking"
    M. Dan Johnston, Pasquale B. Esposito, Vijay Alwar, Stuart W. Demcak, Eric J. Graat, and Robert A. Mase, "Mars Global Surveyor Aerobraking at Mars," 205-224 in Jay W. Middour, Lester L. Sackett, Louis D'Amario, and Dennis V. Byrnes, ed., *Spaceflight Mechanics 1998*, Pt. I (San Diego, Univelt, 1998)

"MGS Aerobraking Broken Wing"
    Daniel T. Lyons, "Mars Global Surveyor: Aerobraking with a Broken Wing," 275-294 in Felix R. Hoots, Bernard Kaufman, Paul J. Cefola, and David B. Spencer, ed., *Astrodynamics 1997*, Pt. I (San Diego: Univelt, 1997)

"MGS Aerobraking Overview"
    Daniel T. Lyons, Joseph G. Beerer, Pasquale Esposito, M. Dan Johnston, and William H. Willcockson, "Mars Global Surveyor: Aerobraking Mission Overview," *Journal of Spacecraft and Rockets* 36 (1999): 307-313

"MGS Mapping OD"
    Stuart Demcak, Eric Graat, Pasquale B. Esposito, and Darren T. Baird, "Mars Global Surveyor Mapping Orbit Determination," 3-24 in Louis A. D'Amario, Lester L. Sackett, Daniel J. Scheeres, and Bobby G. Williams, ed., *Spaceflight Mechanics 2001*, Pt. I (San Diego: Univelt, 2001)

"MGS Nav and Aerobraking"
    Pasquale Esposito, Vijay Alwar, Stuart Demcak, Eric Graat, M. Dan Johnston, and Robert Mase, "Mars Global Surveyor Navigation and Aerobraking at Mars," 1015-1028 in Thomas H. Stengle, ed., *Spaceflight Dynamics 1998*, Pt. II (San Diego: Univelt, 1998)

"MGS OD Uncertainties"
    Eric Carranza, Dah-Ning Yuan, and Alexander S. Konopliv, "Mars Global Surveyor Orbit Determination Uncertainties using High-Resolution Mars Gravity Models," 1633-1650 in David B. Spencer, Calina C. Seybold, Arun K. Misra, and Ronald J. Lisowski, ed., *Astrodynamics 2001*, Pt. II (San Diego: Univelt, 2001)

"MGS Orbit Evolution"
   Pasquale Esposito, Eric Graat, Stuart Demcak, Darren Baird and Vijay Alwar, "Mars Global Surveyor: Mapping Orbit Evolution and Control Throughout One Mars Year," 1615-1632 in David B. Spencer, Calina C. Seybold, Arun K. Misra, and Ronald J. Lisowski, ed., *Astrodynamics 2001*, Pt. II (San Diego: Univelt, 2001)

MM64 Final Report
   Mariner-Mars 1964 Final Project Report, SP-139 (Washington: NASA, 1967)

MM64 Operations Report
   Mariner Mars 1964 Project Report: Mission Operations, TR 32-881 (Pasadena: JPL, June 15, 1966)

MM64 TDA Report
   Nicholas A. Renzetti, *Tracking and Data Acquisition Report: Mariner-Mars 1964 Mission: Volume 1: Near-Earth Trajectory Phase*, TR 33-239 (Pasadena: JPL, January 1, 1965

MM69 Final Report, Vol. I
   Mariner Mars 1969 Final Project Report: Vol. I: Development, Design, and Test, TR 32-1460 (Pasadena: JPL, November 1, 1970)

MM69 Final Report, Vol. II
   Mariner Mars 1969 Final Project Report: Vol. II: Performance, TR 32-1460 (Pasadena: JPL, March 1, 1971)

MM69 Final Report, Vol. III
   Mariner Mars 1969 Final Project Report: Vol. III: Scientific Investigations, TR 32-1460 (Pasadena: JPL, September 15, 1971)

MM69 QR8
   Mariner Mars 1969 Project, *Eighth Quarterly Review*, May 21-22, 1968, F31/B1, JPLDSN

MM69 QR9
   Mariner Mars 1969 Project, *Ninth Quarterly Review*, September 17-18, 1968, JPLDSN

MM71 Final Report, Vol. I
   Mariner Mars 1971 Project Final Report, Vol. I: Project Development through Launch and Trajectory Correction Maneuver, TR 32-1550 (Pasadena: JPL, April 1, 1973)

MM71 Final Report, Vol. IV
   Mariner Mars 1971 Final Project Report, Vol. IV: Science Results, TR 32-1550 (Pasadena: JPL, July 15, 1973)

MM71 Final Report, Vol. V
   Mariner Mars 1971 Project Final Report, Vol. V: Science Experiment Reports, TR 32-1550 (Pasadena: JPL, August 20, 1973)

MM71 OpNav Demo
   George H. Born, Thomas C. Duxbury, William G., Breckenridge, Charles H. Acton, Jr., Srinivas N. Mohan, Navin Jerath, and Hiroshi Ohtakay, *Mariner Mars 1971 Optical Navigation Demonstration Final Report*, TM 33-683 (Pasadena: JPL, April 15, 1974)

Mohan and D'Souza
   Presentation, Srinivas Mohan and Christopher D'Souza, "Magellan: Final Mission Design Review #2: 1989 Mission: Navigation," May 6, 1987, F56, JPL4

Mottinger and Trask
   Neil Mottinger and Donald W. Trask, "Status of DSS Location Solutions for Deep Space Probe Missions: I. Initial Comparisons," 12-22 in JPLSPS 37-48:II

Mowlem
   A. Robin Mowlem, "The Programming System for Orbit Determination at the IBM Space Computing Center," 119-127 in JPL, *Tracking and OD*

Moyer, "DPODP Basis"
   Theodore D. Moyer, "Theoretical Basis for the Double-Precision Orbit Determination Program (DPODP)," 24-27 in JPLSPS 37-38:III

Moyer, "DPODP Basis: Time"
   Theodore D. Moyer, "Theoretical Basis for the DPODP: Time Transformations," 36-38 in JPLSPS 37-39:III

Moyer, *Formulation*
   Theodore D. Moyer, *Formulation for Observed and Computed Values of Deep Space Network Data Types for Navigation* (Pasadena: JPL, October 2000)

Moyer, *Formulation of DPODP*
   Theodore D. Moyer, *Mathematical Formulation of the Double Precision Orbit Determination Program (DPODP)*, TR 32-1527 (Pasadena: JPL, May 15, 1971)

"MRO: Launch to Science Orbit"
   Martin D. Johnston, James E. Graf, Richard W. Zurek, Howard J. Eisen, Benhan Jai, and James K. Erickson, "The Mars Reconnaissance Orbiter Mission: From Launch to the Primary Science Orbit," 334-352 in *2007 IEEE Aerospace Conference*, Pt. 1 (Piscataway, NJ: IEEE, 2007)

"MRO Nav"
   Tung-Han You, Allen Halsell, Dolan Highsmith, Jah Moriba, Stuart Demcak, Earl Higa, Stacia Long, and Shyam Bhaskaran, "Mars Reconnaissance Orbiter Navigation," 158-178 in *A Collection of Technical Papers: AIAA/AAS Astrodynamics Specialist Conference, August 16-19, 2004, Providence, RI*, vol. 1 (Reston, VA: AIAA, 2004)

Mudgway, *Big Dish*
   Douglas J. Mudgway, *Big Dish: Building America's Deep Space Connection to the Planets* (Gainesville: University Press of Florida, 2005)

Mudgway, *Uplink-Downlink*
   Mudgway, *Uplink-Downlink: A History of the Deep Space Network, 1957–1997*, SP-2001-4227 (Washington: NASA, 2001)

Mulhall, "Analysis"
   Brendan D. Mulhall, "Charged-Particle Calibration System Analysis," 13-21 in JPLSPS 37-64:II

Mulhall, "Calibration"
   Brendan D. Mulhall, "In-Flight Ionospheric Calibration Procedures for Mariner Mars 1969 Radio Tracking Data," 66-73 in JPLSPS 37-58:II

Mulhall, "Demonstrations"
   Brendan D. Mulhall, "Navigation Demonstrations With the Mariner Venus-Mercury 1973 Spacecraft Requiring X-Band Receiving Capability at a Second DSN Station," 38-43 in DSNPR:IX

Mulhall and Wimberly
    Brendan D. Mulhall and Ravenel N. Wimberly, "Comparison of Ionospheric Measurements for Evaluation of Techniques Used in Charged Particle Calibration of Radio Tracking Data: Preliminary Results," 58-61 in JPLSPS 37-56:II

Mulholland, "Corrections"
    J. Derral Mulholland, "Corrections to the Lunar Ephemeris," 2-5 in JPLSPS 37-42:IV

Mulholland, "Numerical Studies"
    J. Derral Mulholland, "Numerical Studies of Lunar Motion," Nature 223 (July 19, 1969): 247-249

Mulholland and Sjogren
    J. Derral Mulholland and William L. Sjogren, "Lunar Orbiter Ranging Data: Initial Results," Science 155 (January 6, 1967): 74

Muller, "Polar Motion"
    Paul M. Muller, "Polar Motion and DSN Station Locations," 10-14 in JPLSPS 37-45:III

Muller, "Time"
    Paul M. Muller, "Time and Polar Motion in Early NASA Spacecraft Navigation," 215-219 in Polar Motion

Muller and Sjogren, "Consistency"
    Paul M. Muller and William L. Sjogren, "Consistency of Lunar Orbiter Residuals With Trajectory and Local Gravity Effects," 28-37 in JPLSPS 37-51:II

Muller and Sjogren, "Mascons"
    Paul M. Muller and William L. Sjogren, "Mascons: Lunar Mass Concentrations," 10-16 in JPLSPS 37-53:II

MV67: Final Report
    Mariner-Venus 1967: Final Project Report, SP-190 (Washington: NASA, 1971)

NASA Facts
    NASA Facts, "NASA's Ranger Program," 1962, F13/B 64, Willy Ley

NASA Tech Brief
    NASA, Tech Brief, "Mission Operations and Navigation Toolkit Environment," September 1, 2009, http://www.techbriefs.com/component/content/article/5702 (accessed February 23, 2011)

"Nav and MGS Mission"
    Pasquale Esposito, Vijay Alwar, Stuart Demcak, J. Giorgini, Eric Graat, and M. Don Johnston, "Navigation and the Mars Global Surveyor Mission," 371-376 in Tan-Duc Guyenne, ed., Proceedings of the 12th International Symposium Spaceflight Dynamics, ESA SP-403 (Paris: European Space Agency, 1997)

"Nav Flight Ops for Mars Pathfinder"
    Robin Vaughan, Pieter Kallemeyn, David A. Spencer, Jr., and Robert D. Braun, "Navigation Flight Operations for Mars Pathfinder," 643-662 in Jay W. Middour, Lester L. Sackett, Louis D'Amario, and Dennis V. Byrnes, ed., Spaceflight Mechanics 1998, Pt. 1 (San Diego: Univelt, 1998)

"Nav for New Millennium"
   Joseph E. Riedel, Shyam Bhaskaran, Stephen P. Synnott, Shailen D. Desai, Willard E. Bollman, Philip J. Dumont, C. Allen Halsell, Dan Han, Brian M. Kennedy, George W. Null, William M. Owen, Jr., Robert A. Werner, and Bobby G. Williams, "Navigation for the New Millennium: Autonomous Navigation for Deep Space 1," 303-320 in Tan-Duc Guyenne, ed., *Proceedings of the 12th International Symposium Space Flight Dynamics*, ESA SP-403 (Paris: European Space Agency, 1997)

"Navigating CONTOUR"
   Eric Carranza, Anthony H. Taylor, Bobby G. Williams, Dongsuk Han, Cliff E. Helfrich, Ramachand Bhat, and Jamin S. Greenbaum, "Navigating CONTOUR using the Noncoherent Transceiver Technique," 1473-1492 in Daniel J. Scheeres, Mark E. Pittelkau, Ronald J. Proulx, and L. Alberto Cangahuala, ed., *Spaceflight Mechanics 2003*, Pt. II (San Diego: Univelt, 2003)

"Navigation for NEAR Shoemaker"
   Bobby Williams, Peter Antreasian, John Bordi, Eric Carranza, Stephen Chesley, Clifford Helfrich, James K. Miller, William M. Owen, Jr., and Tseng-Chang Wang, "Navigation for NEAR Shoemaker: The First Mission to Orbit an Asteroid," 973-988 in David B. Spencer, Calina C. Seybold, Arun K. Misra, and Ronald J. Lisowski, ed., *Astrodynamics 2001*, Pt. II (San Diego: Univelt, 2001)

"Nav Software"
   Steve Flanagan, Theodore R. Drain, Todd Ely, Tomas Martin-Mur, "Navigation and Mission Analysis Software for the Next Generation of JPL Missions," 16th International Symposium on Space Flight Dynamics, December 3-7, 2001, Pasadena, CA, http://trs-new.jpl.nasa.gov/dspace/bitstream/2014/36810/1/01-1301.pdf (accessed February 23, 2011)

Nead, "Reminiscences"
   Melba W. Nead, "Reminiscences of California Institute of Technology Guggenheim Aeronautical Laboratory, GALCIT No. 1 later JPL," memorandum from Nead to Kyky Chapman, November 5, 1991, C 16-7, JPLHC

"NEAR Encounter Mathilde"
   Daniel J. Scheeres, David W. Dunham, Robert W. Farquhar, Clifford E. Helfrich, James V. McAdams, William M. Owen, Jr., Stephen P. Synnott, Bobby G. Williams, Peter J. Wolff, and Donald K. Yeomans, "Mission Design and Navigation of NEAR's Encounter with Asteroid 253 Mathilde," 1157-1174 in Jay W. Middour, Lester L. Sackett, Louis A. D'Amario, and Dennis V. Byrnes, ed., *Spaceflight Mechanics 1998*, Pt. II (San Diego: Univelt, 1998)

"NEAR OpNav at Eros"
   William M. Owen, Jr., Tseng-Chang Wang, Ann Harch, Maureen Bell, and Colin Peterson, "NEAR Optical Navigation at Eros," 1075-1088 in David B. Spencer, Calina C. Seybold, Arun K. Misra, and Ronald J. Lisowski, ed., *Astrodynamics 2001*, Pt. II (San Diego: Univelt, 2001)

"Networks Consolidation (42-59)"
   Malvin L. Yeater, David T. Herrman, and George E. Sanner, "Networks Consolidation Program," 107-120 in TDAPR 42-59

"Networks Consolidation (42-63)"
   Edwin C. Gatz, "Networks Consolidation Program System Design," 150-153 in TDAPR 42-63

"Networks Consolidation (42-69)"
   Malvin L. Yeater, David T. Herrman, and Edward B. Luers, "Networks Consolidation Program," 10-13 in TDAPR 42-69

Neufeld
Jacob Neufeld, *The Development of Ballistic Missiles in the United States Air Force, 1945-1960* (Washington: Office of Air Force History, 1990)

Newell
Homer E. Newell, *Beyond the Atmosphere: Early Years of Space Science* (Washington: NASA, 1980)

"New Horizons Pluto Nav"
James K. Miller, Dale Stanbridge, and Bobby G. Williams, "New Horizons Pluto Approach Navigation," 529-540 in Shannon L. Coffey, Andre P. Mazzoleni, K. Kim Luu, and Robert A. Glover, ed., *Spaceflight Mechanics 2004*, Pt. I (San Diego: Univelt, 2004)

"Next 25 Years"
Tomás J. Martin-Mur, Shyam Bhaskaran, Robert J. Cesarone, and Timothy McElrath, "The Next 25 Years of Deep Space Navigation," 397-404 in Michael E. Drews and Robert D. Culp, ed., *Guidance and Control 2008* (San Diego: Univelt, 2008)

Nicks
Oran W. Nicks, *A Review of the Mariner IV Results*, SP-130 (Washington: NASA, 1967)

Null, "Mariner IV Flight Path"
George W. Null, "The Mariner IV Flight Path and Its Determination From Tracking Data," 1-40 in *Mariner IV OD*

Null, "Proposal"
George W. Null, "Proposal for a Determination of the Astronomical Unit and Earth Ephemeris using MIV Return Tracking Data," June 14, 1966, 08 00356 XF, JPLHC

Oberg
James Oberg, "Why the Mars Probe Went Off Course," December 1999, http://sunnyday.mit.edu/accidents/mco-oberg.htm (accessed January 31, 2011)

"OD Performance AutoNav"
Shyam Bhaskaran, Joseph E. Riedel, Shailen D. Desai, Philip J. Dumont, George W. Null, William M. Owen, Jr., Stephen P. Synnott, and Robert A. Werner, "Orbit Determination Performance Evaluation of the Deep Space 1 Autonomous Navigation System," 1295-1314 in Jay W. Middour, Lester L. Sackett, Louis D'Amario, and Dennis V. Byrnes, ed., *Spaceflight Mechanics 1998*, Pt. II (San Diego: Univelt, 1998)

"OD Results"
Ian M. Roundhill, Peter G. Antreasian, Kevin E. Criddle, Rodica Ionasescu, M. Cameron Meek, and Jason R. Stauch, "Orbit Determination Results for the Cassini Titan-A Flyby," 185-202, in David A. Vallado, Michael J. Gabor, and Prasun N. Desai, ed., *Spaceflight Mechanics 2005*, Pt. I (San Diego: Univelt, 2005)

O'Handley, *Card Format*
Douglas A. O'Handley, *Card Format for Optical and Radar Planetary Data*, TR 32-1296 (Pasadena: JPL, May 1, 1968)

O'Handley, "Ephemeris Development"
Douglas A. O'Handley, "Ephemeris Development," 2-2 – 2-12 in *The Mariner VI and VII Flight Paths and Their Determination from Tracking Data*, TR 33-469 (Pasadena: JPL, December 1, 1970)

"O'Keefe," BAAS
David P. Rubincam and Paul D. Lowman, "John Aloysius O'Keefe, 1917-2000," *Bulletin of the American Astronomical Society* 32 (2000): 1683-1684

"O'Keefe," EOS
Paul D. Lowman and David P. Rubincam, "John A. O'Keefe (1916-2000)," *EOS: Transactions of the American Geophysical Union* 82 (January 30, 2001): 55

Olley
Allan Olley, "Just a Beginning: Computers and Celestial Mechanics in the Work of Wallace J. Eckert," Ph.D. dissertation, Institute for History and Philosophy of Science, University of Toronto, 2011

Ondrasik and Rourke, "Advantages"
V. John Ondrasik and Kenneth H. Rourke, "An Analytical Study of the Advantages Which Differenced Tracking Data May Offer for Ameliorating the Effects of Unknown Spacecraft Accelerations," 61-70 in DSNPR:IV

Ondrasik and Rourke, "Application"
V. John Ondrasik and Kenneth H. Rourke, "Application of New Radio Tracking Data Types to Critical Spacecraft Navigation Problems," *JPL Quarterly Technical Review* 1 (Pasadena: JPL January 1972): 116-132

O'Neil and Rudd
William J. O'Neil and Richard P. Rudd, "Introduction," 1-5 in *Viking Navigation*, P 78-38 (Pasadena: JPL, November 15, 1979)

"OpNav for Gaspra"
Robin M. Vaughan, Joseph E. Riedel, R. P. Davis, William M. Owen, Jr., and Stephen P. Synnott, "Optical Navigation for the Galileo Gaspra Encounter," 361-369 in *A Collection of Technical Papers: AIAA/AAS Astrodynamics Conference, August 10-11, 1992, Hilton Head Island, SC* (Reston, VA: AIAA, 1993)

"OpNav for Stardust"
Shyam Bhaskaran, Nickolaos Mastrodemos, Joseph E. Riedel, and Stephen P. Synnott, "Optical Navigation for the Stardust Wild 2 Encounter," 455-460 in Oliver Montenbrook and Bruce Battrick, ed., *Proceedings of the 18th International Symposium on Space Flight Dynamics*, ESA SP-548 (December 2004)

"Optical Observables"
Joseph E. Ball, William G. Breckenridge, Thomas C. Duxbury, and Roger E. Koch, "Optical Observables," 5-1 – 5-26 in *The Mariner VI and VII Flight Paths and Their Determination from Tracking Data*, TR 33-469 (Pasadena: JPL, December 1, 1970)

Orbit Theory
Orbit Theory: Proceedings of the Ninth Symposium in Applied Mathematics (Providence: American Mathematical Society, 1959)

"Orientation of DE200/LE200"
E. Myles Standish, "Orientation of the JPL Ephemerides, DE200/LE200, to the Dynamical Equinox of J2000," *Astronomy and Astrophysics* 114 (October 1982): 297-302

Osterbrock and Seidelmann
Donald E. Osterbrock and P. Kenneth Seidelmann, "Paul Herget, January 30, 1908-August 27, 1981," National Academy of Sciences, *Biographical Memoirs* 57 (1987): 59-86

Peebles
Curtis Peebles, *High Frontier: The U. S. Air Force and the Military Space Program* (Washington: Air Force History and Museums Program, 1997)

"Performance"
Tony H. Taylor, James K. Campbell, Robert A. Jacobson, Benjamin Moultrie, Ralph A. Nichols, Jr., and Joseph E. Riedel, "Performance of Differenced Range Data Types in Voyager Navigation," 40-52 in TDAPR 42-71

Pickering
William H. Pickering, with James H. Wilson, "Countdown to Space Exploration: A Memoir of the Jet Propulsion Laboratory, 1944-1958," 385-422 in Hall, *Essays*

Pioneer: First to Jupiter
Richard O. Fimmel, James Van Allen, and Eric Burgess, *Pioneer: First to Jupiter, Saturn, and Beyond*, SP-446 (Washington: NASA, 1980)

Pioneer IV OD Program
Manfred Eimer and Yasushi Hiroshige, *Evaluation of Pioneer IV Orbit-Determination Program*, TR 34-26 (Pasadena: JPL, February 22, 1960)

Pioneer Odyssey
Richard O. Fimmel, William Swindell, and Eric Burgess, *Pioneer Odyssey*, SP-349 rev. ed. (Washington: NASA, 1977)

Polar Motion
Steven Dick, Dennis McCarthy, and Brian Luzum, ed., *Polar Motion: Historical and Scientific Problems* (San Francisco: Astronomical Society of the Pacific, 2000)

"Precision OD"
Byron D. Tapley, John C. Ries, George W. Davis, Richard J. Eanes, Bob E. Schutz, C. K. Shum, Michael M. Watkins, John A. Marshall, R. Steven Nerem, Barbara H. Putney, Steven M. Klosko, Scott B. Luthcke, Despina E. Pavlis, Ronald G. Williamson, and Nikita P. Zelensky, "Precision Orbit Determination for TOPEX/POSEIDON," *Journal of Geophysical Research* 99 (1994): 24,383–24,404

"Preliminary Results"
Arvydas J. Kliore, Dan L. Cain, Gerald S. Levy, Von R. Eshleman, Gunnar Fjeldbo, and Frank D. Drake, "Preliminary Results of the Mariner IV Occultation Measurement of the Atmosphere of Mars," 258-265 in Harrison Brown, Gordon J. Stanley, Duane O. Muhleman, and Guido Münch, ed., *Proceedings of the Caltech-JPL Lunar and Planetary Conference* (Pasadena: Caltech, 1966)

"Progress"
Claude E. Hildebrand, James S. Border, Frank F. Donivan, Susan G. Finley, Benjamin Moultrie, X X Newhall, Lyle J. Skjerve, Thomas P. Yunck, Frank R. Bletzacker, and Cory B. Smith, "Progress in the Application of VLBI to Interplanetary Navigation," 55-72 in *VLBI Techniques*

"Quasar Experiment: Part I"
Martin A. Slade, Peter F. MacDoran, Irwin I. Shapiro, Donovan J. Spitzmesser, Jack Gubbay, Anthony Legg, David S. Robertson, and Lyle J. Skjerve, "The Mariner 9 Quasar Experiment: Part I," 31-35 in DSNPR:XIX

Quayle
James Danforth Quayle, *Standing Firm: A Vice-Presidential Memoir* (New York: HarperCollins Publishers, 1994)